Shop Manual for

Automatic Transmissions
and Transaxles

TODAY'S TECHNICIAN

Shop Manual for

Automatic Transmissions and Transaxles

Jack Erjavec

Columbus State Community College
Columbus, Ohio

Delmar Publishers

1945 - 1995
50 years

I(T)P   An International Thomson Publishing Company

Albany • Bonn • Boston • Cincinnati • Detroit • London • Madrid • Melbourne
Mexico City • New York • Pacific Grove • Paris • San Francisco • Singapore • Tokyo
Toronto • Washington

NOTICE TO THE READER

COVER PHOTO: Courtesy of Stacey Riggert

PHOTO SEQUENCES: Photography by Rod Dixon Associates, Jeff Hinckley, and Michael A. Gallitelli (Metroland Photo)

Portions of materials contained herein have been reprinted with permission of General Motors Corporation, Service Technology Group.

DELMAR STAFF

Senior Administrative Editor: Vernon Anthony
Developmental Editor: Catherine Eads
Project Editor: Eleanor Isenhart
Production Coordinator: Karen Smith
Art/Design Coordinator: Heather Brown

COPYRIGHT © 1995
By Delmar Publishers
an International Thomson Publishing Company
The ITP logo is a trademark under license

Printed in the United States of America

For information, contact:

Delmar Publishers
3 Columbia Circle, Box 15015
Albany, New York 12212-5015

International Thomson Publishing Europe
Berkshire House 168-173
High Holborn
London, WC1V7AA
England

Thomas Nelson Australia
102 Dodds Street
South Melbourne, 3205
Victoria, Australia

Nelson Canada
1120 Birchmont Road
Scarborough, Ontario
Canada M1K 5G4

International Thomson Editors
Campos Eliseos 385, Piso 7
Col Polanco
11560 Mexico DF Mexico

International Thomson Publishing GmbH
Königswinterer Strasse 418
53227 Bonn
Germany

International Thomson Publishing Asia
221 Henderson Road
#05-10 Henderson Building
Singapore 0315

International Thomson Publishing Japan
Hirakawacho Kyowa Building, 3F
2-2-1 Hirakawacho
Chiyoda-ku, Tokyo 102
Japan

1 2 3 4 5 6 7 8 9 10 XXX 01 00 99 98 97 96 95

Library of Congress Cataloging-in-Publication Data

Erjavec, Jack.
    Automatic transmissions and transaxles / Jack Erjavec.
       p. cm. -- (Today's technician)
    Includes index.
    Contents: 1. Classroom manual -- 2. Shop manual.
    ISBN 0-8273-6190-4 (set)
    1. Automobiles -- Transmission devices, Automatic -- Maintenance and
repair. 2. Automobiles -- Transaxles -- Maintenance and repair.
I. Title. II. Series.
TL263.E75 1995
629.24'46 -- dc20
                               94-39394
                                 CIP

CONTENTS

Photo Sequences

PREFACE

Unlike yesterday's mechanic, the technician of today and for the future must know the underlying theory of all automotive systems and be able to service and maintain those systems. Today's technician must also know how these individual systems interact with each other. Standards and expectations have been set for today's technician, and these must be met in order to keep the world's automobiles running efficiently and safely.

The *Today's Technician* series, by Delmar Publishers, features textbooks that cover all mechanical and electrical systems of automobiles and light trucks. Principal titles correspond with the eight major areas of ASE (National Institute for Automotive Service Excellence) certification. Additional titles include remedial skills and theories common to all of the certification areas and advanced or specialized subject areas that reflect the latest technological trends.

Each title is divided into two manuals: a Classroom Manual and a Shop Manual. Dividing the material into two manuals provides the reader with the information needed to begin a successful career as an automotive technician without interrupting the learning process by mixing cognitive and performance-based learning objectives.

Each Classroom Manual contains the principles of operation for each system and subsystem. It also discusses the design variations used by different manufacturers. The Classroom Manual is organized to build upon basic facts and theories. The primary objective of this manual is to allow the reader to gain an understanding of how each system and subsystem operates. This understanding is necessary to diagnose the complex automobile systems.

The understanding acquired by using the Classroom Manual is required for competence in the skill areas covered in the Shop Manual. All of the high priority skills, as identified by ASE, are explained in the Shop Manual. The Shop Manual also includes step-by-step instructions for diagnostic and repair procedures. Photo Sequences are used to illustrate many of the common service procedures. Other common procedures are listed and are accompanied with fine-line drawings and photographs that allow the reader to visualize and conceptualize the finest details of the procedure. The Shop Manual also contains the reasons for performing the procedures, as well as when that particular service is appropriate.

The two manuals are designed to be used together and are arranged in corresponding chapters. Not only are the chapters in the manuals linked together, the contents of the chapters are also linked. Both manuals contain clear and thoughtfully selected illustrations. Many of the illustrations are original drawings or photos prepared for inclusion in this series. This means that the art is a vital part of each manual.

The page layout is designed to include information that would otherwise break up the flow of information presented to the reader. The main body of the text includes all of the "need-to-know" information and illustrations. In the side margins are many of the special features of the series. Items such as definitions of new terms, common trade jargon, tools lists, and cross-references are placed in the margin, out of the normal flow of information so as not to interrupt the thought process of the reader. Each manual in this series is organized in a like manner and contains the same features.

Jack Erjavec, Series Advisor

Classroom Manual

To stress the importance of safe work habits, the Classroom Manual dedicates one full chapter to safety. Included in this chapter are common safety practices, safety equipment, and safe handling of hazardous materials and wastes. This includes information on MSDS sheets and OSHA regulations. Other features of this manual include:

Cognitive Objectives

These objectives define the contents of the chapter and define what the student should have learned upon completion of the chapter.
Each topic is divided into small units to promote easier understanding and learning.

Marginal Notes

Page numbers for cross-referencing appear in the margin. Some of the common terms used for components, and other bits of information, also appear in the margin. This provides an understanding of the language of the trade and helps when conversing with an experienced technician.

Cautions and Warnings

Throughout the text, cautions are given to alert the reader to potentially hazardous materials or unsafe conditions. Warnings are also given to advise the student of things that can go wrong if instructions are not followed or if a nonacceptable part or tool is used.

References to the Shop Manual

Reference to the appropriate page in the Shop Manual is given whenever necessary. Although the chapters of the two manuals are synchronized, material covered in other chapters of the Shop Manual may be fundamental to the topic discussed in the Classroom Manual.

A Bit of History

This feature gives the student a sense of the evolution of the automobile. This feature not only contains nice-to-know information, but also should spark some interest in the subject matter.

Summaries

Each chapter concludes with summary statements that contain the important topics of the chapter. These are designed to help the reader review the contents.

Terms to Know

A list of new terms appears next to the Summary. Definitions for these terms can be found in the Glossary at the end of the manual.

Review Questions

Short answer essay, fill-in-the-blank, and multiple-choice type questions follow each chapter. These questions are designed to accurately assess the student's competence in the stated objectives at the beginning of the chapter.

Shop Manual

To stress the importance of safe work habits, the Shop Manual also dedicates one full chapter to safety. Other important features of this manual include:

Performance Objectives

These objectives define the contents of the chapter and define what the student should have learned upon completion of the chapter. These objectives also correspond with the list of required tasks for ASE certification. *Each ASE task is addressed.*

Although this textbook is not designed to simply prepare someone for the certification exams, it is organized around the ASE task list. These tasks are defined generically when the procedure is commonly followed and specifically when the procedure is unique for specific vehicle models. Imported and domestic model automobiles and light trucks are included in the procedures.

Photo Sequences

Many procedures are illustrated in detailed Photo Sequences. These detailed photographs show the students what to expect when they perform particular procedures. They also can provide a student a familiarity with a system or type of equipment, which the school may not have.

Marginal Notes

New terms are pulled out and defined. Common trade jargon also appears in the margin and gives some of the common terms used for components. This allows the reader to speak and understand the language of the trade, especially when conversing with an experienced technician.

Cautions and Warnings

Throughout the text, cautions are given to alert the reader to potentially hazardous materials or unsafe conditions. Warnings are also given to advise the student of things that can go wrong if instructions are not followed or if a nonacceptable part or tool is used.

References to the Classroom Manual

Reference to the appropriate page in the Classroom Manual is given whenever necessary. Although the chapters of the two manuals are synchronized, material covered in other chapters of the Classroom Manual may be fundamental to the topic discussed in the Shop Manual.

Customer Care

This feature highlights those little things a technician can do or say to enhance customer relations.

Tools Lists

Each chapter begins with a list of the Basic Tools needed to perform the tasks included in the chapter. Whenever a Special Tool is required to complete a task, it is listed in the margin next to the procedure.

Service Tips

Whenever a special procedure is appropriate, it is described in the text. These tips are generally those things commonly done by experienced technicians.

Case Studies

Case Studies concentrate on the ability to properly diagnose the systems. Each chapter ends with a case study in which a vehicle has a problem, and the logic used by a technician to solve the problem is explained.

Terms to Know

Terms in this list can be found in the Glossary at the end of the manual.

Diagnostic Chart

Chapters include detailed diagnostic charts linked with the appropriate ASE task. These charts list common problems and most probable causes. They also list a page reference in the Classroom Manual for better understanding of the system's operation and a page reference in the Shop Manual for details on the procedure necessary for correcting the problem.

ASE Style Review Questions

Each chapter contains ASE style review questions that reflect the performance objectives listed at the beginning of the chapter. These questions can be used to review the chapter as well as to prepare for the ASE certification exam.

Instructor's Guide

The Instructor's Guide is provided free of charge as part of the *Today's Technician Series* of automotive technology textbooks. It contains Lecture Outlines, Answers to Review Questions, Pretest and Test Bank including ASE style questions.

Classroom Manager

The complete ancillary package is designed to aid the instructor with classroom preparation and provide tools to measure student performance. For an affordable price, this comprehensive package contains:

Instructor's Guide
200 Transparency Masters
Answers to Review Questions

Lecture Outlines and Lecture Notes
Printed and Computerized Test Bank
Laboratory Worksheets and Practicals

Reviewers

The author would like to thank Brooke Mossgrove and Steve Levin, Columbus State Community College, for their contributions to this project. Thanks guys! Thanks also to the following reviewers:

Chane Bush
Southern Alberta Institute of Technology
Calgary, Alberta, CANADA

Henry Chiulli
New England Institute of Technology
Warwick, RI

Richard Fox
Montcalm Community College
Sidney, MI

Stevon D. Gregory
Oklahoma State University at Okmulgee
Morris, OK

Charles Hollar
Liberal Area VoTech
Liberal, KS

Thomas G. Jenkins
MOTECH Educational Center
Madison Heights, MI

Shelby Laborde
Avoyelles Technical Institute
Cotton Port, LA

Gene Zielinski
Indiana State College
Gary, IN

Basic Tools and Procedures

Upon completion and review of this chapter, you should be able to:

❑ List the basic units of measure for length and volume in the metric and USCS systems.

❑ Convert measurements between the US Customary and the international system of units.

❑ Identify and describe the purpose of commonly used power tools.

❑ Identify the major measuring instruments and devices used by technicians.

❑ Explain what these instruments and devices measure and how to use them.

❑ Read a vernier scale.

❑ Explain the procedure for using and measuring with a micrometer.

❑ Describe the measurements normally taken by a technician while working on an automatic transmission.

Servicing modern automotive drive train systems requires the use of various tools. Many of these tools are common hand and power tools. Other tools are very specialized and are intended only for specific repairs and procedures on specific transmissions and/or vehicles. The hand tools that should be found in a basic mechanic's tool set are identified and discussed, as are power and lifting tools and commonly used measuring instruments. Since most tools are available in a variety of sizes, a brief discussion of the common measuring systems is necessary.

Measuring Systems

Two different systems of weights and measures are currently used in the United States — the United States Customary System (USCS) and the international system (SI). The USCS units of measurement were brought here by the original English settlers. The US is slowly changing over to the metric system. During the changeover, cars produced in the US are being made with both English and metric fasteners and specifications. Most of the world outside the US uses the metric system.

The basic unit of linear measurement in the USCS is the inch. The inch is commonly divided into small fractional or decimal increments to provide a way to define the exact length of an object. The basic unit of linear measurement in the metric system is the meter. A meter is slightly longer than 36 inches or a yard. Exact measurements are based on decimal divisions of the meter, such as 10 mm, 11 mm, 12 mm, and so on. The abbreviation "mm" means millimeter, or one-thousandth of a meter.

Because some vehicles have metric fasteners, some have USCS, and others have both, automotive technicians must have both English and metric tools. Vehicle specifications are normally listed in both meters and inches; therefore, measuring tools, such as micrometers and dial indicators, can be based on either system.

Following are some common equivalents between the two systems.

1 meter (m) = 39.37 inches (in.)
1 centimeter (cm) = 0.01 m = 0.3937 in.
1 millimeter (mm) = 0.001 m = 0.03937 in.
1 inch = 2.54 cm = 0.0254 m
1 inch = 25.4 mm = 0.0254 m

United States Customary System

In the United States Customary System, linear measurements are measured by inches, feet, and yards. The inch can be broken down into fractions, such as 1/64, 1/32, 1/16, 1/8, 1/4, and 1/2 of

The USCS system is also known as the English system.

The SI system is normally called the metric system.

A meter is 1/10,000,000 of the distance from the North Pole to the Equator, or 39.37 inches.

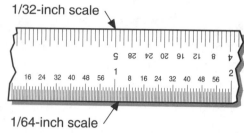

Figure 1-1 An inch is normally broken down into these fractional increments.

an inch (Figure 1-1). The inch is also commonly broken down by decimals. When an inch is divided into tenths, each part is a tenth of an inch (0.1 in.). Tenths of an inch can be further divided by ten into hundredths of an inch (0.01 in.). The division after hundredths is thousandths (0.001 in.), which is followed by ten-thousandths (0.0001 in.).

Metrics

In the metric system, the basic measurement unit of length is the meter. For exact measurements, the meter is divided into units of ten. The first division is called a decimeter (dm). The second division is the centimeter (cm) and the third and most commonly used division is the millimeter (mm). One dm is equal to 0.1 meter, 1 cm equals 0.01 m, and 1 mm equals 0.001 m.

☑ **SERVICE TIP:** A great variety of fasteners are used by the automotive industry, each one designed for a specific purpose and for specific conditions (Figure 1-2). Using an incorrect fastener or a fastener of inferior quality (Figure 1-3) can result in early failure and even injury to the driver and passengers. Some precautions to observe when replacing fasteners are:

Grade 2 (GM 200-M) Grade 5 (GM 280-M) Grade 7 (GM 290-M) Grade 8 (GM 300-M)

Customary (inch) bolts - Identification marks correspond to bolt strength - Increasing numbers represent increasing strength.

Metric Bolts - Identification class numbers correspond to bolt strength - Increasing numbers represent increasing strength.

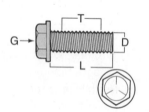
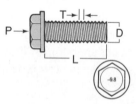

G–Grade marking (bolt strength)
L–Length, (inches)
T–Thread Pitch (thread/inch)
D–Nominal diameter (inches)

P–Property class (bolt strength)
L–Length (millimeters)
T–Thread pitch (thread width crest to crest mm)
D–Nominal diameter (millimeters)

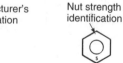

Manufacturer's identification

Nut strength identification

Posidriv screw head

Identifcation marks (4)

Figure 1-2 Proper nomenclature for English (USCS) and metric bolts (Reprinted with the permission of Ford Motor Co.)

Figure 1-3 Metric bolt and nut identification (Courtesy of General Motors Corp.)

1. Always use the same diameter, length, grade, and type of fastener as were originally used by the vehicle manufacturer.
2. Never thread a fastener of one thread type to a fastener of a different thread type.
3. Always use the same number of fasteners as were originally used by the manufacturer of the vehicle.
4. Always observe the manufacturer's recommendations for tightening sequence, tightening steps, and torque values.
5. Always use the correct washers, pins, and locks as specified by the manufacturer.
6. Always replace stretched fasteners or fasteners with damaged threads.
7. Never use a cotter pin more than once.

Common Special Tools

A cotter pin is a device used to hold a bolt in position and prevent it from loosening.

Automatic transmission technicians use a variety of tools. As you progress through the chapters of this manual, you will be introduced to the special tools required to complete a particular task. These tools will be listed in the margin, next to the procedure, and included in the procedure. At the beginning of each chapter, a list of basic tools is given in the margin.

Although every technician's tool box contains many different tools (Figure 1-4), there are certain hand tools that are a must for general work and there are other tools that are used for specific purposes. These tools are described in the following paragraphs.

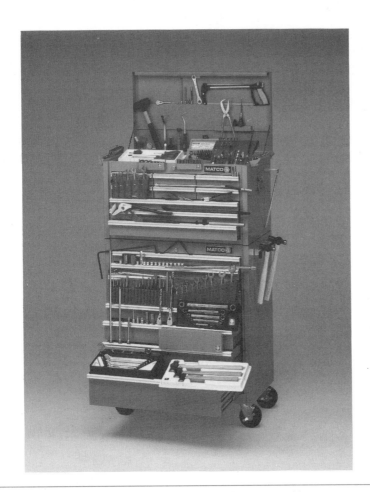

Figure 1-4 A basic mechanic's tool set (Courtesy of Matco Tools)

Figure 1-5 A six-point flare nut wrench (Courtesy of Matco Tools)

Line Wrenches

Flare nut wrenches are commonly called line or tubing wrenches.

To loosen or tighten line or tubing fittings, flare nut wrenches (Figure 1-5) should be used rather than open-end wrenches. Using an open-end wrench will tend to round the corners of the nut. These nuts are typically made of soft metal and can distort easily. Line wrenches surround the nut and provide a better grip on the fitting. These wrenches should always be used when loosening or tightening transmission cooler lines.

WARNING: Metric and USCS wrenches are not interchangeable. For example, a 9/16-inch wrench is 0.02 inch larger than a 14-millimeter nut. If the 9/16-inch wrench is used to turn or hold a 14-millimeter nut, the wrench will probably slip. This may cause the points of the bolt head or nut to round off and can possibly cause skinned knuckles.

Torque Wrenches

Classroom Manual
Chapter 1, page 3

Torque wrenches measure how tight a nut or bolt is. Many of a car's nuts and bolts should be tightened to a certain amount and have a torque specification that is expressed in foot-pounds (USCS) or Newton-meters (metric). A foot-pound is the work or pressure accomplished by a force of 1 pound through a distance of 1 foot. A Newton-meter is the work or pressure accomplished by a force of 1 kilogram through a distance of 1 meter.

Torque wrenches come with drives that correspond with sockets: 1/4, 3/8, and 1/2 inch. There are four types of torque wrenches: dial-type, breakover-type, torsion bar-type (Figure 1-6), and the digital readout type. For most drive train work, a dial, torsion bar, or digital readout type is recommended. These have a scale that can be read and can be used to measure turning effort, as well as tightening bolts. With the breakover type, you must dial in the desired torque. The wrench makes an audible click when you have reached the correct force.

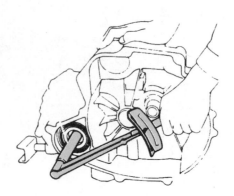

Figure 1-6 Using a torque wrench to measure turning effort (Courtesy of Nissan Motor Co., Ltd.)

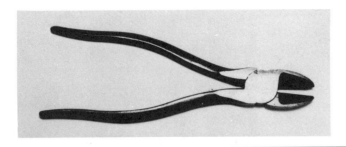

Figure 1-7 Diagonal side cutting pliers

Pliers

Two most commonly used pliers are a pair of interlocking jaw pliers that are about eight or nine inches long and a pair of diagonal cutters about seven inches long, which can be used to cut wire and remove cotter pins. Other designs may also be required while servicing a vehicle. These include: slip-joint, needle nose, duck bill, adjustable joint (channel locks) (water pump pliers), and offset needle nose pliers.

Your tool set should include a variety of pliers, including at least one pair of slip-joint pliers, two or three sizes of adjustable joint pliers, a pair of needle nose pliers, and a pair of diagonal cutters (Figure 1-7). It is also recommended that you have a pair of vise-grip pliers to hold parts while grinding or to use as a "third hand."

Punches and Chisels

Most tool sets include a variety of drift punches and starter punches. Drift punches are used to remove drift and roll pins. Some drifts are made of brass; these should be used whenever you are concerned about possible damage to the pin or surface surrounding the pin. Tapered punches are used to line up bolt holes. Starter or center punches are used to make an indent before drilling to prevent the drill bit from wandering. A variety of chisels is also recommended; these should include flat, cape, round-nose cape, and diamond point chisels.

Files

A set of files should also be included in your tool set. Files are used to remove metal and to deburr parts. Files are available in many different shapes: round, half-round, flat, crossing, knife, square, and triangular. The most commonly used files are the half-round and flat, with either single-cut or double-cut designs. A single-cut file has its cutting grooves lined up diagonally across the face of the file. The cutting grooves of a double-cut file run diagonally in both directions across the face. Double-cut files are considered first-cut or roughening files because they can remove large amounts of metal. Single-cut files are considered finishing files because they remove small amounts of metal.

Taps and Dies

Problems are often caused by defective fasteners or damaged threads in the bore of an assembly. Fasteners can be replaced or their threads restored with a die. A tap can cut and restore the threads in a bore. It is recommended that you have two sets of taps and dies: one USCS and one metric.

Special Transmission Tools

Many tools are designed for a specific purpose. An example of a special tool is a gear and bearing puller. Many gears and bearings have a slight interference fit when they are installed on a shaft or

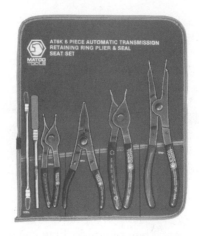

Figure 1-8 Retaining ring plier and seal tool kit for transmission service (Courtesy of Matco Tools)

in a housing. This press-fit prevents the parts from moving on each other. The removal of these gears and bearings must be done carefully to prevent damage to the gears, bearings, or shafts. Prying or hammering can break or damage the parts. A puller with the proper jaws and adapters should be used to remove gears and bearings. Using the proper puller, the force required to remove a gear or bearing can be applied with a steady motion.

Often, an automatic transmission technician will run into many different styles and sizes of retaining rings that hold subassemblies together or keep them in a fixed location. Using the correct tool to remove and install these rings is the only safe way to work with them. All automatic transmission technicians should have an assortment of retaining ring pliers (Figure 1-8).

Other commonly used special tools are the various designs of bushing and seal drivers and pullers. Pullers are either a threaded or slide hammer-type tool. Always make sure you use the correct tool for the job. Bushings and seals are easily damaged if the wrong tool or procedure is used. Car manufacturers and specialty tool companies work closely together to design and manufacture special tools required to repair cars. Most of these special tools are listed in the appropriate service manuals and are part of each manufacturer's Essential Tool Kit. All special tools required for the procedures discussed in this manual will be displayed in the margin.

Power Tools

Power tools save time during assembly or disassembly and allow the technician to use much less energy. Although many power tools will become part of your tool box, many are supplied by the shop. Power tools are either pneumatic or electric. Pneumatic tools are typically used by technicians because they have more torque, weigh less, and require less maintenance than electric power tools. However, electric power tools tend to cost less than the pneumatics. Electric power tools can be plugged into most electric wall sockets or have rechargeable batteries, but to use a pneumatic tool, you must have an air compressor and an air storage tank.

 CAUTION: Carelessness or mishandling of power tools can cause serious injury. Make sure you know how to operate a tool before using it.

Impact Wrench

An impact wrench (Figure 1-9) uses compressed air or electricity to hammer or impact a nut or bolt loose or tight. Light-duty impact wrenches are available in three drive sizes, 1/4, 3/8, and 1/2 inch, and two heavy-duty sizes, 3/4 and 1 inch.

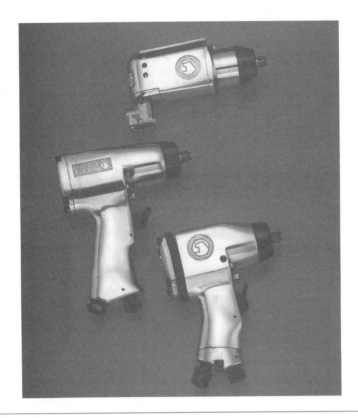

Figure 1-9 Assortment of air impact wrenches (Courtesy of Matco Tools)

WARNING: Impact wrenches should not be used to tighten critical parts or parts that may be damaged by the hammering force of the wrench.

CAUTION: The sockets designed for impact wrenches are constructed of thicker and softer steel to withstand the force of the impact. Ordinary sockets must not be used with impact wrenches. They are made of very hard metal, will crack or shatter because of the force, and can cause injury.

Air Ratchet

Air ratchets are often used during disassembly or reassembly work to save time. Because the ratchet turns the socket without an impact force, these wrenches can be used on most parts and with ordinary sockets. Air ratchets usually have a 3/8-inch drive. Air ratchets are not torque sensitive; therefore, to properly tighten a bolt, a torque wrench should be used on all fasteners after snugging them up with an air ratchet.

Lifting Tools

Lifting tools are necessary tools for most transmission repair procedures. These tools are typically provided by the shop and are not the property of a technician. Correct operating and safety procedures should always be followed when using lifting tools.

Jacks

Jacks are used to raise a vehicle off the ground and are available in two basic designs and in a variety of sizes. The most common jack is the hydraulic floor jack, which is classified by the weight it can lift: 1-1/2, 2, 2-1/2 tons, and so on. These jacks are controlled by moving the handle up and down. The other design of portable floor jack uses compressed air. Pneumatic jacks are operated by controlling air pressure at the jack.

Transmission Jacks

To reduce the risk of injury and the amount of work a technician must do while removing transmissions from a vehicle, special jacks are specifically designed to raise and lower transmissions from vehicles. These jacks are appropriately called transmission jacks (Figure 1-10). Transmission jacks allow a technician to tilt transmissions in many directions without the fear of the unit falling and without the strain of manually doing the same.

Safety Stands

Safety stands are commonly called jack stands.

When a vehicle is raised by a jack, it should be supported by safety stands (Figure 1-11). Never work under a vehicle with only a jack supporting it. Always use safety stands. Hydraulic seals in the jack can let go and allow the vehicle to drop. Service manuals note the proper locations for jacking and supporting a vehicle while it is raised from the ground. Always follow those guidelines.

Hydraulic and Electric Lifts

The hydraulic floor lift is the safest lifting tool and it is able to raise the vehicle high enough to allow you to walk and work under it. Various safety features prevent a hydraulic lift from dropping if a seal leaks or if air pressure is lost. Before lifting a vehicle, make sure the lift is correctly positioned on the vehicle's specified lifting points.

Engine Hoist

Engine hoists are often referred to as "cherry pickers."

The engine hoist allows you to lift an engine from a car. Hydraulic pressure converts power to a mechanical advantage and lifts the engine from the car. After the engine has been removed, the engine should be mounted to an engine stand and not left dangling on the engine hoist. On some vehicles, it may be necessary to lift the engine and transmission as an assembly. The two units are separated after they are removed from the vehicle.

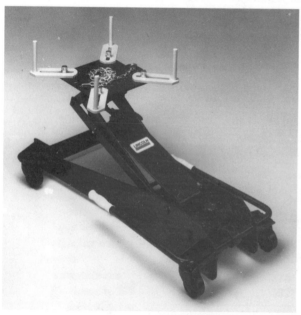

Figure 1-10 Hydraulically operated transmission jack (Courtesy of Blackhawk Automotive, Inc.)

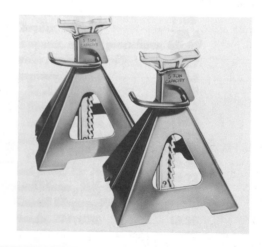

Figure 1-11 Safety stands are used to support a vehicle after it has been jacked up. (Courtesy of Lincoln, St. Louis)

Shop Manuals

Shop manuals are a necessary tool for servicing automatic transmissions. They are needed to obtain the desired specifications and for specific service procedures. Service manuals also provide drawings and photographs that show where and how to perform certain procedures on the particular car you are working on. Service manuals are also very valuable during diagnostics, because the operation of each specific transmission is outlined in the manuals. It is very difficult to troubleshoot a transmission if you don't know how it is supposed to work. Special tools or instruments are listed and shown, when they are required. Precautions are also given to prevent injury or damage to parts.

Most automobile manufacturers publish a service manual or set of manuals for each model and year of their cars. These manuals provide the best and most complete information for those cars. The most commonly used specifications and procedures are compiled in comprehensive service manuals. Various editions are available, covering different ranges of model years for both domestic and imported cars. Specialized shop manuals are also produced for special areas, such as transmissions. Although similar to manufacturers' manuals in many ways, they do not provide as much information or detail as do the manufacturers' manuals.

Although the manuals from different publishers vary in presentation and arrangement of topics, all service manuals are easy to use, after you become familiar with their organization. Most shop manuals are divided into a number of sections, each covering different aspects of the vehicle (Figure 1-12). The beginning sections commonly provide vehicle identification and basic maintenance infor-

Shop manuals are often referred to as service manuals or by the name of the publisher; i.e., Motors, Mitchells, or Chiltons.

Classroom Manual
Chapter 1, page 13

TABLE OF CONTENTS	SECTION
GENERAL INFORMATION	
General Information	0A
Maintenance and Lubrication	0B
HEATING AND AIR CONDITIONING	
Heating and Ventilation	1A
Manual Air Conditioning	1B
V-5 Compressor - Unit Repair	1D3
BUMPERS AND SHEET METAL	
Bumpers	2B
Sheet Metal	2C
STEERING, SUSPENSION, WHEELS AND TIRES	
Diagnosis	3
Wheel Alignment	3A
Power Rack and Pinion	3B1
Power Steering Pump	3B3
Steering Wheel and Columns	3B4
Front Suspension	3C
Rear Suspension	3D
Wheels and Tires	3E
FRONT DRIVE AXLE	4D
BRAKES	5
ENGINE	
Engine Diagnosis	6
General Engine Mechanical	6A
2.0L L4 Engine	6A2
2.8L V6 Engine	6A3
Engine Cooling	6B
Engine Fuel	6C
Engine Electrical	6D
Driveability & Emission	6E
Driveability & Emission - T.B.I.	6E2
Driveability & Emissions - P.F.I	6E3
Engine Exhaust	6F
TRANSAXLE	
Automatic – General Service	7A
Automatic – Service	7A1
125 Hydraulic Diagnosis	
125C – Unit Repair	
Manual 5 Speed	7B2
Manual 5 Speed	7B3
Clutch	7C
ELECTRICAL	
Electrical Diagnosis	8A
Chassis Electrical	8B
Instrument Panel and Gages	8C
Windshield Washers and Wipers	8E
ACCESSORIES	9
BODY SERVICE	10
ELECTRICAL WIRING DIAGRAMS	REAR OF BOOK

Figure 1-12 Typical index for a service manual (Courtesy of Chevrolet Motor Division, General Motors Corp.)

mation. The remaining sections deal with each different vehicle system in detail, including diagnostic, service, and overhaul procedures. Each section has an index, indicating more specific areas of information.

To use a shop manual:

1. Select the appropriate manual for the vehicle being serviced.
2. Use the table of contents to locate the section that applies to the work being done.
3. Use the index at the front of that section to locate the required information.
4. Carefully read the information and study the applicable illustrations and diagrams.
5. Follow all of the required steps and procedures given for that service operation.
6. Adhere to all of the given specifications and perform all measurement and adjustment procedures with accuracy and precision.

Throughout this book, you will be told to refer to the appropriate service manual to find the correct procedures and specifications. Although the various systems of all automobiles function in much the same way, there are many variations in design. Each design has its own set of repair and diagnostic procedures. Therefore, it is important that you always follow the recommendations of the manufacturer to identify and repair problems.

Measuring Tools

Many of the procedures discussed in this manual require exact measurements of parts and clearances. Accurate measurements require the use of precision measuring devices, which are designed to measure things in very small increments. Measuring tools are delicate instruments and should be handled with great care. Never strike, pry, drop, or force these tools. Also make sure you clean them before and after every use.

✓ **SERVICE TIP:** It's a good idea to check all measuring tools against known good equipment or tool standards. This will ensure that they are operating properly and are capable of accurate measurement.

There are many different measuring devices used by automotive technicians. This chapter will only cover those that are commonly used to service automatic transmissions.

Micrometers

A micrometer is used to measure the outside diameter of an object and the inside diameter of a bore. The procedures for calibrating and for reading an outside micrometer are the same as for an inside micrometer. The major components and markings of a micrometer include the frame, anvil, spindle, locknut, sleeve, sleeve numbers, sleeve long line, thimble marks, thimble, and ratchet. Micrometers are calibrated in either inch or metric graduations and are available in a range of sizes.

To measure small objects with an outside micrometer, open the jaws of the tool and slip the object between the spindle and the anvil. While holding the object against the anvil, turn the thimble, using your thumb and forefinger, until the spindle contacts the object. Use only enough pressure on the thimble to allow the object to just fit between the anvil and the spindle. The object should slip through with only a very slight resistance. When a satisfactory feel is reached, lock the micrometer. Since each graduation on the sleeve represents 0.025 inch, begin reading the measurement by counting the visible lines on the sleeve and multiply that number by 0.025. The graduations on the thimble assembly define the area between the lines on the sleeve; therefore, the number indicated on the thimble should be added to the measurement shown on the sleeve. This sum is the outside measurement of the object. Photo Sequence 1 guides you through the correct procedure for measuring with and reading a micrometer.

To measure larger objects, hold the frame of the micrometer and slip it over the object. Continue to slip the micrometer over the object until you feel a very light resistance, while at the same

Photo Sequence 1
Typical Procedure for Using a Micrometer

P1-1 Micrometers can be used to measure the diameter of many objects. By measuring the stem of a valve in two places, the wear of the stem can be determined.

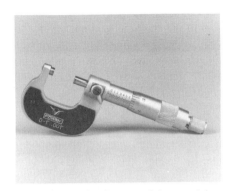

P1-2 Because the diameter of the stem is less than one inch, a 0-to-1-inch micrometer is used.

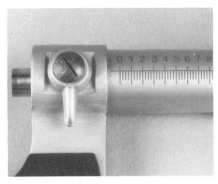

P1-3 The graduations on the sleeve each represent 0.025 inch. To read a measurement on a micrometer, begin by counting the visible lines on the sleeve by 0.025.

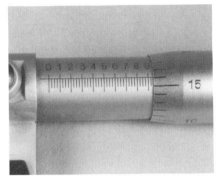

P1-4 The graduations on the thimble assembly define the area between the lines on the sleeve. The number indicated on the thimble is added to the measurement shown on the sleeve.

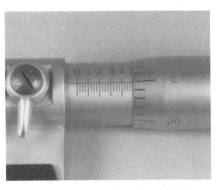

P1-5 Micrometer reading of 0.500 inch.

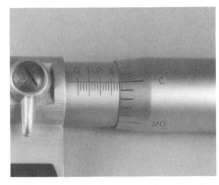

P1-6 Micrometer reading of 0.375 inch.

P1-7 Normally, little stem wear is evident directly below the keeper grooves. To measure the diameter of the stem at that point, close the micrometer around the stem.

P1-8 To get an accurate reading, slowly close the micrometer until a slight drag is felt while passing the valve in and out of the micrometer.

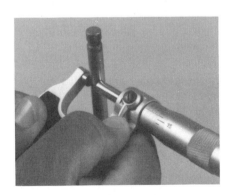

P1-9 To prevent the reading from changing while you move the micrometer away from the stem, use your thumb to activate the lock lever.

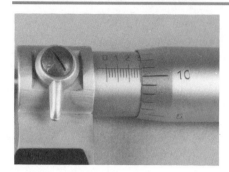

P1-10 This reading (0.311 inch) represents the diameter of the valve stem at the top of the wear area.

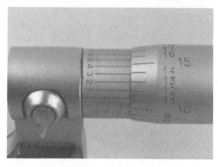

P1-11 Some micrometers are able to measure in 0.0001 inch. Use this type of micrometer if the specifications call for this type of accuracy. Note that the exact diameter of the valve stem is 0.3112 inch.

P1-12 Most valve stem wear occurs above the valve head. The diameter here should also be measured. The difference between the diameter of the valve stem just below the keepers and just above the valve head represents the amount of valve stem wear.

time rocking the tool from side to side to make certain the spindle cannot be closed any further. Then, lock the micrometer and take a measurement reading.

Inside micrometers also have a micrometer body, thimble, sleeve, and ratchet. They have interchangeable spindles of various lengths to use in holes of different sizes. To use this type micrometer, select a spindle that will just fit into the bore when it is threaded to the thimble and sleeve. Then, extend the thimble until the spindle touches one side of the bore and the thimble end touches the other side. Read the measurement in the same way as you would read an outside micrometer. It is good practice to take measurements at two or three different locations as a check for bore out-of-roundness and taper.

> ☑ **SERVICE TIP:** Like all tools, measuring tools should only be used for the purpose for which they were designed. Some instruments are not accurate enough for very precise measurements, others are too accurate to be practical for less critical measurements.

Bore Gauges

Bore gauges are commonly called dial bore gauges.

Bore gauges are used to measure bore taper and bore out-of-roundness. These gauges are equipped with various lengths of spindles which must be selected for the bore being measured, in the same way as an inside micrometer. After the gauge has been inserted into the bore, any variations in diameter will be displayed on the gauge's indicator as the gauge is moved within the bore.

To check the taper of a bore, insert the gauge and zero the indicator. Then slide the gauge down the bore. The taper of the bore will show on the indicator. To check for out-of-roundness, insert the gauge into the bore. Set the indicator to zero, then turn the gauge in the bore. The indicator will show the amount of out-of-roundness.

Telescoping Gauges

Telescoping gauges are sometimes called span gauges.

Small hole gauges are sometimes called ball gauges.

Telescoping gauges are used to measure the size of a hole. However, these gauges do not actually measure, they are used to transfer the size of the bore to another measuring tool. Telescoping gauges are inserted into a bore and allowed to expand or telescope. When the gauge is setting level in the bore, the gauge is locked. The gauge is then removed and its length measured with an outside micrometer. This measurement represents the diameter of the bore.

Small-hole gauges work in the same way as telescoping gauges, except they are used on smaller holes. The gauge is inserted into the bore and expanded by turning the knob on the end of the gauge until the gauge makes contact with the sides of the bore. The gauge is then removed and measured with an outside micrometer (Figure 1-13).

Calipers and Dividers

Calipers and dividers also transfer lengths to another measuring instrument. These are used to measure inside and outside diameters. To take a measurement with these tools, adjust the legs of the tool to fit the inside or outside diameter of the object. Then lay the caliper or divider over a rule and read the length on the rule.

A vernier caliper is a measuring tool. It can make inside, outside, or depth measurements. It is marked in both USCS and metric divisions called a vernier scale. A vernier scale consists of a stationary scale and a movable scale, in this case the vernier bar to the vernier plate. The length is read from the vernier scale.

Dial Indicator

A dial indicator is used to measure variations or movements of an object. Dial indicators are normally calibrated in 0.001-inch increments. Metric indicators are also available and measure in 0.01-mm increments. Common uses of a dial indicator include gear backlash, runout, and endplay (Figure 1-14). A dial indicator uses a spring-loaded rod that contacts the part being measured. Once the rod is in contact and the indicator is locked into position, the dial gauge is set to zero. All changes or movements will show on the indicator.

Feeler Gauge

A feeler gauge is a thin strip of metal of a known and closely controlled thickness. A feeler gauge set is a collection of several of these metal strips, each of which has a different thickness. A steel feeler gauge set usually contains strips of 0.002- to 0.010-inch thicknesses (in steps of 0.001 inch) and strips

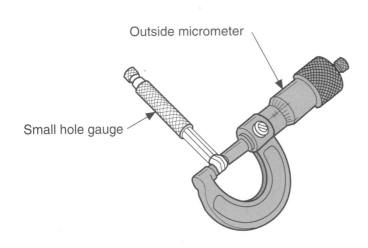

Outside micrometer

Small hole gauge

Figure 1-13 Measuring a ball gauge with a micrometer to determine the diameter of a small bore

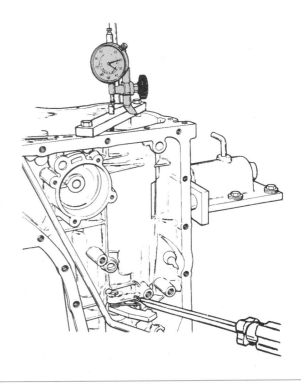

Figure 1-14 A dial indicator being used to measure shaft movement (Courtesy of Chevrolet Motor Division, General Motors Corp.)

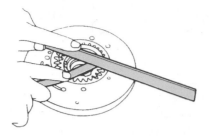

Figure 1-15 Using a feeler gauge and straight edge to check the flatness of a transmission oil pump (Courtesy of Nissan Motor Co., Ltd.)

of 0.012- to 0.024-inch thicknesses (in steps of 0.002). A feeler gauge can be used by itself to measure side clearances, endplay, and other similar procedures. A feeler gauge set can also be used with a straight edge to measure bore alignments and the flatness of a surface (Figure 1-15).

Basic Gear Adjustments

While a transmission is in operation, the gears, shafts, and bearings are subjected to loads and vibrations. Because of this, the drive train must normally be adjusted for the proper fit between parts. These adjustments require the use of precision measuring tools. There are three basic adjustments that are made when reassembling a unit or when a problem suggests that readjustment is necessary. Adjusting the clearance or play between two gears in mesh is referred to as adjusting the backlash. Endplay adjustments limit the amount of end-to-end movement of a shaft. Preload is an adjustment made to put a load on an assembly to offset the loads the assembly will face during operation.

Special Tools
Dial indicator
Dial indicator mount

Backlash in Gears

Backlash (Figure 1-16) is the clearance between two gears in mesh. Excessive backlash can be caused by worn gear teeth, improper meshing of teeth, or bearings that do not support the gears properly. Excessive backlash can result in severe impact on the gear teeth from sudden stops or directional changes of the gears, which can cause broken gear teeth and gears. Insufficient backlash causes excessive overload wear on the gear teeth and could cause premature gear failure.

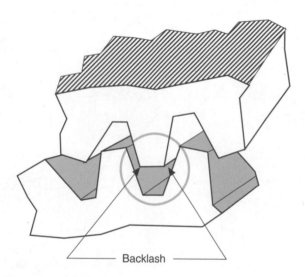

Backlash

Figure 1-16 Backlash is the clearance between two gears in mesh.

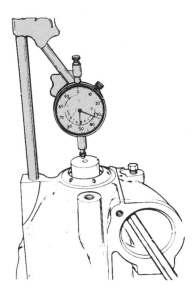

Figure 1-17 Measuring shaft endplay with a dial indicator (Courtesy of Chevrolet Motor Division, General Motors Corp.)

Backlash is measured with a dial indicator mounted so that its stem is in line with the rotation of the gear and perpendicular to the angle of the teeth. The gear is moved in both directions while the other gear it meshes with is held. The amount of movement on the dial indicator equals the amount of backlash present. The proper placement of shims on a gear shaft is the normal procedure for making backlash adjustments.

Endplay in Gears and Shafts

Endplay refers to the measurable axial or end-to-end looseness of a bearing. Endplay is always measured in an unloaded condition. To check endplay, a dial indicator is mounted against the outside of a gear or the end of a shaft (Figure 1-17). The gear or shaft is then pried in both directions and the readings noted. The difference between the two readings is the amount of endplay. Shims or adjusting nuts are widely used to adjust endplay.

Classroom Manual
Chapter 1, page 7

Preloading of Gear Trains

When normal operating loads are great, gear trains are often preloaded to reduce the deflection of parts. The amount of preload is specified in service manuals and must be correct for the design of the bearings and the strength of the parts. If bearings are excessively preloaded, they will heat up and fail. When bearings are set too loose, the shaft will wear rapidly due to the great amounts of deflection it will experience. Gear trains are preloaded by shims, thrust washers, adjusting nuts, or by using double-race bearings. Preload adjustments are normally checked by measuring turning effort with a torque wrench.

Summary

❑ The USCS measuring system is based on the inch; fine measurements are expressed in fractions of an inch or decimals.

❑ The metric system is based on the meter; fine measurements are expressed in tenths, hundredths, and thousandths.

❑ One millimeter is equal to 0.001 meter or 0.03937 inch.

❏ Whenever bolts are replaced, they should be replaced with exactly the same size, grade, and type as was installed by the manufacturer.

❏ Torque wrenches are used to tighten fasteners to a specified torque.

❏ Basic tool sets should include a variety of standard and Phillips screwdrivers, as well as an assortment of Torx-type screwdrivers.

❏ Many special tools are available for special purposes. These tools are not normally part of a basic tool set but are purchased on an as-need basis.

❏ Special care should be taken whenever using power tools such as impact wrenches, gear pullers, and jacks.

❏ One of the most important tools for a technician is a service manual.

❏ Micrometers are used to measure the diameter of an object. Outside micrometers measure the outside diameter, while inside micrometers measure the inside diameter.

❏ Dial indicators are used to measure movement and are commonly used to measure the backlash and endplay of a set of gears.

❏ Gear backlash is a statement of how tightly the teeth of two gears mesh.

❏ Endplay refers to the measurable axial or end-to-end looseness of a shaft or bearing.

Terms to Know

Backlash	Meter	Safety stands
Ball gauge	Micrometer	Telescoping gauge
Bore gauge	Millimeter	Torque wrench
Cotter pin	Pneumatic tools	USCS
Dial indicator	Preload	Vernier caliper
Endplay	Press-fit	

ASE Style Review Questions

1. While discussing measuring systems:
   Technician A says the USCS measuring system is based on the inch and fine measurements are expressed in fractions of an inch or decimals.
   Technician B says the metric system is based on the meter and fine measurements are expressed in tenths, hundredths, and thousandths.
   Who is correct?
   A. A only   **C.** Both A and B
   B. B only   **D.** Neither A nor B

2. While discussing the metric system:
   Technician A says one millimeter is equal to 0.001 meter.
   Technician B says one meter is equal to 0.03937 inch.
   Who is correct?
   A. A only   **C.** Both A and B
   B. B only   **D.** Neither A nor B

3. While discussing automotive fasteners:
   Technician A says bolt sizes are listed by their appropriate wrench size.
   Technician B says whenever bolts are replaced, they should be replaced with exactly the same size and type as was installed by the manufacturer.
   Who is correct?
   A. A only   **C.** Both A and B
   B. B only   **D.** Neither A nor B

4. While discussing the purpose of a torque wrench:
   Technician A says they are used to tighten fasteners to a specified torque.
   Technician B says they are used for added leverage while loosening or tightening a bolt.
   Who is correct?
   A. A only   **C.** Both A and B
   B. B only   **D.** Neither A nor B

5. While discussing shaft endplay:
   Technician A says shaft endplay is measured with a dial indicator.
   Technician B says shaft endplay can be corrected by the addition or subtraction of shims.
   Who is correct?
   A. A only   **C.** Both A and B
   B. B only   **D.** Neither A nor B

6. While discussing the purpose of micrometers:
   Technician A says micrometers are used to measure the diameter of an object.
   Technician B says outside micrometers are used to measure the outside diameter of an object, while inside micrometers are used to measure the inside diameter.
   Who is correct?
   A. A only   **C.** Both A and B
   B. B only   **D.** Neither A nor B

7. While discussing dial indicators:
   Technician A says dial indicators are commonly used to measure the backlash of a set of gears.
   Technician B says dial indicators are commonly used to measure the endplay of a shaft.
   Who is correct?
   A. A only   **C.** Both A and B
   B. B only   **D.** Neither A nor B

8. While discussing gear backlash:
   Technician A says that gear backlash is a statement of how much radial movement a gear shaft has.
   Technician B says gear backlash is a statement of how tightly the teeth of two gears mesh.
   Who is correct?
   A. A only   **C.** Both A and B
   B. B only   **D.** Neither A nor B

9. While discussing the uses of pliers:
   Technician A says pliers are commonly used to hold items.
   Technician B says pliers should never be used to tighten or loosen nuts and bolts.
   Who is correct?
   A. A only   **C.** Both A and B
   B. B only   **D.** Neither A nor B

10. While discussing technicians' tools:
    Technician A says one of the most important tools for a technician is a service manual.
    Technician B says technicians must constantly update their tools in order to work on newer vehicles.
    Who is correct?
    A. A only   **C.** Both A and B
    B. B only   **D.** Neither A nor B

Safety

Upon completion and review of this chapter, you should be able to:

❏ Discuss how to ensure a safe work environment in a shop.

❏ Properly lift heavy objects.

❏ Extinguish the common variety of fires.

❏ Inspect and use tools safely.

❏ Properly work around batteries.

❏ Discuss basic safety rules and describe how common sense dictates these rules.

Safe Work Practices

Classroom Manual
Chapter 2, page 26

Exhaust contains an odorless, colorless, and deadly gas called carbon monoxide. This poisonous gas gives very little warning to the victim and can kill in just a few minutes.

R-12 is the type of refrigerant used on most cars; it is commonly referred to as freon. Newer vehicles are using R-134A refrigerant.

There are times that a technician must work around a running engine, although this should be avoided whenever possible. Certain precautions should always be followed when it is necessary to work around a running engine. When the engine is running, make sure the parking brakes are applied and use wheel chocks at the wheels to reduce the chances of the vehicle slipping into gear and running over you or someone else. Also, as a precaution, stand to the side of the vehicle when it is running. By standing in front, you are a good target for the moving vehicle.

Always keep your hands and clothing well out of the way around moving engine parts. Be especially careful around electric cooling fans as they may come on without notice and shear a finger or hand right off.

An engine should never be run in a shop without properly ventilating the exhaust fumes. Exhaust gases may contain large amounts of carbon monoxide, so always vent the exhaust outside or connect the exhaust pipes to a ventilating system (Figure 2-1).

Automotive air conditioning systems are filled with R-12 or another type of refrigerant. These refrigerants will cool anything around them as they escape from the system. They will freeze your eyes, hands, or any part of you. To prevent frostbite, always be careful when working around air conditioning lines and components, and avoid contact with leaking freon. All leaks should be immediately repaired or the refrigerant should be removed from the system with a refrigerant reclaimer. This will avoid personal injury and damage to the environment. If freon is burned by the engine or by contact with an open flame or extreme heat, phosgene gas is formed.

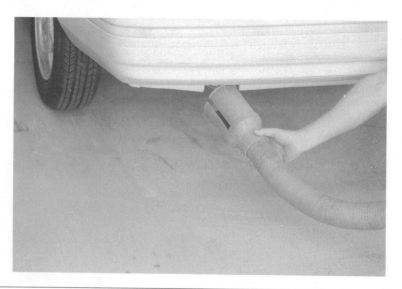

Figure 2-1 When running an engine in the shop, always connect the car's exhaust to the shop's exhaust ventilation system.

SERVICE TIP: R-12 is being phased out and will not be used as the refrigerant in automotive air conditioning systems. A new refrigerant, R-134A, is being introduced. R-134A is less harmful to the atmosphere, but it requires the same safe-handling practices as R-12.

CAUTION: Phosgene gas is poisonous and will make you sick or fatally ill.

WARNING: R-12 should never be released to the atmosphere, it should always be captured and reclaimed by special equipment whenever an A/C system is opened. It has been determined that R-12 has an adverse affect on the earth's ozone layer.

Lifting Heavy Objects

When lifting a heavy object like a transmission, use a hoist or have someone else help you. If you must work alone, *always* lift heavy objects with your legs, not your back. Bend down with your legs, not your back, and securely hold the object you are lifting. Then stand up, keeping the object close to you (Figure 2-2). Trying to "muscle" something with your arms or back can result in severe damage to your back and may end your career and limit what you do the rest of your life!

Before lifting a heavy object, position yourself at the angle the object should be lifted at and with your feet close to it. Then bend your knees, with your back straight, and lower yourself enough to firmly grab the object. Begin to lift by straightening your knees and keeping your elbows as straight as possible and the object close to your body. Continue lifting by straightening your legs and keeping your back straight. To place the object onto a workbench, turn your entire body. Never twist at the hips. Lower the object to the bench by bending your knees and place one end on the bench. While keeping your back straight, slide the rest of the object onto the bench.

Safe Work Areas

Familiarize yourself with the layout of the shop. Find out where fire extinguishers, first aid kits, eyewash stations (Figure 2-3), and other safety items are in the shop. Take time to read the oper-

Classroom Manual
Chapter 2, page 26

Figure 2-2 Begin to lift a heavy object by keeping it close to you and by straightening your knees and keeping your arms as straight as possible.

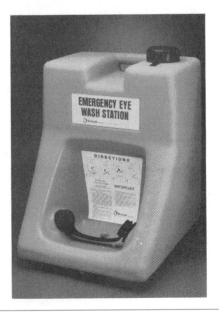

Figure 2-3 A typical emergency eyewash station (Courtesy of Brodhead-Garrett)

ating instructions for the fire extinguishers and the type of fire they are meant to be used on. Be familiar with the instructions given on the eyewash stations. Make sure that you know the route to the exit in case of a fire. Find out whether there are certain stalls that are reserved for special jobs. Pay attention to all of the warning signs around the shop. No smoking signs, special instructions for shop tools and equipment, danger zones, and so on are all there to help the shop run smoothly and safely. Always follow the shop's safety rules.

Housekeeping

Classroom Manual
Chapter 2, page 27

There are two areas of housekeeping that you are responsible for: your work area and the rest of the shop. A clean, organized work area will help you to be a better technician. If the work area is clean and organized, chances are that your work will be the same. Housekeeping within the shop is a safety consideration. A cluttered shop is a dangerous one.

When you raise a vehicle up with a jack, set it down on safety stands and remove the jack. Don't leave a jack handle sticking out for someone to trip over. When you are not using your creeper, shove it back under the vehicle or prop it against a wall. Don't leave it out so someone can step on it. Don't leave air hoses in walkways. Make sure exits aren't blocked with equipment. Keep them clear at all times. Keep the shop floor and workbenches clean and tidy. If you see something out of place, pick it up and put it away.

If you spill oil or drop grease on the floor, wipe it up. Oil on the floor can cause serious accidents and injuries. Automatic transmission fluid is especially slippery, so clean up all spills immediately. Never initially clean oil spills by hosing the floor down and allowing the oil and other materials to enter the floor drain. Oil, like many other materials found in an automotive shop, should be disposed of properly. Dirty and oily rags should be stored in a closed metal container to avoid catching fire.

Do not operate shop tools or equipment that are in an unsafe condition. Electrical cords and connectors must be in good condition. Extension cords should not pose a hazard by being strung across walkways. Never drive cars over electrical cords; this could cause a short circuit. Also, never overload an electrical circuit.

Bench grinding wheels and wire brushes should be replaced if they are defective. The bench grinding machine should be equipped with tool stands and protective shields. If they are not equipped this way, *don't* use them. Floor jacks and hoists must be in safe operating condition and should not be used above their rated capacity. The same applies to mechanical and hydraulic presses, drills, and drill presses. Bring all unsafe conditions to the attention of your instructor or shop foreman.

Fire Hazards and Prevention

Classroom Manual
Chapter 2, page 28

In case of a fire, you should know the location of the fire extinguishers and fire alarms in the shop and you should know how to use them. You should also be aware of the different types of fires and the fire extinguishers used to put out these types of fires.

Basically, there are four types of fires: Class A fires are those in which wood, paper, and other ordinary materials are burning; Class B fires are those involving flammable liquids, such as gasoline, diesel fuel, paint, grease, oil, and other similar liquids; and Class C fires are electrical fires. Class D fires are a unique type of fire, for the material burning is a metal. An example of a Class D fire is a burning "mag" wheel. The magnesium used in the construction of the wheel is a flammable metal and will burn brightly when subjected to high heat.

Fire Extinguishers

Classroom Manual
Chapter 2, page 29

Using the wrong type of extinguisher may cause the fire to grow, instead of being put out. All extinguishers are marked with a symbol or letter to signify what class of fire they were intended for (see

Table 2-1). Fire extinguishers that use pressurized water should only be used on Class A fires. Soda-acid extinguishers are also used to put out Class A fires. Foam-type extinguishers are used on Class A and B fires; the foam should be sprayed directly on the fire. Those extinguishers filled with a dry chemical should be aimed at the base of a Class B or C fire, so as to suffocate the fire. Some extinguishers that use a multipurpose dry chemical are also suitable for three classes of fires — Classes A, B, and C. Carbon dioxide is also used in some fire extinguishers that are designed to put out Class B and C fires. Halon gas extinguishers can also be used on Class B and C fires. Class D fires require the use of a special dry powder-type extinguisher. Some extinguishers are designed to fight more than one class of fire and have multiple ratings such as A & B, B & C, or A, B, C, & D. You should locate each fire extinguisher in the shop and know its ratings before you need one. You need to be able to react immediately when a fire starts.

Table 2-1 GUIDE TO EXTINGUISHER SELECTION

	Class of Fire	Typical Fuel Involved	Type of Extinguisher
Class **A** Fires (green)	**For Ordinary Combustibles** Put out a Class A fire by lowering its temperature or by coating the burning combustibles.	Wood Paper Cloth Rubber Plastics Rubbish Upholstery	Water[*1] Foam[*] Multipurpose dry chemical[4]
Class **B** Fires (red)	**For Flammable Liquids** Put out a Class B fire by smoldering it. Use an extinguisher that gives a blanketed flame-interrupting effect; cover whole flaming liquid surface.	Gasoline Oil Grease Paint Lighter fluid	Foam[*] Carbon dioxide[5] Halogenated agent[6] Standard dry chemical[3] Purple K dry chemical[3] Multipurpose dry chemical[4]
Class **C** Fires (blue)	**For Electrical Equipment** Put out a Class C fire by shutting off power as quickly as possible and by always using a nonconducting extinguishing agent to prevent electric shock.	Motors Appliances Wiring Fuse boxes Switchboards	Carbon dioxide[5] Halogenated agent[6] Standard dry chemical[2] Purple K dry chemical[3] Multipurpose dry chemical[4]
Class **D** Fires (yellow)	**For Combustibles Metals** Put out a Class D fire of metal chips, turnings, or shavings by smothering or coating with a specially designed extinguishing agent.	Aluminum Magnesium Potassium Sodium Titanium Zirconium	Dry powder extinguishers and agents only

*Cartridge-operated water, foam, and soda-acid types of extinguishers are no longer manufactured. These extinguishers should be removed from service when they become due for their next hydrostatic pressure test.

Notes:

(1) Freeze in low temperatures unlesstreated with antifreeze solution, usually weighs over 20 pounds, and is heavier than any other extinguisher mentioned.

(2) Also called ordinary or regular dry chemical. (solution bicarbonate)

(3) Has the greatest initial fire-stopping power of the extinguishers mentioned for class B fires. Be sure to clean residue immediately after using the extinguisher so sprayed surfaces will not be damaged. (potassium bicarbonate)

(4) The only extinguishers that fight A, B, and C classes of fires. However, they should not be used on fires in liquefied fat or oil of appreciable depth. Be sure to clean residue immediately after using the extinguisher so sprayed surfaces will not be damaged. (ammonium phosphates)

(5) Use with caution in unventilated, confined spaces.

(6) May cause injury to the operator if the extinguisher agent (a gas) or the gases produced when the agent is applied to a fire is inhaled.

If there is not a fire extinguisher handy, a blanket or fender cover may be used to smother the flames. You must be careful when doing this because the heat of the fire may burn you and the blanket. If the fire is too great to smother, move everyone away from the fire and call the local fire department. A simple under-the-hood fire can cause the total destruction of the car and the building and can take some lives. You must be able to respond quickly and precisely to avoid a disaster.

Hand Tool Safety

Classroom Manual
Chapter 2, page 28

Hand tools should always be kept clean and should be used only for the purpose for which they were designed. Use the right tool for the right job. If you use the wrong tool, you will damage the part, the tool, or yourself. Keep your hand tools clean. Oily tools can slip out of your hand while you are using them, causing broken fingers or at least cut or skinned knuckles. Your tools should also be inspected for cracks, broken parts, or other dangerous conditions before you use them. The following list includes examples of what to do when using certain tools:

1. Always pull on a wrench handle.
2. Always use the correct size of wrench.
3. Use a box-end or socket wrench whenever possible.
4. Use an adjustable wrench only when it is absolutely necessary. Pull the wrench so that the force of the pull is on the nonadjustable jaw.
5. When using an air impact wrench, always use impact sockets.
6. Never use wrenches or sockets that have cracks or breaks.
7. Never use a wrench or pliers as a hammer.
8. Never use pliers to loosen or tighten a nut. Use the correct wrench.
9. Always be sure to strike an object with the full face of the hammer head.
10. Always wear safety glasses when using a hammer and/or chisel.
11. Never strike two hammer heads together.
12. Never use screwdrivers as chisels.

If the tool that you are using is pointed or sharp, always aim it away from you. Knives, chisels, and scrapers must be used in a motion that will keep the point or blade moving away from your body. When you are handing a pointed or sharp tool to someone else, always give the handle of the tool to the other person, keeping the point or sharp end away from them.

When using power tools, be sure that the equipment, parts, and other people are away from the tool. All electric power tools must be grounded. Always wear safety glasses when using power tools. Check all hose connections when using tools powered by compressed air.

When using compressed air, be very careful and never play or clown with it. Never aim an air nozzle at someone or yourself. Pressurized air can rupture an eardrum, cause blindness, destroy lungs, or cause skin damage. The dirt and metal particles that can get blown around by the air can cause other problems. When using compressed air, wear safety glasses and be extremely careful.

Equipment Safety

Defective and misused equipment is one of the most common hazards in a shop. Defective electrical cords, ungrounded or unprotected equipment, and improper use are some of the common causes of accidents in a shop. Never operate a machine or piece of equipment unless you have been instructed on the use of the machine, have inspected the machine for potential safety hazards, and are wearing the appropriate safety gear.

Many shop procedures require that you raise the car off the floor. Always make sure that the vehicle is properly placed on the lift before raising it and always use the safety locks to prevent the lift from coming down. A floor jack is often used to raise a vehicle. When positioning a floor jack

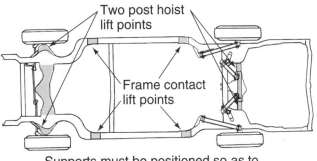

Supports must be positioned so as to distribute load and support car in a stable manner.

Figure 2-4 Recommended hoist lift points for a typical car (Courtesy of Oldsmobile Division, General Motors Corp.)

under a vehicle, make sure it is under a frame member or a strong part of the vehicle. In many unibody vehicles, proper lift points are very critical.

 WARNING: Check your shop manual before lifting a vehicle. Failure to use the correct lifting points is not only potentially dangerous but may damage the sheet metal of the vehicle. It is wise to refer to a shop manual before lifting any vehicle to identify the proper locations for lifting (Figure 2-4).

SERVICE TIP: To ensure safe operation of a lift, here is a list of suggestions to follow:

1. Inspect each lift daily.
2. Never operate a lift if it doesn't work right or if it has broken or damaged parts. Repairs should be made with exact replacement parts.
3. Never overload your lift. The rated capacity of the lift is noted on the manufacturer's nameplate. Never exceed that rating.
4. Always make sure the vehicle is properly positioned before raising the lift.
5. Never raise a vehicle with someone in it.
6. Always keep the lift area clean and free of hazards and obstructions.
7. Before moving a vehicle over the lift, position the arms and supports to allow for free movement of the vehicle over the lift. Never drive over or hit the lift arms, adapters, or supports. This may damage the lift or the vehicle.
8. Carefully load the vehicle onto the lift and align the lift arms and contact pads with the specified lift points on the vehicle. Raise the lift until it barely supports the vehicle, then check the contact area of the lift.
9. Always lock the lift into position while working under it.
10. Before lowering the lift, make sure all tools and other equipment are removed from under the vehicle. Also make sure no one is standing under or near the vehicle as it descends.

Battery Safety

When possible, you should disconnect the battery of a car before you disconnect any electrical wire or component. This prevents the possibility of a fire or electrical shock. It also eliminates the possibility of an accidental short, which can ruin the car's electrical system. This is especially true of newer cars that are equipped with many electronic and computerized controls. Any electrical arcing can cause damage to the components.

Classroom Manual
Chapter 2, page 30

To properly disconnect the battery, you should disconnect the negative or ground cable first, then disconnect the positive cable. Since electrical circuits require a ground to be complete, by removing the ground cable you eliminate the possibility of a circuit accidentally becoming completed. When reconnecting the battery, connect the positive cable first, then the negative.

CAUTION: Never smoke or cause sparks around a battery. The slightest increase in heat can cause the battery to explode.

CUSTOMER CARE: To keep your customer happy, always record the stations set on the radio before disconnecting the battery. The absence of power will remove the stations from the radio's memory. Also be sure to reset those stations and the clock after the battery has been reconnected.

Jump Starting

To avoid hurting yourself or the car's electrical system, carefully follow these steps when using jumper cables (Figure 2-5) or a booster battery to help start an engine. If the battery is a sealed battery equipped with a charge indicator lamp, do not follow this procedure if the battery lamp indicates a need for replacing the battery.

1. Remove the vent caps of both batteries. (If the battery is a sealed type, do not attempt to pry off the top.)
2. Cover the vent holes with a wet rag or cloth.
3. Make sure that both cars are not touching each other.
4. Turn off all electrical accessories on both cars.
5. Connect the positive sides of both batteries with the positive booster cable.
6. Connect the negative booster cable to the negative post of the booster battery.
7. Connect the other end of the booster cable to a known good ground on the car being "jumped."
8. Start the engine of the booster car.
9. Start the engine of the dead car.
10. First disconnect the negative booster cable of the jumped car, then disconnect the cable from the booster battery.
11. Disconnect the positive cable.

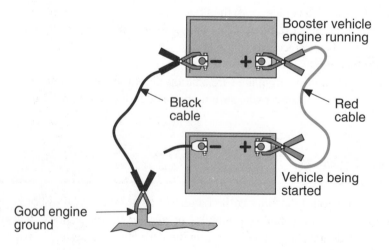

Figure 2-5 Proper set-up and connections for jump starting a vehicle with a low battery

NOTICE

CONTENTS SENSITIVE
TO
STATIC ELECTRICITY

HANDLE IN ACCORDANCE WITH STATIC CONTROL
PROCEDURES GM9107P AND GM9108P,
OR GM DIVISIONAL SERVICE MANUALS.

Figure 2-6 Some replacement parts have this packaging label and require special handling. (Courtesy of Buick Motor Division, General Motors Corp.)

Electrical System Repairs

When replacing electrical parts or repairing connectors and wires, always disconnect the power from the circuit being worked on. The best way to turn off the power is to disconnect the negative cable from the battery. Not only will this practice keep you safe, it will also help prevent damage to the electrical, especially the electronic, systems of the vehicle.

Some electronic replacement parts are very sensitive to static electricity. These parts will be labeled as such. An example of the label is shown (Figure 2-6). Whenever you are handling a part that is sensitive to static, you should follow these guidelines to reduce any possible electrostatic charge buildup on your body and the electronic part:

1. Don't open the package until it is time to install the component.
2. Before removing the part from the package, ground the package to a known good ground on the car.
3. Always touch a known good ground before handling the part. This should be repeated while handling the part and more frequently after sliding across the seat, sitting down from a standing position, or walking a distance.
4. Never touch the electrical terminals of the component.

Static is a form of electricity caused by friction.

Accidents

In case of an accident, you should be familiar with the shop's procedures for dealing with an emergency. You should also know the location of the nearest phone and the procedure for calling outside. Also make sure you are aware of the location and contents of the shop's first-aid kit. There should be an eyewash station in the shop so that you can rinse your eyes thoroughly should you get acid or some other irritant into them. Find out if there is a resident nurse in the shop or at the school, and find out where the nurse's office is located. If there are specific first-aid rules in your school or shop, make sure you are aware of them and follow them.

Some first-aid rules apply to all circumstances and are normally included in everyone's rules. If someone is overcome by carbon monoxide, get him or her fresh air immediately. Burns should be cooled immediately by rinsing them with water. Whenever there is severe bleeding from a wound, try to stop the bleeding by applying pressure with clean gauze on or around the wound, and get medical help. Never move someone who may have broken bones unless the person's life is otherwise endangered. Moving that person may cause additional injury. Call for medical assistance.

Classroom Manual
Chapter 2, page 31

Common Sense Safety Rules

There are many things that you should or should not do while working on an engine or a car. Most of these are based on common sense. Following are some common sense safety rules:

1. Think about what you are doing at all times.
2. Double-check all your work to ensure that nothing is undone.
3. Keep your tools and equipment clean and in their proper place when you are not using them.
4. Check all equipment for possible safety hazards before using it.
5. Wipe off all excessive grease from your hands to prevent your tools from slipping out.
6. Never play around in the shop.
7. Always set the parking brake of the vehicle you are working on when the engine is running.
8. Never run an engine in a shop without proper ventilation and adequate means of getting rid of exhaust gases.
9. Never wear jewelry while working on a car.
10. Wear the correct clothing for the trade and make sure that no part of it dangles freely.
11. Protect your feet with appropriate safety shoes or boots.
12. Whenever there is a possibility of dirt or metal getting in your eyes, wear safety glasses.
13. Wear safety glasses whenever you are working with chemicals.
14. Individuals who wear prescription glasses should use glasses with safety lenses and side shields.
15. Use protective clothing and equipment where needed.
16. Never use compressed air to blow dirt off you or your clothes.
17. Never point compressed air at anyone.
18. Do not use compressed air to blow away dirt from brake parts since cancer-causing asbestos dust may be inhaled as a result.
19. Never carry screwdrivers, punches, or other sharp objects in your pockets. You could injure yourself or damage the car you are working on.
20. Immediately clean up any spills.
21. Keep tools and equipment out of normal walk paths.
22. Always use the correct tool in the correct way.
23. Keep hands, tools, and clothing away from a rotating engine cooling fan.
24. Be careful around electric cooling fans; they may turn on at any time even when the engine is off.
25. Do not stand in the plane of rotating parts, such as fans and grinding wheels.
26. When disconnecting electrical components or wires, disconnect the negative cable of the battery first.
27. Use rubber gloves and an apron as well as a face mask for handling batteries.
28. Never stand directly in line with a cooling fan, as the blades may come loose and cause injury.
29. Always use jack stands to support a car that has been raised with a jack.
30. Keep jack handles out of the way, so that others cannot accidentally bump into them.
31. Use a creeper when working under a vehicle.
32. Always support the car with properly placed jack stands before working under it.
33. When placing a floor jack under the vehicle, make sure it is under a frame member or strong part, such as a rear axle housing.
34. Take extra care when moving a vehicle in and out of the shop.
35. Never pour gasoline into an engine in an attempt to get it started.
36. Never continue to work in gas-soaked clothing.
37. Know the location of the shop's fire extinguisher and eye flush containers.
38. Report all safety hazards to the person responsible for that area.

39. Use wheel chocks to block the wheels of the car while you are working on it and it is running.

40. Do not use drugs or alcohol before or during your work.

41. Do not work when you are overtired or depressed. These conditions can reduce your concentration and get you or someone else hurt.

42. If you are unfamiliar with the operation of any equipment in the shop, ask a knowledgeable person to demonstrate its proper use to you before you attempt to use it.

These basic rules should become a part of your life. Following these rules can extend your life or career, as well as the lives and property of others. Working on cars can be hazardous, but it doesn't have to be. You can prevent accidents by simply taking the time to think and to be careful.

Summary

❏ It is everyone's job to periodically check for safety hazards in a shop.

❏ Carbon monoxide is a deadly gas that is emitted by engines. Exhaust fumes should always be vented outside the shop.

❏ The refrigerant in automotive air conditioning systems produces a poisonous gas, phosgene gas, when it is introduced to heat.

❏ Heavy objects should be lifted with your legs not your back.

❏ You are responsible for two areas of housekeeping: your work area and the rest of the shop.

❏ Different types of fires require different types of fire extinguishers. You should know these differences and how to operate the extinguishers.

❏ Extreme care should be taken whenever doing electrical repairs, especially with electronic parts.

❏ Always use the correct tool, in the correct way, for the job.

❏ Common sense usually will dictate whether a practice or behavior is safe or not. Think before you act.

Terms to Know

Carbon monoxide	Class D fire	Lift points
Class A fire	Electrolyte	Phosgene gas
Class B fire	Fire extinguishers	Static
Class C fire	Freon	

ASE Style Review Questions

1. While discussing jacks:
   Technician A says to remove the jack from under a vehicle after the safety stands are in place.
   Technician B says to shove the creeper back under the car when it is not being used.
   Who is correct?
   A. A only   **C.** Both A and B
   B. B only   **D.** Neither A nor B

2. While discussing wrenches:
   Technician A says to always push on the handle of a wrench when tightening a bolt.
   Technician B says to never use an incorrectly sized wrench or socket on a bolt or nut.
   Who is correct?
   A. A only   **C.** Both A and B
   B. B only   **D.** Neither A nor B

3. While discussing the safe use of hammers:
Technician A says to be sure to strike an object with the full face of the hammer head.
Technician B says safety glasses should be worn when using a hammer and/or chisel.
Who is correct?
A. A only
B. B only
C. Both A and B
D. Neither A nor B

4. While discussing why exhaust fumes should be vented outdoors or drawn into a ventilation/filtration system:
Technician A says exhaust gases contain large amounts of carbon monoxide.
Technician B says carbon dioxide is an odorless, colorless, and deadly gas.
Who is correct?
A. A only
B. B only
C. Both A and B
D. Neither A nor B

5. While discussing wrenches:
Technician A says that a box-end or socket wrench should be used whenever it is possible.
Technician B says that when using an adjustable wrench, the wrench should be pulled so that the force of the pull is on the adjustable jaw.
Who is correct?
A. A only
B. B only
C. Both A and B
D. Neither A nor B

6. While discussing simple first-aid procedures:
Technician A says that if someone is overcome by carbon monoxide, get him or her fresh air immediately.
Technician B says burns should be cooled immediately by rinsing them with water.
Who is correct?
A. A only
B. B only
C. Both A and B
D. Neither A nor B

7. While discussing the use of an air impact wrench:
Technician A says that impact sockets can be used with an air impact wrench.
Technician B says sockets that have cracks or breaks should never be used with an air impact wrench.
Who is correct?
A. A only
B. B only
C. Both A and B
D. Neither A nor B

8. While discussing working safely around automotive air conditioning systems:
Technician A says the refrigerants used in these systems will heat anything that is around when it escapes from the system.
Technician B says that if the refrigerant is exposed to flame or heat, a poisonous gas is formed.
Who is correct?
A. A only
B. B only
C. Both A and B
D. Neither A nor B

9. While discussing the car's electrical system:
Technician A says you should always disconnect the negative or ground battery cable first, then disconnect the positive cable.
Technician B says you should always connect the positive battery cable first, then the negative.
Who is correct?
A. A only
B. B only
C. Both A and B
D. Neither A nor B

10. While discussing the uses of wrenches and pliers:
Technician A says that a wrench should never be used as a hammer.
Technician B says pliers should never be used to loosen or tighten a nut.
Who is correct?
A. A only
B. B only
C. Both A and B
D. Neither A nor B

Diagnosis, Maintenance, and Basic Adjustments

Upon completion and review of this chapter, you should be able to:

❏ Listen to the driver's complaint, road test the vehicle, then determine the needed repairs.

❏ Diagnose unusual fluid usage, level, and condition problems.

❏ Replace automatic transmission fluid and filters.

❏ Diagnose noise and vibration problems.

❏ Diagnose electronic, mechanical, and vacuum control systems.

❏ Inspect, replace, and align power train mounts.

❏ Inspect, adjust, and replace vacuum modulator, lines, and hoses.

❏ Inspect, adjust, and replace manual valve shift linkage.

❏ Inspect, adjust, and replace cables or linkages for throttle valve (T.V.) kickdown and accelerator pedal.

❏ Inspect and replace external seals and gaskets while the transmission is in the vehicle.

❏ Inspect, repair, and replace extension housing.

❏ Inspect and replace speedometer drive gear, driven gear, and retainers while the transmission is in the vehicle.

❏ Inspect and replace parking pawl, shaft, spring, and retainer while the transmission is in the vehicle.

❏ Adjust bands.

Basic Tools

Basic mechanic's tool set

Appropriate service manual

Some of the procedures noted in this chapter are for both in-vehicle and out-of-vehicle service. If there is an SM reference in the margin, go to that page for more details.

Automatic transmissions are used in many rear-wheel-drive and four-wheel-drive vehicles. Automatic transaxles are most used on front-wheel-drive vehicles. The major components of a transaxle are the same as those in a transmission, except the transaxle assembly includes the final drive and differential gears. An automatic transmission/transaxle shifts automatically without the driver moving a gearshift or pressing a clutch pedal.

Because of the many similarities between a transmission and a transaxle, most of the diagnostic and service procedures are similar. Therefore, all references to a transmission apply equally to a transaxle unless otherwise noted. This rule also holds true for the questions on the certification test for automatic transmissions and transaxles.

 SERVICE TIP: Whenever you are diagnosing or repairing a transaxle or transmission, make sure you refer to the appropriate service manual before you begin.

Transmissions are strong and typically troublefree units that require little maintenance. Maintenance normally includes fluid checks and scheduled linkage adjustments and oil changes.

Identification

Prior to beginning any service or repair work, be sure you know exactly which transmission you are working on. This will ensure that you are following the correct procedures and specifications and are installing the correct parts. Proper identification can be difficult because transmissions cannot be accurately identified by the way they look. The only exception to this is the shape of the oil pan, which can be used as positive identification on some transmissions (Figure 3-1).

The only positive way to identify the exact design of the transmission is by its identification numbers. Transmission identification numbers are found either as stamped numbers in the case or on a metal tag held by a bolt head. Use a service manual to decipher the identification number. Most identification numbers include the model, manufacturer, and assembly date. Whenever you

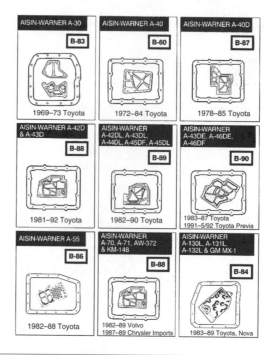

Figure 3-1 The shape of the oil pan can help identify the exact model of some transmissions. (Courtesy of Automatic Transmission Parts, Inc.)

work with a transmission with a metal ID tag, make sure the tag is put back on the transmission so that the next technician will be able to properly identify the transmission.

Some examples of the location of and the information contained on the identification tags of common transmissions follow:

On Chrysler FWD vehicles, the transaxle can be identified by a transaxle identification number (TIN) stamped on a boss, located on the transaxle housing just above the oil pan flange. On RWD vehicles, the TIN is stamped on a pad on the left side of the transmission case's oil pan flange. In addition to the TIN, each transmission carries an assembly part number that must be referenced when ordering transaxle replacement parts. Transmission operation requirements are different for each vehicle and engine combination. Some internal parts will differ between models. Always refer to the seven-digit part number for positive transmission identification when replacing parts.

Late-model Ford RWD vehicles can be identified by an identification code letter found on the lower line of the vehicle certificate label under "TR." This label is attached to the left (driver's) side door lock post. A RWD transmission can be identified by a metal tag attached to the transmission by the lower extension housing retaining bolt. The tag on a transaxle is attached to the valve body cover and gives the transmission model number, line shift code, build date code, and assembly and serial numbers (Figure 3-2).

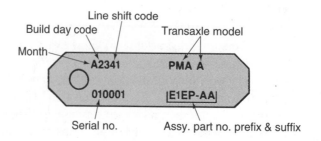

Figure 3-2 A typical identification tag for a Ford Motor Company automatic transmission (Reprinted with the permission of Ford Motor Co.)

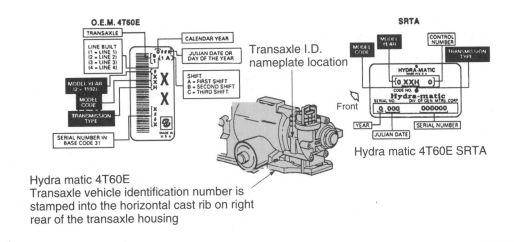

Transaxle I.D. nameplate location

Front

Hydra matic 4T60E SRTA

Hydra matic 4T60E
Transaxle vehicle identification number is
stamped into the horizontal cast rib on right
rear of the transaxle housing

Figure 3-3 Location of and information contained on a General Motors' transaxle identification plate (Courtesy of the Hydra-Matic Division of General Motors Corp.)

Early Ford RWD models can be identified by an identification tag located under the lower intermediate servo cover bolt or attached to the lower extension housing retaining bolt. A number appearing after the suffix indicates internal parts in the transmission have been changed after initial production startup. The top line of the tag shows the transmission model number and build date code.

The transaxle in General Motors FWD vehicles can be identified by an identification number stamped on a plate attached to the rear face of the transaxle (Figure 3-3). Transaxle model is printed on the Service Parts Identification label located inside the vehicle.

The transmission in early RWD General Motors products can be identified by the production number, which is located on the ID plate attached to the right side of the case near the modulator. The production number consists of a year code, a two-character model code and a build-date code.

The transmission in late-model RWD GM products can be identified by a letter code contained in the identification number. The ID number is stamped on the transmission case above the oil pan rail on the right rear side. The identification number contains information that must be used when ordering replacement parts.

The transaxle in Hondas and Acuras can be identified by an identification number stamped on a metal pad on top of the transmission. The first two characters indicate the transmission model.

Aisin Warner transmissions, which are used by Jeep, Isuzu, and Volvo, can be identified by a plate attached to a side of the transmission case. The plate shows the transmission model number and serial number.

The transmission used in some models from Audi, Porsche, and Volkswagen can be identified by numbers cast into the top rear of the case. The transaxle model code is identified by figures stamped into the torque converter housing. These figures consist of the model code and build-date code. Some models will also have code letters and the date of manufacture stamped into the machined flat on the bellhousing rim. The valve body may also be stamped on the machined boss on the valve body. The valve body identification tag is secured with valve body mounting screws. A torque converter code letter is stamped on the side of the attaching lug.

Models that use Borg Warner transmissions have an identification number stamped on a plate attached to the torque converter housing near the throttle cable and distributor.

Jatco transmissions, which are manufactured by the Japan Automatic Transmission Company, are used by Chrysler, Mitsubishi, Mazda, and Nissan. The model may be identified by a stamped metal plate attached to the right side of the transmission case. The model code appears on the second line and the serial number is on the bottom line.

Some Mazda, Mercedes Benz, Toyota, and Subaru (Gunma) transmissions are best identified by the eleventh character in the VIN, which is located at the top left of the instrument panel and on the transaxle flange on the exhaust side of the engine or the driver's door post.

BMW, Peugeot, and some models of Volvo use ZF transmissions, which have an identification plate fixed to the left side of the transmission case. The lower left series of numbers on the plate indicate the number of gears, type of controls, type of gears, and torque capacity.

Diagnostics

An automatic transmission receives engine power through a torque converter, which is indirectly attached to the engine's crankshaft. Hydraulic pressure in the converter allows power to flow from the torque converter to the transmission's input shaft. The only time a torque converter is mechanically connected to the transmission is during torque converter lockup.

The input shaft drives a planetary gear set, which provides the different forward gears, a neutral position, and reverse. Power flow through the gears is controlled by multiple-disc clutches, one-way valves, and friction bands (Figure 3-4). These hold a member of the gear set when hydraulic pressure is applied to them. By holding different members of the planetary gear set, different gear ratios are possible.

Hydraulic pressure is routed to the correct holding element by the transmission's valve body, which contains many hydraulic valves and controls the pressure of the hydraulic fluid and its direction.

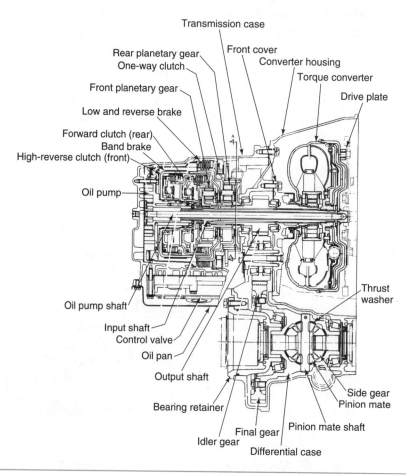

Figure 3-4 Basic components of a transaxle. Note the location of the planetary gear set and apply and holding devices. (Courtesy of Nissan Motor Co., Ltd.)

Because of the complexities involved in the operation of an automatic transmission, diagnostics can be quite complicated. This is especially true if the technician does not have a thorough understanding of the operation of a normally working transmission.

Before beginning to rebuild or repair a transmission, make sure it has a problem and requires repairs. Gathering as much information as you can to describe the problem will help you decide if the problem is a transmission problem.

Automatic transmission problems are usually caused by one or more of the following conditions: poor engine performance, problems in the hydraulic system, mechanical malfunctions, or improper adjustments. Diagnosis of these problems should begin by checking the condition of the fluid and its level, conducting a thorough visual inspection, and by checking the various linkage adjustments.

Poor engine performance can have a drastic affect on transmission operation. Low engine vacuum will cause a vacuum modulator to sense a load condition when it actually is not present. This will cause delayed and harsh shifts. Delayed shifts can also result from the action of the T.V. assembly, if the engine runs so badly that the throttle pedal must be frequently pushed to the floor to keep it running or to get it going.

Engine performance can also affect torque converter clutch operation. If the engine is running too poorly to maintain a constant speed, the converter clutch will engage and disengage at higher speeds. The customer complaint may be that the converter chatters; however, the problem may be the result of engine misses.

If the vehicle has an engine performance problem, the cause should be found and corrected before any conclusions on the transmission are made. A quick way to identify if the engine is causing shifting problems is to connect a vacuum gauge to the engine and take a reading while the engine is running. The gauge should be connected to intake manifold vacuum, anywhere below the throttle plates. A normal vacuum gauge reading is steady and at 17 in. Hg. The rougher the engine runs, the more the gauge readings will fluctuate. The lower the vacuum readings, the more severe the problem.

Diagnosing a problem should follow a systematic procedure to eliminate every possible cause that can be corrected without removing the transmission.

Fluid Check

Your diagnosis should begin with a fluid level check. To check the ATF level, make sure the vehicle is on a level surface. Before removing the dipstick, wipe all dirt off of the protective disc and the dipstick handle. On most automobiles, the ATF level can be checked accurately only when the transmission is at operating temperature and the engine is running. Remove the dipstick and wipe it clean with a lintfree cloth or paper towel. Reinsert the dipstick, remove it again, and note the reading. Markings on a dipstick indicate ADD levels, and on some models, FULL levels for cool, warm, or hot fluid (Figures 3-5 and 3-6).

Classroom Manual
Chapter 3, page 50

If the dipstick has a HOT level marked on it, this mark should be used only when the dipstick is hot to the touch.

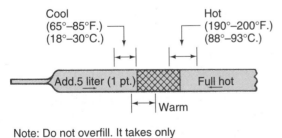

Figure 3-5 Typical dipstick markings for an automatic transmission (Courtesy of the Hydra-Matic Division of General Motors Corp.)

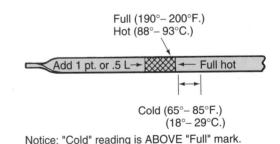

Figure 3-6 Typical dipstick markings for an automatic transaxle (Courtesy of the Hydra-Matic Division of General Motors Corp.)

33

✔️ **SERVICE TIP:** The temperature of the transmission or transaxle is a vital factor when checking ATF levels. Fluid level checking can be misleading because some transmissions have bimetallic elements that block fluid flow to the transmission cooler until a predetermined temperature is reached. The bimetal element keeps the fluid inside the transmission hot.

If the fluid level is low and/or off the crosshatch section of the dipstick, the problem could be external fluid leaks. Check the transmission case, oil pan, and cooler lines for evidence of leaks.

Low fluid levels can cause a variety of problems. Air can be drawn into the oil pump's inlet circuit and mixed with the fluid. This will result in aerated fluid, which causes slow pressure buildup and low pressures, which will cause slippage between shifts. Air in the pressure regulator valve will cause a buzzing noise when the valve tries to regulate pump pressure.

Excessively high fluid levels can also cause aeration. As the planetary gears rotate in high fluid levels, air can be forced into the fluid. Aerated fluid can foam, overheat, and oxidize. All of these problems can interfere with normal valve, clutch, and servo operation. Foaming may be evident by fluid leakage from the transmission's vent.

⬤ **CUSTOMER CARE:** Customers should be made aware that fluid level and condition should be checked at least every six months. Temperature fluctuations from summer to winter can cause a thermal breakdown of ATF. Even high-quality fluids can experience breakdown as a result of these frequent and extreme temperature changes. It is said that nearly 90% of all transmission failures are caused by fluid breakdown or oxidation.

The condition of the fluid should be checked while checking the fluid level. Examine the fluid carefully. The normal color of ATF is pink or red. If the fluid has a dark brownish or blackish color and/or a burned odor, the fluid has been overheated. A milky color indicates that engine coolant has been leaking into the transmission's cooler in the radiator. If there is any question about the condition of the fluid, drain out a sample for closer inspection.

After checking the ATF level and color, wipe the dipstick on absorbent white paper and look at the stain left by the fluid. Dark particles are normally band and/or clutch material, while silvery metal particles are normally caused by the wearing of the transmission's metal parts. If the dipstick cannot be wiped clean, it is probably covered with varnish, which results from fluid oxidation. Varnish will cause the spool valves to stick, causing improper shifting speeds. Varnish or other heavy deposits indicate the need to change the transmission's fluid and filter.

✔️ **SERVICE TIP:** Contaminated fluid can sometimes be felt better than seen. Place a few drops of fluid between two fingers and rub them together. If the fluid feels dirty or gritty, it is contaminated with burned frictional material.

Fluid Changes

Special Tools

Large drain pan

Clean white rags

Inch-pound torque
    wrench

⬤ **CUSTOMER CARE:** Abusive driving can overheat a transmission and cause fluid oxidization and breakdown. Inform your customers that they should always stay within the recommended towing load for their vehicle. Also tell them to avoid excessive rocking back and forth when they are stuck in snow or mud.

The transmission's fluid and filter should be changed whenever there is an indication of oxidation or contamination. Photo Sequence 2 shows the correct procedure for performing an automatic transmission fluid and filter change. Periodic fluid and filter changes are also part of the preventative program for most vehicles. The frequency of this service depends on the conditions that the transmission normally operates under. Severe usage requires that the fluid and filter be changed every 15,000 miles. Severe usage is defined as:

1. More than 50% operation in heavy city traffic during hot weather above 90° F.
2. Police, taxi, commercial-type operation and trailer towing.

The mileage interval that a manufacturer will recommend a fluid and filter change will also depend upon the type of transmission. For example, some General Motors transmissions use aluminum valves in their valve body. Since aluminum is softer than steel, aluminum valves are less tolerant of abrasives and dirt in the fluid. Therefore, to prolong the life of the valves, more frequent fluid changes are recommended by General Motors.

CUSTOMER CARE: Older transmissions did not require as frequent fluid and filter changes as do the later-model vehicles. Customers should be made aware of the recommended frequency and the reason for this change. In older vehicles, both the engine and transmission ran cooler, which extended the life of transmission fluid. These transmissions operated at 175° F and the transmission fluid lasted about 100,000 miles before oxidizing. However, fluid life is halved with every 20° its temperature rises above 175° F. For example, fluid life expectancy drops to 50,000 miles at 195° F and just 25,000 miles at 215° F, which is the temperature at which most newer transmissions run.

CAUTION: Be careful when draining the transmission fluid. It can be very hot and will tend to adhere to your skin, causing severe burns.

Change the fluid only when the engine and transmission are at normal operating temperatures. On most transmissions, you must remove the oil pan to drain the fluid. Some transmission pans on recent vehicles include a drain plug. A filter or screen is normally attached to the bottom of the valve body. Filters are made of paper or fabric and held in place by screws, clips, or bolts (Figure 3-7). Filters should be replaced, not cleaned.

To drain and refill a typical transmission, the vehicle must be raised on a hoist. After the vehicle is safely in position, place a drain pan with a large opening under the transmission's oil pan. Then loosen the pan bolts and tap the pan at one corner to break it loose. Fluid will begin to drain from around the pan. When all fluid is drained, remove the oil pan.

After draining, carefully remove the pan. There will be some fluid left in the pan. Be prepared to dump it into the drain pan. Check the bottom of the pan for deposits and metal particles. Slight contamination, blackish deposits from clutches and bands, is normal. Other contaminants should be of concern. Clean the oil pan and its magnet.

Classroom Manual
Chapter 3, page 51

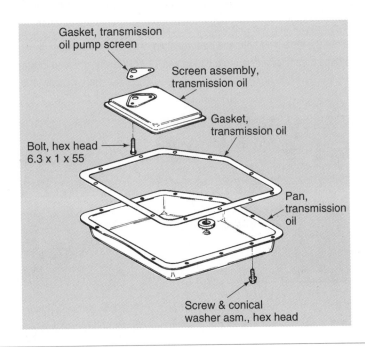

Figure 3-7 Oil filters and pans are attached to the transmission case by screws, bolts, and/or clips. (Courtesy of the Hydra-Matic Division of General Motors Corp.)

Photo Sequence 2
Typical Procedure for Performing an Automatic Transmission Fluid and Filter Change

P2-1 To begin the procedure for changing automatic transmission fluid and filter, raise the car on a lift. Before working under the car, make sure it is safely positioned on the lift.

P2-2 Locate the transmission pan. Remove any part that may interfere with the removal of the pan.

P2-3 Position the oil drain pan under the transmission pan. A large diameter drain pan helps prevent spills.

P2-4 Loosen all of the pan bolts and remove all but three. This will cause fluid to flow out around the pan into the drain pan.

P2-5 Supporting the pan with one hand, remove the remaining bolts and pour the fluid from the transmission pan into the drain pan.

P2-6 Carefully inspect the transmission pan and the residue in it. The condition of the fluid is often an indication of the condition of the transmission and serves as a valuable diagnostic tool.

P2-7 Remove the old pan gasket and wipe the transmission pan clean with a lint-free rag.

P2-8 Unbolt the fluid filter from the transmission's valve body. Keep the drain pan under the transmission while doing this. Additional fluid may leak out of the filter or valve body.

P2-9 Before continuing, compare the old filter and pan gasket with the new ones.

P2-10 Install the new filter onto the valve body and tighten the attaching bolts to proper specifications. Then lay the new pan gasket over the sealing surface of the pan and move the pan into place.

P2-11 Install the attaching bolts and tighten them to the recommended specifications. Read the specifications for installing the filter and pan carefully. Some transmissions require torque specifications of inch-pounds rather than the typical foot-pounds.

P2-12 With the pan tightened, lower the car. Start the engine. With the parking brake applied and your foot on the brake pedal. Move the shift lever through the gears. This allows the fluid to circulate throughout the entire transmission. After the engine reaches normal operating temperature, check the transmission fluid level and correct it if necessary.

Remove the filter and inspect it. Use a magnet to determine if metal particles are steel or aluminum. Steel particles indicate severe internal transmission wear or damage. If the metal particles are aluminum, they may be part of the torque-converter stator. Some torque converters use phenolic plastic stators; therefore, metal particles found in these transmissions must be from the transmission itself. Filters are always replaced, whereas screens are cleaned. Screens are removed in the same way as filters. Clean a screen with fresh solvent and a stiff brush. Remove any traces of the old pan gasket on the case housing. Then, install a new filter and gasket on the bottom of the valve body and tighten the retaining bolts to the specified torque.

⚠ **WARNING:** Make sure you check the specifications. The required torque is often given in inch-pounds. You can easily break the bolts or damage something if you tighten the bolts to foot-pounds.

Remove any traces of the old pan gasket from the oil pan. Make sure the mounting flange of the oil pan is not distorted and bent. Then, reinstall the pan using the sealant recommended by the manufacturer. Normally RTV sealant is recommended. Tighten the pan retaining bolts to the specified torque.

Pour a little less than the required amount of fluid into the transaxle through the dipstick tube. Always use the recommended type of ATF. The wrong fluid will alter the shifting characteristics of the transmission. For example, if type F fluid is used in a transmission designed for Dexron-type fluid, the shifting will be harsher.

Start the engine and allow it to idle for at least one minute. Then, with the parking and service brakes applied, move the gear selector lever momentarily to each position, ending in park. Recheck the fluid level and add a sufficient amount of fluid to bring the level to about 1/8-inch below the ADD mark.

Run the engine until it reaches normal operating temperature. Then recheck the fluid level. It should be in the HOT region on the dipstick. Make sure the dipstick is fully seated into the dipstick tube opening. This will prevent dirt from entering into the transmission.

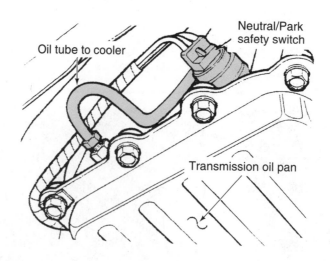

Figure 3-8 Oil cooler lines and electrical switches are common sources of fluid leaks. (Courtesy of Chrysler Corp.)

Fluid Leaks

Continue your diagnostics by conducting a quick and careful visual inspection. Check all drive train parts for looseness and leaks. If the transmission fluid was low or there was no fluid, raise the vehicle and carefully inspect the transmission for signs of leakage. Leaks are often caused by defective gaskets or seals. Common sources of leaks are the oil pan seal, rear cover and final drive cover (on transaxles), extension housing, speedometer drive gear assembly, and electrical switches mounted into the housing (Figure 3-8). The housing itself may have a porosity problem, allowing fluid to seep through the metal. Case porosity may be repaired using an epoxy-type sealer.

Oil Pan

A common cause of fluid leakage is the seal of the oil pan to the transmission housing. If there are signs of leakage around the rim of the pan, retorquing the pan bolts may correct the problem. If tightening the pan does not correct the problem, the pan must be removed and a new gasket installed. Make sure the sealing surface of the pan's rim is flat and capable of providing a seal before reinstalling it.

Torque Converter

Torque converter problems can be caused by a leaking converter (Figure 3-9). This type of problem may be the cause of customer complaints of slippage and a lack of power. To check the converter for leaks, remove the converter access cover and examine the area around the torque converter shell. An engine oil leak may be falsely diagnosed as a converter leak. The color of engine oil is different than transmission fluid and may help identify the true source of the leak. However, if the oil or fluid has absorbed much dirt, both will look the same. An engine leak typically leaves an oil film on the front of the converter shell, whereas a converter leak will cause the entire shell to be wet. If the transmission's oil pump seal is leaking, only the back side of the shell will be wet. If the converter is leaking or damaged, it should be replaced.

Extension Housing

An oil leak stemming from the mating surfaces of the extension housing and the transmission case may be caused by loose bolts. To correct this problem, tighten the bolts to the specified torque. Also check for signs of leakage at the rear of the extension housing. Fluid leaks from the seal of the

Most transaxles vent through the dipstick handle or tube.

Case porosity is caused by tiny holes that are formed by trapped air bubbles during the casting process.

Some transmissions do not use a gasket, rather the pan is sealed with RTV.

Classroom Manual
Chapter 3, page 51

Shop Manual
Chapter 5, page 98

Classroom Manual
Chapter 3, page 39

Classroom Manual
Chapter 3, page 42

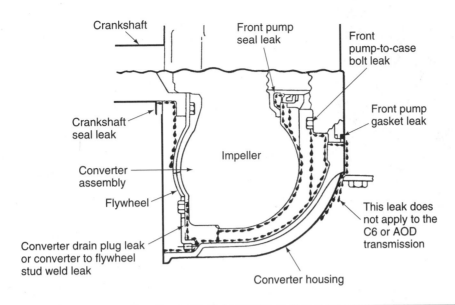

Crankshaft

Front pump seal leak

Front pump-to-case bolt leak

Crankshaft seal leak

Front pump gasket leak

Converter assembly

Impeller

Flywheel

This leak does not apply to the C6 or AOD transmission

Converter drain plug leak or converter to flywheel stud weld leak

Converter housing

Figure 3-9 By determining the direction of fluid travel, the cause of a fluid leak around the torque converter can be identified. (Reprinted with the permission of Ford Motor Co.)

extension housing can be corrected with the transmission in the car. Often, the cause for the leakage is a worn extension housing bushing, which supports the sliding yoke of the drive shaft. When the drive shaft is installed, the clearance between the sliding yoke and the bushing should be minimal. If the clearance is satisfactory, a new oil seal will correct the leak. If the clearance is excessive, the repair requires that a new seal and a new bushing be installed. If the seal is faulty, the transmission vent should be checked for blockage.

Rear Oil Seal and Bushing Replacement

Procedures for the replacement of the rear oil seal and bushing vary little with each car model. General procedures for the replacement of the oil seals and bushings follow.

To replace the rear seal:

1. Remove the drive shaft.
2. Remove the old seal from the extension housing (Figure 3-10).
3. Lubricate the lip of the seal, then install the new seal in the extension housing.
4. Install the drive shaft.

To replace the rear bushing and seal:

1. Remove the drive shaft from the car. Inspect the yoke for wear and replace it if necessary.
2. Insert the appropriate puller tool into the extension housing until it grips the front side of the bushing.
3. Pull the seal and bushing from the housing.
4. Drive a new bushing into the extension housing.
5. Install a new seal in the housing (Figure 3-11).
6. Install the drive shaft.

Special Tools

Slide hammer

Seal remover tool

Bushing removing tool

Seal driver

Bushing driver

Speedometer Drive

The vehicle's speedometer can be purely electronic, which requires no mechanical hookup to the transmission, or it can be driven off the output shaft via a cable. An oil leak at the speedometer cable can be corrected by replacing the O-ring seal. While replacing the seal, inspect the drive gear for chips and missing teeth. Always lubricate the O-ring and gear prior to installation.

Classroom Manual
Chapter 3, page 45

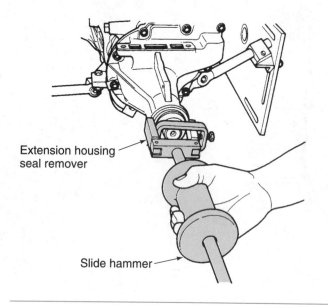

Extension housing seal remover

Slide hammer

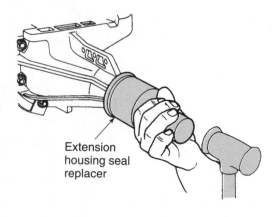

Extension housing seal replacer

Figure 3-11 Rear seals can be easily installed into the extension housing with the proper driver and a hammer. (Reprinted with the permission of Ford Motor Co.)

Figure 3-10 Extension housing rear seals can be removed with a slide hammer and a special removal tool. (Reprinted with the permission of Ford Motor Co.)

When replacing the speedometer drive seal, first clean the top of the speedometer cable retainer. Then remove the hold-down screw, which keeps the retainer in its bore. Carefully pull up on the speedometer cable, pulling the speedometer retainer and drive gear assembly from its bore (Figure 3-12). Carefully remove the old seal.

To reinstall the retainer, lightly grease and install the O-ring onto the retainer. Gently tap the retainer and gear assembly into its bore while lining the groove in the retainer with the screw hole in the side of the clutch housing case. Install the hold-down screw and tighten it in place.

Electrical Connections

Shop Manual
Chapter 5, page 48

A visual inspection of the transmission should include a careful check of all electrical wires, connectors, and components. This inspection is especially important for electronically controlled transmissions and for transmissions that have a lockup torque converter. All faulty connectors, wires, and components should be repaired or replaced before continuing your diagnosis of the transmission.

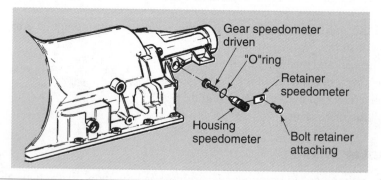

Gear speedometer driven

"O"ring

Retainer speedometer

Housing speedometer

Bolt retainer attaching

Figure 3-12 Speedometer drive gears can be removed from the transmission case by removing the retaining bolt and pulling the assembly out of the case. (Courtesy of the Hydra-Matic Division of General Motors Corp.)

Check all electrical connections to the transmission. Faulty connectors or wires can cause harsh or delayed and missed shifts. On transaxles, the connectors can normally be inspected through the engine compartment, whereas they can only be seen from under the vehicle on longitudinally mounted transmissions. To check the connectors, release the locking tabs and disconnect them, one at a time, from the transmission. Carefully examine them for signs of corrosion, distortion, moisture, and transmission fluid. A connector or wiring harness may deteriorate if ATF reaches it. Also check the connector at the transmission. Using a small mirror and flashlight may help you get a good look at the inside of the connectors. Inspect the entire transmission wiring harness for tears and other damage. Road debris can damage the wiring and connectors mounted underneath the vehicle.

Because the operation of the engine and transmission is integrated through the control computer, a faulty engine sensor or connector may affect the operation of both the engine and the transmission. The various sensors and their locations can be identified by referring to the appropriate service manual. The engine control sensors that are the most likely to cause shifting problems are the throttle-position sensor, MAP sensor, and vehicle speed sensor.

Throttle position sensors (Figure 3-13) are typically located on the side of the carburetor or fuel injection throttle plate assembly. Remove the electrical connector from the sensor and inspect both ends for signs of corrosion and damage. Poor contact can cause the transmission to miss shifts. Also inspect the wiring harness to the TPS for evidence of damage.

A MAP sensor (Figure 3-14) is usually located on top of the intake manifold or mounted to the firewall, where a vacuum hose connects it to the intake manifold. Check both ends of its three-pronged connector and wiring for corrosion and damage. Also check the condition of the vacuum hose. A vacuum leak can cause harsh shifting.

A vehicle speed sensor is often used on late-model cars to monitor speed. The function of this sensor is similar to a speedometer except it operates electronically rather than mechanically. Proper shift points cannot occur if the signal from the speed sensor is faulty. The sensor's connections and wiring should be inspected for signs of damage and corrosion. These sensors are located near the output shaft of the transmission.

Checking the Transaxle Mounts

The engine and transmission mounts on FWD cars are important to the operation of the transaxle. Any engine movement may change the effective length of the shift and throttle cables and there-

Even the slightest amount of corrosion can affect the output of a sensor. Increased resistance will always change the voltage signal of a circuit.

A throttle position sensor is most often referred to as a TPS.

Classroom Manual
Chapter 3, page 52

A MAP sensor is a manifold absolute pressure sensor, which senses engine vacuum.

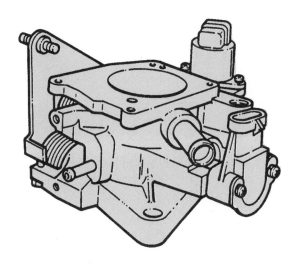

Figure 3-13 Typical throttle position sensor (TPS) (Courtesy of the Buick Motor Division of General Motors Corp.)

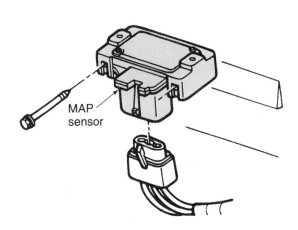

MAP sensor

Figure 3-14 Typical manifold absolute pressure (MAP) sensor (Courtesy of the Buick Motor Division of General Motors Corp.)

Some manufacturers require that the mount bolts be discarded and new ones installed during reassembly.

Classroom Manual
Chapter 3, page 48

fore may affect the engagement of the gears. Delayed or missed shifts may result from linkage changes as the engine pivots on its mounts. Problems with transmission mounts may also affect the operation of a transmission on a RWD vehicle, but this type of problem will be less detrimental than the same type of problem on a FWD vehicle. Many shifting and vibration problems can be caused by worn, loose, or broken engine and transmission mounts. Visually inspect the mounts for looseness and cracks. To get a better look at the condition of the mounts, pull up and push down on the transaxle case while watching the mount. If the mount's rubber separates from the metal plate or if the case moves up but not down, replace the mount. If there is movement between the metal plate and its attaching point on the frame, tighten the attaching bolts to an appropriate torque.

Then, from the driver's seat, apply the foot brake, set the parking brake, and start the engine. Put the transmission into a gear and gradually increase the engine speed to about 1500–2000 rpm. Watch the torque reaction of the engine on its mounts. If the engine's reaction to the torque appears to be excessive, broken or worn drive train mounts may be the cause.

If it is necessary to replace the transaxle mount, make sure you follow the manufacturer's recommendations for maintaining the alignment of the drive line. Failure to do this may result in poor gear shifting, vibrations, and/or broken cables. Some manufacturers recommend that a holding fixture or special bolt be used to keep the unit in its proper location.

When removing the transaxle mount, begin by disconnecting the battery's negative cable. Disconnect any electrical connectors that may be located around the mount. It may be necessary to move some accessories, such as the horn, in order to service the mount without damaging some other assembly. Be sure to label any wires you remove to facilitate reassembly.

Install the engine support fixture (Figure 3-15) and attach it to an engine hoist. Lift the engine just enough to take the pressure off of the mounts. Remove the bolts attaching the transaxle mount to the frame and the mounting bracket, then remove the mount.

To install the new mount, position the transaxle mount in its correct location on the frame and tighten its attaching bolts to the proper torque. Install the bolts that attach the mount to the transaxle bracket. Prior to tightening these bolts, check the alignment of the mount. Once you have confirmed that the alignment is correct, tighten all loosened bolts to their specified torque.

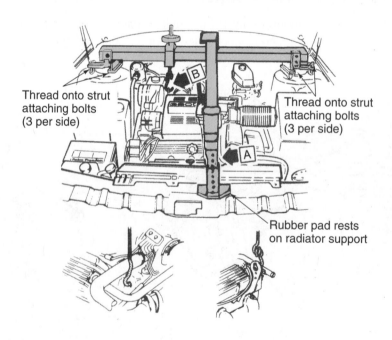

Figure 3-15 A typical engine support fixture in place to remove and/or install engine and transaxle mounts (Courtesy of the Buick Motor Division of General Motors Corp.)

Remove the engine hoist fixture from the engine and reinstall all accessories and wires that may have been removed earlier.

> **SERVICE TIP:** In the procedure for removing the mount, some manufacturers recommend the use of a special alignment bolt, which is installed in an engine mount. This bolt serves as an indicator of power train alignment. If excessive effort is required to remove the alignment bolt, the power train must be shifted to allow for proper alignment.

Road Testing the Vehicle

All transmission complaints should be verified by road testing the vehicle and duplicating the customer's complaint. A knowledge of the exact conditions that cause the problem and a thorough understanding of transmissions will allow you to accurately diagnose problems. Many problems that appear to be transmission problems may be caused by problems in the engine, drive shaft, U-joint or CV-joint, wheel bearings, wheel/tire imbalance, or other conditions.

Make sure these are not the cause of the problem before you begin to diagnose and repair a transmission. Diagnosis becomes easy if you think about what is happening in the transmission when the problem occurs. If there is a shifting problem, think about the parts that are being engaged and what these parts are attempting to do.

> **SERVICE TIP:** Always refer to your service manual to identify the particulars of the transmission you are diagnosing. It is also helpful to check for any Technical Service Bulletins that may be related to the customer's complaint.

Noises

Abnormal transmission noises and vibrations can be caused by faulty bearings, damaged gears, worn or damaged clutches and bands, or a bad oil pump, as well as contaminated fluid or improper fluid levels. Vibrations can also be caused by torque converter problems. All noises and vibrations that occur during the road test should be noted, as well as when they occur.

Transaxles have unique noises associated with them because of their construction. These noises can result from problems in the differential or drive axles. Use the following guide to determine if a noise is caused by the transaxle or by other parts in the drive train.

A knock at low speeds:

1. Worn drive axle CV joints
2. Worn side gear hub counterbore

Noise most pronounced on turns:

1. Differential gear noise
2. Dry, worn CV joints

Clunk on acceleration or deceleration:

1. Loose engine mounts
2. Worn differential pinion shaft or side gear hub counterbore in the case
3. Worn or damaged drive axle inboard CV joints

Clicking noise in turns:

1. Worn or damaged outboard CV joint

Road testing a vehicle to exactly duplicate the symptoms is essential to proper diagnosis. During the a road test, the transmission should be operated in all possible modes and its operation noted. The customer's complaint should be verified and any other transmission malfunctions should be noted. Your observations during the road test, your understanding of the operation of a transmission, and the information given in service manuals will help you identify the cause of any transmission problem. If conducted properly, road testing is diagnosis through a process of elimination.

Range		Gear ratio	Clutch		Low & reverse brake	Lockup	Band servo		One-way clutch	Parking pawl
			High-reverse clutch (Front)	Forward clutch (Rear)			Operation	Release		
Park										on
Reverse		2.364	on		on					
Neutral										
Drive	D₁ Low	2.826		on					on	
	D₂ Second	1.543		on			on			
	D₃ Top (3rd)	1.000	on	on		on	(on)	on		
2	2₁ Low	2.826		on					on	
	2₂ Second	1.543		on			on			
1	1₁ Low	2.826		on	on				on	
	1₂ Second	1.543		on			on			

Figure 3-16 Typical band and clutch application chart. This chart should be referred to during a road test and when determining the cause of any shifting problems. (Courtesy of Nissan Motors Co., Ltd.)

Before beginning your road test, find and duplicate from a service manual the chart (Figure 3-16) that shows the band and clutch application for different gear selector positions. Using these charts will greatly simplify your diagnosis of automatic transmission problems. It is also wise to have a notebook or piece of paper to jot down notes about the operation of the transmission. By doing this, you won't have to worry about remembering all of the details, nor will you have to decide if you remembered them correctly. Some manufacturers recommend the use of a symptoms chart and/or checklist that is provided in their service manuals (Figure 3-17).

Prior to road testing a vehicle, always check the transmission fluid level and condition of oil and correct any engine performance problems. Inspect the transmission for signs of fluid leakage. If leakage is evident, wipe off the leaking oil and dust. Then, note the location of the leaks in your notebook or on the symptoms chart.

Begin the road test with a drive at normal speeds to warm the engine and transmission. If a problem appears only when starting and/or when the engine and transmission is cold, record this symptom on the chart or in your notebook.

After the engine and transmission are warmed up, place the shift selector into the DRIVE position and allow the transmission to shift through all of its normal shifts. Any questionable transmission behavior should be noted.

During a road test, check for proper gear engagement as you move the selector lever to each gear position, including park. There should be no hesitation or roughness as the gears are engaging. Check for proper operation in all forward ranges, especially the 1–2, 2–3, 3–4 upshifts, and converter lockup during light throttle operation. These shifts should be smooth and positive and occur at the correct speeds. These same shifts should feel firmer under medium-to-heavy throttle pressures. Transmissions equipped with lockup torque converters should be brought to the spec-

DATE _____ / _____ / _____

TRANSMISSION/TRANSAXLE CONCERN CHECK SHEET

(* — INFORMATION REQUIRED FOR TECHNICAL ASSISTANCE)

SERVICE ADVISOR

*VIN _____   *MILEAGE _____   R.O.#_____

*MODEL YEAR _____   *VEHICLE MODEL _____   *ENGINE_____

*TRANS. MODEL _____   *TRANS. SERIAL # _____

*CUSTOMER'S CONCERN

_____

_____

_____

_____

CHECK THE ITEMS THAT DESCRIBE THE CONCERN —

WHAT:	WHEN:	OCCURS:	USUALLY NOTICED:
_ NO POWER	_ VEHICLE WARM	_ ALWAYS	_ IDLING
_ SHIFTING	_ VEHICLE COLD	_ INTERMITTENT	_ ACCELERATING
_ SLIPS	_ ALWAYS	_ SELDOM	_ COASTING
_ NOISE	_ NOT SURE	_ FIRST TIME	_ BRAKING
_ SHUDDER			AT _____ MPH

TECHNICIAN

PRELIMINARY CHECK PROCEDURES	NOTE FINDINGS
INSPECT	
• FLUID LEVEL & CONDITION	_____
• ENGINE PERFORMANCE — VACUUM & ECM CODES	_____
• TV CABLE AND/OR MODULATOR VACUUM	_____
• MANUAL LINKAGE ADJUSTMENT	NOTE: DUPLICATE THE CONDITIONS UNDER WHICH
• ROAD TEST TO VERIFY CONCERN *	CUSTOMER'S CONCERN WAS OBSERVED

*PROPOSED OR COMPLETED REPAIRS

ON-CAR   BENCH

_____

_____

_____

_____

(OVER)

---

*TRANS. TEMPERATURE _____ HOT _____ COLD

*VACUUM READINGS AT MODULATOR
(180C, 250C, 350C, 400, 440-T4)

READING AT MODULATOR _____ IN. HG. (ENGINE AT HOT IDLE, TRANS. IN DRIVE)

CHECK FOR VACUUM RESPONSE DURING ACCELERATOR MOVEMENT

PRESSURE TEST

*TRANS. LINE PRESSURES

	MINIMUM		MAXIMUM	
	SPEC. (FROM MANUAL)	ACTUAL	SPEC. (FROM MANUAL)	ACTUAL
P	_____	_____	_____	_____
R	_____	_____	_____	_____
N	_____	_____	_____	_____
D	_____	_____	_____	_____
D	_____	_____	_____	_____
2	_____	_____	_____	_____
1	_____	_____	_____	_____

*FINDINGS BASED ON ROAD TEST

CHECK ITEMS FOUND ON ROAD TEST THAT DESCRIBE THE CUSTOMER'S COMMENTS ABOUT:

ROAD TEST

GARAGE SHIFT FEEL	UPSHIFTS	DOWNSHIFTS	TORQUE CONVERTER CLUTCH
_ ENGINE STOPS	_ EARLY	_ BUSYNESS	_ BUSYNESS
_ HARSH	_ HARSH	_ HARSH	_ HARSH
_ DELAYED	_ DELAYED	_ DELAYED	_ NO RELEASE
_ NO DRIVE	_ SLIPS	_ SLIPS	_ SHUDDER
	_ NO UPSHIFT	_ NO DOWNSHIFT	_ EARLY APPLY
			_ NO APPLY
			_ LATE APPLY

CONCERNS OCCUR WHEN/DURING:

✔	GEAR	RANGE
		P - N
	1st	D
	2nd	
	3rd	
	4th	
	1st	D
	2nd	
	3rd	
	1st	2
	2nd	
	1st	1
	REVERSE	R

DURING

_ 1-2 UPSHIFT
_ 2-3 UPSHIFT
_ 3-4 UPSHIFT
_ 4-3 DOWNSHIFT
_ 3-2 DOWNSHIFT
_ 2-1 DOWNSHIFT

NOISE

TYPE	WHEN NOTICED
_ BUZZ	_ ALWAYS
_ WHINE	_ LOAD SENSITIVE
_ CLUNK	_ STEERING SENSITIVE
	AT _____ MPH
	IN _____ GEAR

THROTTLE POSITION	PITCH	LEVEL
_ LIGHT	_ LOW	_ LIGHT
_ MEDIUM	_ MEDIUM	_ MEDIUM
_ HEAVY	_ HIGH	_ HEAVY
_ W.O.T.		

Figure 3-17 Typical road test check list recommended by manufacturer to identify the cause of shifting problems (Courtesy of the Hydra-Matic Division of General Motors Corp.)

ified lockup speed and their engagement noted. Again, record the operation of the transmission in these different modes in your notebook or on the diagnostic chart.

Force the transmission to "kickdown" and record the quality of this shift and the speed at which it downshifts. These will be compared later to the specifications. Manual downshifts should also be made at a variety of speeds. The reaction of the transmission should be noted as should all abnormal noises, and the gears and speeds at which they occur.

After the road test, check the transmission for signs of leakage. Any new leaks and their probable cause should be noted. Then compare your written notes from the road test to the information given in the service manual to identify the cause of the malfunction. The service manual usually has a diagnostic chart to aid you in this process.

SERVICE TIP: The following problems and their causes are given as examples. The actual causes of these types of problems will vary with the different models of transmissions. Always refer to the appropriate band and clutch application chart while diagnosing shifting problems. Using the chart will allow you to identify the cause of the shifting problems through the process of elimination.

Normally, if the shift for all forward gears is delayed, a slipping front or forward clutch is indicated. A slipping rear clutch is indicated when there is a delay or slip when the transmission shifts into any two forward gears. However, if there is a shift delay while the transmission is in DRIVE and it shifts into third gear, either the front or rear clutch may be slipping. To determine which clutch is defective, look at the behavior of the transmission when one of the two clutches is not applied. For example, if the rear clutch is not engaged while the transmission is in reverse and there is no slippage in reverse, the problem of the slippage in third gear is caused by the rear clutch.

The reverse/high clutch is sometimes called the direct clutch.

If the reverse/high clutch is slipping, there is a delay only during shifts into reverse and from second to third gear. Intermediate clutch or band problems will result in delayed shifting from first to second gear. If there is slippage in first gear when the gear selector is in the DRIVE position, but not when first gear is manually selected, the one-way or overrunning clutch may be the cause.

It is important to remember that delayed shifts or slippage may also be caused by leaking hydraulic circuits or sticking spool valves in the valve body. Since the application of bands and clutches is controlled by the hydraulic system, improper pressures will cause shifting problems. Other components of the transmission can also contribute to shifting problems. For example, on transmissions equipped with a vacuum modulator, if upshifts do not occur at the specified speeds or do not occur at all, either the modulator is faulty or the vacuum supply line is leaking.

Vacuum Modulator

Classroom Manual
Chapter 3, page 45

Diagnosing a vacuum modulator begins with checking the vacuum at the line or hose to the modulator (Figure 3-18). The modulator should be receiving engine manifold vacuum. If it does, there are no vacuum leaks in the line to the modulator. Check the modulator itself for leaks with a hand-held vacuum pump (Figure 3-19). The modulator should be able to hold approximately 18 in. Hg. If transmission fluid is found when you disconnect the line at the modulator, the vacuum diaphragm in the modulator is leaking and the modulator should be replaced. If the vacuum source, vacuum lines, and vacuum modulator are in good condition but shift characteristics indicate a vacuum modulator problem, the modulator may need adjustment.

Most modulators must be removed to be adjusted. However, there are some that have an external adjustment. This adjustment allows for fine tuning of modulator action. To remove a vacuum modulator from the transmission, loosen the retaining clamp and bolt. Some units are screwed into the transmission case. While pulling the modulator out of the housing, be careful not to lose the modulator actuating pin, which may fall out as the modulator is removed. Use a hand-held vacuum pump with a vacuum gauge and the recommended gauge pins to adjust the modulator according to specifications.

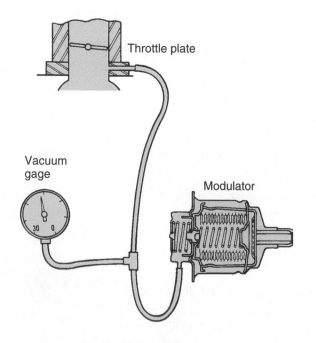

Figure 3-18 The correct hookup for connecting a vacuum gauge to a vacuum modulator (Courtesy of the Hydra-Matic Division of General Motors Corp.)

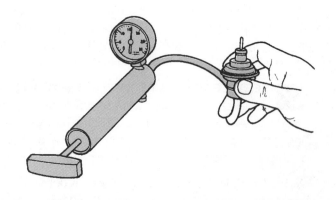

Figure 3-19 The vacuum modulator can be checked for leakage and action by activating it with a hand-held vacuum pump and observing the vacuum gauge and the action of the modulator. (Reprinted with the permission of Ford Motor Co..)

Linkages

Many transmission problems are caused by improper adjustment of the linkages. All transmissions have either a cable or rod-type gear selector linkage. Some transmissions also have a throttle valve linkage, while others use an electric switch connected to the throttle to control forced downshifts.

Normal operation of a neutral safety switch provides a quick check for the adjustment of the gear selector linkage. To do this, move the selector lever slowly until it clicks into the PARK position. Turn the ignition key to the "Start" position. If the starter operates, the PARK position is correct. After checking the PARK position, move the lever slowly toward the NEUTRAL position until the lever drops at the end of the N stop in the selector gate. If the starter also operates at this point, the gearshift linkage is properly adjusted. This quick test also tests the adjustment of the neutral safety switch. If the engine does not start in either or both of these positions, the neutral safety switch or the gear selector linkage needs adjustment or repair.

> ■ **CAUTION:** Since you must work under the vehicle to adjust most shift linkages, make sure you properly raise and support the vehicle before working under it. Also, wear safety glasses or goggles while working under the vehicle.

The gear selector linkage is often referred to as the manual linkage because it controls the manual shift valve.

Neutral Safety Switch

Most neutral safety switches are combinations of a neutral safety switch and a backup lamp switch. Others have separate units for these two distinct functions. When the reverse light switch is not part of the neutral safety switch, it seldom needs adjustment, as it is directly activated by the transmission (Figure 3-20) or the gear selector linkage. A neutral safety switch allows current to pass from the starter switch to the starter when the lever is placed in the "P" or "N" position. Some neutral safety switches are nonadjustable. If these prevent starting in P and/or N and the gear selector linkage is correctly adjusted, they should be replaced.

A voltmeter can be used to check the switch for voltage when the ignition key is turned to the START position with the shift lever in P or N. If there is no voltage, the switch should be adjusted or replaced.

To adjust a typical neutral safety switch:

1. Place the shift lever in neutral.
2. Loosen the attaching bolts for the switch.
3. Using an aligning pin, move the switch until the pin falls into the hole in its rotor.
4. Tighten the attaching bolts.
5. Recheck the switch for continuity. If no voltage is present, replace the switch.

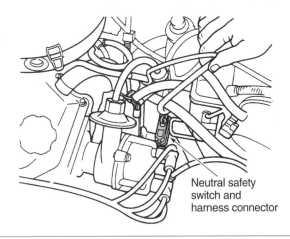

Neutral safety switch and harness connector

Figure 3-20 Typical location of a neutral safety switch on a transaxle (Reprinted with the permission of Ford Motor Co.)

Gear Section Linkage

A worn or misadjusted gear selection linkage will affect transmission operation. The transmission's manual shift valve must completely engage the selected gear (Figure 3-21). Partial manual shift valve engagement will not allow the proper amount of fluid pressure to reach the rest of the valve body. If the linkage is misadjusted, poor gear engagement, slipping, and excessive wear can result. The gear selector linkage should be adjusted so the manual shift valve detent position in the transmission matches the selector level detent and position indicator.

To check the adjustment of the linkage, move the shift lever from the PARK position to the lowest DRIVE gear. Detents should be felt at each of these positions. If the detent cannot be felt in either of these positions, the linkage needs to be adjusted. While moving the shift lever, pay attention to the gear position indicator. Although the indicator will move with an adjustment of the linkage, the pointer may need to be adjusted so that it shows the exact gear after the linkage has been adjusted.

Linkage Adjustment

CAUTION: Always set the parking brake before moving the gear selector through its positions.

Classroom Manual
Chapter 3, page 46

To adjust a typical floor-mounted gear selector linkage:

1. Place the shift lever into the DRIVE position.
2. Loosen the locknuts and move the shift lever until DRIVE is properly aligned and the vehicle is in the "D" range.
3. Tighten the locknut (Figure 3-22).

To adjust a typical cable-type linkage:

1. Place the shift lever into the PARK position.
2. Loosen the clamp bolt on the shift cable bracket.
3. Make sure the preload adjustment spring engages the fork on the transaxle bracket.
4. By hand, pull the shift lever to the front detent position (PARK), then tighten the clamp bolt. The shift linkage should now be properly adjusted.

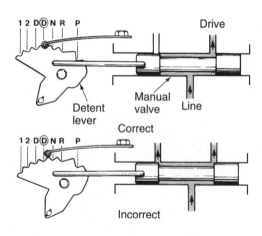

Figure 3-21 Incorrect linkage adjustments may cause the manual shift valve to be positioned improperly in its bore and cause slipping during gear changes. (Courtesy of the Hydra-Matic Division of General Motors Corp.)

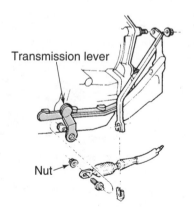

Figure 3-22 Gearshift linkages normally attach directly to the shift lever at the transmission and are retained by a locknut. (Reprinted with the permission of Ford Motor Co.)

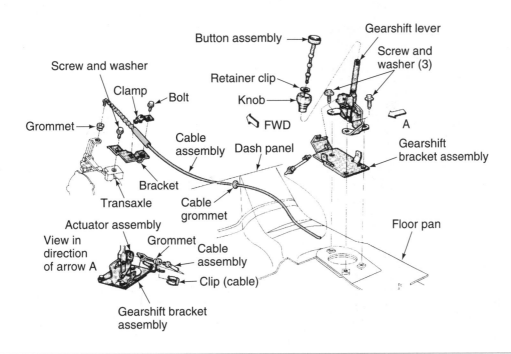

Figure 3-23 If the shift cable cannot be properly adjusted, the cable is distorted or some of the various grommets or brackets are worn and should be replaced. (Courtesy of Chrysler Corp.)

To adjust a typical rod-type linkage:

1. Loosen or disconnect the shift rod at the shift lever bracket.
2. Place the gear selector into PARK and the manual shift valve lever into the PARK detent position.
3. With both levers in position, tighten the clamp on the sliding adjustment to maintain their relationship. On the threaded type of linkage adjustment, lengthen or shorten the connection as needed. On some vehicles, you may need to adjust the neutral safety switch after resetting the linkage.

After adjusting any type of shift linkage, recheck it for detents throughout its range. Make sure a positive detent is felt when the shift lever is placed into the PARK position, as a safety measure. If you are unable to make an adjustment, the levers' grommets may be badly worn or damaged and should be replaced (Figure 3-23). When it is necessary to disassemble the linkage from the levers, the plastic grommets used to retain the cable or rod should be replaced. Use a prying tool to force the cable or rod from the grommet, then cut out the old grommet. Pliers can be used to snap the new grommets into the levers and the cable or rod into the levers.

Throttle Valve Linkages

The throttle valve cable connects the movement of the throttle pedal movement to the throttle valve in the transmission's valve body. On some transmissions, the throttle linkage may control both the downshift valve and the throttle valve. Others use a vacuum modulator to control the throttle valve and a throttle linkage to control the downshift valve. Late-model transmissions may not have a throttle cable. Instead, they rely on electronic sensors and switches to monitor engine load and throttle plate opening. The action of the throttle valve produces throttle pressure (Figure 3-24). Throttle pressure is used as an indication of engine load and influences the speed at which automatic shifts will take place.

A misadjusted T.V. linkage may also result in a throttle pressure that is too low in relation to the amount the throttle plates are open, causing early upshifts. Too high of a throttle pressure can cause harsh and delayed upshifts and part and wide-open throttle downshifts will occur earlier

The throttle valve linkage is called the T.V. linkage.

Classroom Manual
Chapter 3, page 46

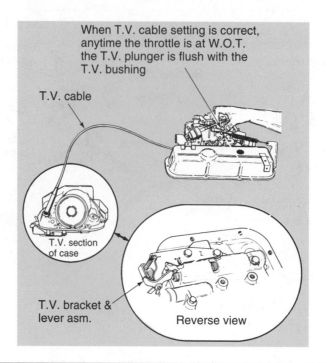

When T.V. cable setting is correct, anytime the throttle is at W.O.T. the T.V. plunger is flush with the T.V. bushing

T.V. cable

T.V. section of case

T.V. bracket & lever asm.

Reverse view

Figure 3-24 The movement of the throttle plate causes the T.V. valve to move. (Courtesy of the Hydra-Matic Division of General Motors Corp.)

than normal. When adjusting the T.V. and downshift linkages, always follow the manufacturer's recommended procedures. An adjustment as small as a half-turn can make a big difference in shift timing and feel.

SERVICE TIP: It is important that you check the service manual before proceeding to make adjustments to the throttle or downshift linkages. Some of these linkage systems are not adjustable and if you loosen them in an attempt to adjust them, you may break them.

To adjust a typical throttle cable:

1. Run the engine until it has reached normal operating temperature.
2. Loosen the cable mounting bracket or swivel lock screw.
3. Position the bracket so that both bracket alignment tabs are touching the transaxle, then tighten the lock screw to the recommended torque.
4. Release the cross-lock on the cable assembly.
5. Make sure the cable is free to slide all the way toward the engine against its stop, after the cross-lock or locking tab is released (Figure 3-25).
6. Move the throttle control lever clockwise against its internal stop, then press the cross-lock downward into its locked position.
7. At this point, the cable is adjusted and the backlash in the cable is removed.

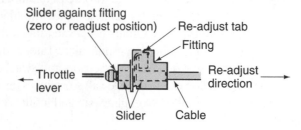

Slider against fitting
(zero or readjust position)

Re-adjust tab

Fitting

Throttle lever

Re-adjust direction

Slider

Cable

Figure 3-25 Typical adjusting mechanism for a T.V. cable (Courtesy of the Chevrolet Motor Division of General Motors Corp.)

8. Check the cable for free movement by moving the throttle lever counterclockwise and slowly release it to confirm it will return fully clockwise.

9. No lubrication is required for any component of the throttle cable system.

To adjust a typical throttle lever rod:

1. Run engine until it has reached normal operating temperature.

2. Loosen the linkage's swivel lock screw.

3. Make sure the swivel is free to slide along the flat end of the throttle rod. Disassemble and clean or repair parts to assure free movement.

4. Hold the transaxle's throttle lever firmly toward the engine and against its internal stop, then tighten the swivel lock screw.

5. The rod is now adjusted and any backlash in the linkage should be taken up by the preload spring.

To adjust a typical downshift linkage:

1. Run the engine until it has reached normal operating temperature.

2. Put the transmission in neutral with the parking brake set and allow the engine to run at its normal idle speed.

3. Using the specified amount of pressure, press down on the downshift rod.

4. Rotate the adjustment screw to obtain the specified clearance between the screw and the throttle arm.

Kickdown Switch Adjustment

Most late-model transmissions are not equipped with downshift linkages, rather they use a kickdown switch typically located at the upper post of the throttle pedal (Figure 3-26). Movement of the throttle pedal to the wide-open position signals to the transmission that the driver desires a forced downshift.

To check the operation of the switch, fully depress the throttle pedal and listen for a click that should be heard just before the pedal reaches its travel stop. If the click is not heard, loosen the lock-nut and extend the switch until the pedal lever makes contact with the switch. If the pedal contacts the switch too early, the transmission may downshift during part-throttle operation.

If you feel and hear the click of the switch but the transmission still doesn't kickdown, use an ohmmeter to check the continuity of the switch when it is depressed. An open switch will prevent forced downshifting, whereas a shorted switch can cause upshift problems. Defective switches should be replaced.

Electrical Controls

The neutral safety switch and the kickdown switch are two examples of the switches used on today's transmissions. There are many more possible switches connected to a transmission and

Classroom Manual
Chapter 3, page 48

A neutral safety switch is called an inhibitor switch by some manufacturers.

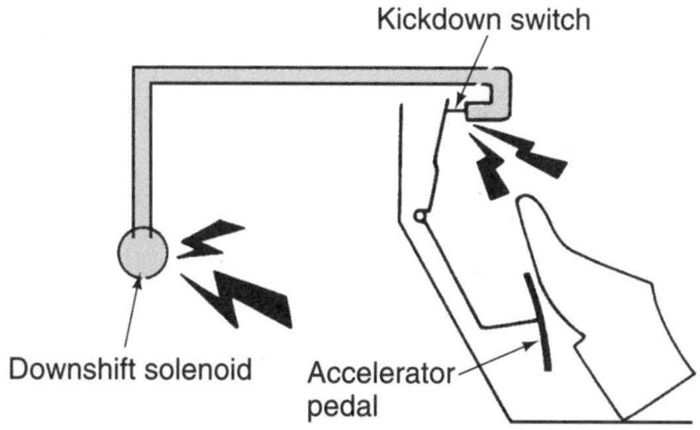

Figure 3-26 Action of a kickdown switch (Courtesy of Nissan Motors Co., Ltd.)

Switches that supply a ground for the circuit are called grounding switches.

Classroom Manual
Chapter 3, page 52

It has been said that 70% of all electrical problems are caused by faulty connections.

they can serve many different purposes. However, all of these switches either complete or open an electrical circuit. By completing a circuit, these switches either provide a ground for the circuit or connect two wires together. To open the circuit, switches either disconnect the ground from the circuit or disconnect two wires.

Testing Switches

Prior to the use of electronic controls and their associated sensors, grounding switches were often used to control the engagement of lockup torque converters. These switches are controlled by governor pressure and supply the ground for the lockup solenoid when a particular speed is reached. Different lockup speeds are possible through the use of different spring tensions on the switch. To engage the clutch earlier, a weaker spring is used and when the lockup should occur later, a stronger spring is used. Grounding switches are also used to indicate or sense what gear the transmission is in.

Switches that connect or disconnect two wires are used to allow or disallow an electrical component to operate when the transmission is in a particular gear. These switches are typically activated by the presence of oil pressure. These switches can be either normally open or normally closed switches. Normally open switches do not complete the circuit until oil pressure is applied to them (Figure 3-27), whereas normally closed switches open the circuit when oil pressure is applied.

An ohmmeter can be used to identify the type of switch being used and can be used to test the operation of the switch. There should be continuity when the switch is closed and no continuity when the switch is open. This type of switch can be activated by applying air pressure to the part that would normally be exposed to oil pressure (Figure 3-28).

⚠️ **WARNING:** Always disconnect an electrical switch or component from the circuit or from a power source before testing it with an ohmmeter. Failure to do so will damage the meter.

When applying air pressure to these switches, check them for leaks. Although a malfunctioning electrical switch will probably not cause a shifting problem, it can if it leaks. If the switch leaks off the applied pressure in a hydraulic circuit to a holding device, the holding member may not be able to function properly.

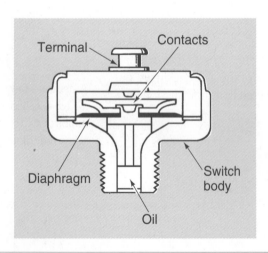

Figure 3-27 Typical normally open oil pressure switch. When oil pressure is present, the diaphragm moves up and closes the switch. (Courtesy of the Hydra-Matic Division of General Motors Corp.)

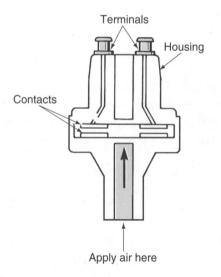

Figure 3-28 Typical normally open, two-terminal oil pressure switch. To check, apply air to the bottom oil passage and check for continuity across the terminals. (Courtesy of the Hydra-Matic Division of General Motors Corp.)

Speed Sensors

Electronically controlled transmissions rely on electrical signals from a speed sensor to control shift timing (Figure 3-29). The use of this type of sensor negates the need for hydraulic signals from a governor. A faulty speed sensor can cause improper shifting, as can malfunctioning electrical solenoids or other components and corroded or loose connections.

The speed sensor used in many late-model transmissions is a permanent magnetic (PM) generator. The operation of a PM generator can be tested with a DVOM set to measure ac voltage. Raise the vehicle on a lift. Allow the wheels to be suspended and free to rotate freely. Connect the meter to the speed sensor. Start the engine and put the transmission in gear. Slowly increase the engine's speed until the vehicle is at approximately 20 mph, then measure the voltage at the speed sensor. Slowly increase the engine's speed and observe the voltmeter. The voltage should increase smoothly and precisely with an increase in speed.

⚠️ **WARNING:** While conducting this procedure, it is possible to damage the CV joints if the drive wheels are allowed to dangle in the air. Place safety stands under the front suspension arms to maintain proper drive axle operating angles.

A speed sensor can also be tested with it out of the vehicle. Connect an ohmmeter across the sensor's terminals. The desired resistance readings across the sensor will vary with each and every

Shop Manual
Chapter 5, page 117

Classroom Manual
Chapter 3, page 45

A DVOM is a digital volt-ohmmeter.

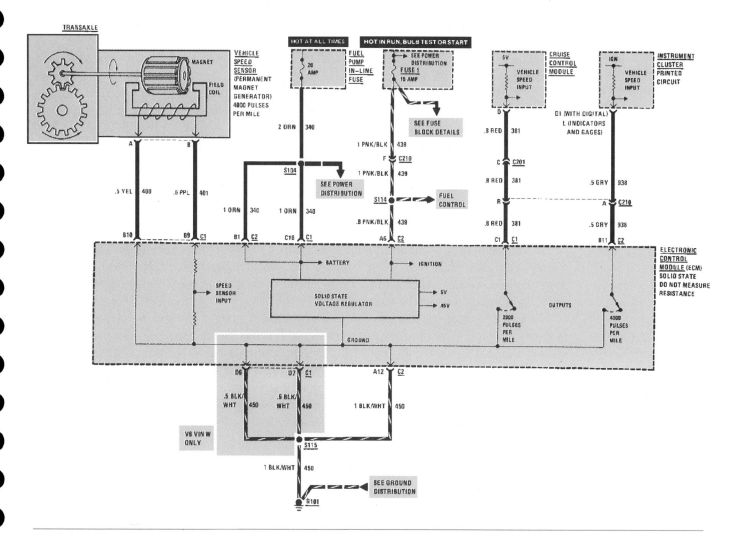

Figure 3-29 Typical circuitry for a vehicle speed sensor (Courtesy of the Chevrolet Motor Division of General Motors Corp.)

individual sensor, but there should be continuity across the leads. If there is no continuity, the sensor is open and should be replaced. Reposition the leads of the meter so that one lead is on the sensor's case and connect the other to a terminal. There should be no continuity in this position. If there is any measurable amount of resistance, the sensor is shorted.

Speed sensor operation can also be monitored by scan tools and digital oscilloscopes. Both of these instruments allow you to watch the activity of the sensor as vehicle speed changes. If the readings on either of these instruments are not steady and do not smoothly change with a change in speed, suspect a faulty connector, wiring harness, or sensor.

Electronic Defaults

The default mode of Chrysler's transmission is commonly known as the "Limp-In" mode. This accurately describes the operation of the transmission during this mode.

Classroom Manual
Chapter 3, page 41

Some electronically controlled transmissions will exhibit strange characteristics if the control computer senses a problem within its system. These transmissions will go into a default mode of operation in which the computer will disregard all input signals. During this period of time, the transmission may lock itself into a single forward gear or bypass all electronic controls. While diagnosing a problem in an electronically controlled transmission, always refer to the appropriate service manual to identify the normal default operation of the transmission. By not recognizing that the transmission is operating in default, you could spend time tracing the wrong problem.

When the computer senses something wrong, the A-604 transmission from Chrysler Corporation will go into a default mode of operation. While in this mode, the transaxle will operate in second gear only with second gear starts and reverse and park. The transaxle goes into the default mode because a device, usually an input device, has sent faulty or erroneous signals to the transaxle's control computer.

Whenever the computer sees a potential problem that may increase wear and/or damage to the transmission, the system also defaults to limp-in. Minor slipping can be sensed by the computer through its input and output sensors. This slipping will cause premature wear and may cause the computer to move into its default mode. A totally burnt clutch assembly will cause limp-in operation, as will some internal pressure leaks, which may not be apparent until pressure tests are run.

The Chrysler transaxle relies on the input from many different sensors. If a sensor does not send a proper signal or sends an erroneous signal, the computer sees it and attempts to prevent transaxle damage. The system can be totally diagnosed with a scan tool, a service manual, and pressure tests.

Chrysler and other manufacturers have transmissions that use solenoids to control fluid pressures and to apply certain holding devices. To test the solenoid valves, remove the electrical connector and measure the resistance of the solenoid coil between the connector and ground. The solenoids can also be tested by using a small jumper wire and applying current to the disconnected solenoid valve. You should be able to hear or feel the opening and closing of the solenoid. A sticking or faulty solenoid should be replaced.

The first production transmissions that were shifted by a computer were introduced by Toyota in the early eighties. Nearly all of the manufacturers now offer transmissions that are computer shifted. Diagnosing these units requires a different approach than diagnosing hydraulically controlled units. One of the first tasks during diagnosis is to determine if the shifting problem is caused by the transmission or by electronics. To determine this, the transmission must be observed to see if it responds to commands given by the computer. Identifying whether the problem is the transmission or electrical will determine what steps need to be followed to diagnose the cause of the problem.

An electronically controlled transmission will work only as well as the commands it receives from the computer, even if the hydraulic and mechanical parts of the transmission are fine. A quick way to identify the cause of the problem is to activate the shift solenoids manually by connecting a jumper wire to them. Prior to doing this, the wiring to the solenoids should be studied to determine if the computer activates them by supplying voltage to them or by completing the ground circuit. You also need to know what gear certain solenoids are activated in. This information can be found in the service manual.

Using a Toyota A140E four-speed automatic transmission as an example, we find that two solenoids are responsible for all shifting. For first gear, the #1 solenoid is on. In second gear, both solenoids are on. Third gear operation occurs when solenoid #2 is on and #1 is off, and fourth gear is activated when both solenoids are off. By examining the wiring diagram for the vehicle, we find that the white wire in the transmission wiring harness is connected to solenoid #1 and the black wire is connected to solenoid #2.

To begin testing of the solenoids, disconnect the transmission wiring harness to the solenoids. Then connect a jumper wire to a 12-volt source and to the #1 solenoid. Start the engine and move the gear selector to DRIVE. The transmission should be in first gear. Then begin to accelerate. At an appropriate time, connect the 12-volt power supply to solenoid #2 as well. This should cause a shift into second gear. When a shift into third gear is appropriate, disconnect the jumper wire to solenoid #1; an upshift should occur. Then disconnect the other jumper wire. This should cause the transmission to shift into fourth gear. Repeat this sequence many times during light, medium, and full throttle with the transmission hot and cold.

If the transmission consistently responds to these commands, the cause of any shifting problems is not the transmission. It is undoubtedly an electrical problem. If the transmission did not respond, the cause of the shifting problems is the transmission.

If the computer controls the ground side of the solenoids, the jumper wires should be connected to a known good ground rather than a 12-volt source. Some transmissions, such as Chrysler's A-604, regulate the solenoids on a duty cycle and should not be tested with a jumper wire. These units should be monitored with a scan tool.

Band Adjustment

If a transmission problem still exists after the shift linkage and throttle pressure cable and rod have been adjusted and all electrical switches and sensors checked, the bands of the transmission may need adjustment. Photo Sequence 3 shows a typical procedure for adjusting a transmission band.

On some transmissions, slippage during shifting can be corrected by adjusting the holding bands. To help identify if a band adjustment will correct the problem, refer to the written results of your road test. Compare your results with the Clutch and Band Application Chart in the service manual. If slippage occurs when there is a gear change that requires the holding by a band, the problem may be corrected by tightening the band.

On some vehicles, the bands can be adjusted externally with a torque wrench. On others, the transmission fluid must be drained and the oil pan removed. Loosen the band adjusting bolt locknut and back it off approximately five turns (Figure 3-30). Use a calibrated inch-pound torque

An automatic transmission tune-up usually includes fluid and filter change, linkage checks, and adjustment of the bands.

Classroom Manual
Chapter 3, page 51

Shop Manual
Chapter 5, page 124

Special Tools

Large drain pan

Inch-pound torque wrench

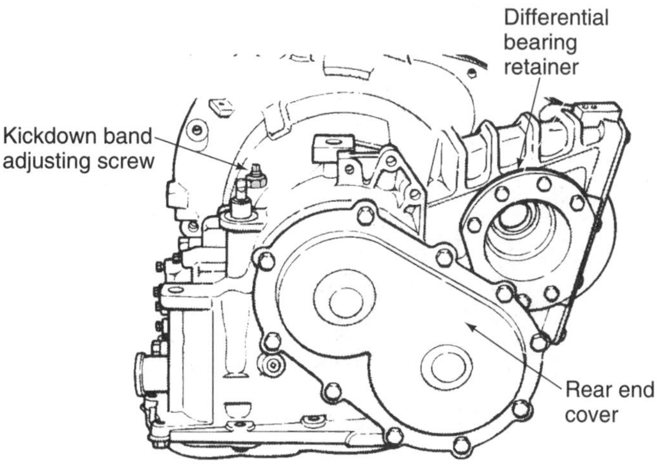

Figure 3-30 Location of external band adjusting screw (Courtesy of Chrysler Corp.)

Photo Sequence 3
Typical Procedure for Adjusting Transmission Bands

P3-1 Make sure you properly identify the transmission before proceeding to make band adjustments.

P3-2 Locate band adjusting screw.

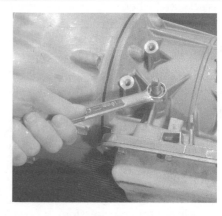

P3-3 Loosen adjusting screw locknut. If the locknut has a fluid seal, do not reuse the locknut. Install a new nut.

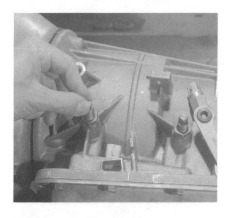

P3-4 Loosen the adjusting screw so that the band can relax around the drum and all tension is off of the adjusting screw.

P3-5 Tighten the adjusting screw to the specified torque.

P3-6 Back off the adjusting screw the exact number of turns that are specified in the service manual.

P3-7 Position a wrench on the adjusting screw and over the locknut so that you can tighten the locknut without moving the adjusting screw.

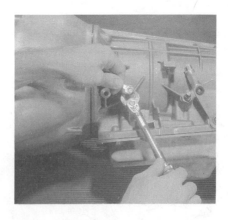

P3-8 Hold the adjusting screw in position and tighten the locknut to the specified torque.

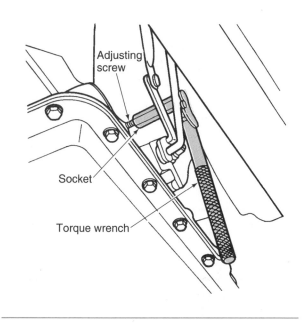

Figure 3-31 Bands are typically adjusted to a specific inch-pound torque setting. (Reprinted with the permission of Ford Motor Co.)

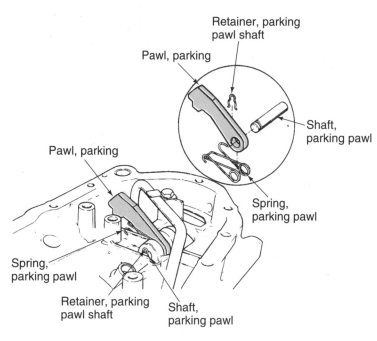

Figure 3-32 The entire parking pawl and gear assembly should be carefully inspected for wear or defects. (Courtesy of the Chevrolet Motor Division of General Motors Corp.)

wrench to tighten the adjusting bolt to the specified torque (Figure 3-31). Then, back off the adjusting screw the specified number of turns and tighten the adjusting bolt locknut while holding the adjusting stem stationary. Reinstall the oil pan with a new gasket and refill the transmission with fluid. If the transmission problem still exists, an oil pressure test or transmission teardown must be done.

> ▲ **WARNING:** Do not excessively back off the adjusting stem as the anchor block may fall out of place, making it necessary to remove and disassemble the transmission to fit it back in place.

Parking Pawl

Anytime you have the oil pan off, you should inspect the transmission parts that are exposed. This is especially true of the parking pawl assembly (Figure 3-32). This component is typically not hydraulically activated, rather the gear shift linkage moves the pawl into position to lock the output shaft of the transmission. Unless the customer's complaint indicates a problem with the parking mechanism, no test will detect a problem here.

Check the pawl assembly for excessive wear and other damage. Also check to see how firmly the pawl is in place when the gear selector is shifted into the PARK mode. If the pawl can be easily moved out, it should be repaired or replaced.

Classroom Manual
Chapter 3, page 38

C A S E   S T U D Y

A customer with a late-model Ford pickup equipped with an E4OD transmission had complained that every time he hit a bump, the transmission would downshift. Then, when driven on smooth surfaces, the transmission would begin to cycle on its own between third and fourth gears.

The technician originally thought the problem was caused by a bad fourth gear clutch that couldn't hold the planetary gearset in overdrive or by a problem in the lockup torque converter circuit. Beginning the diagnosis with a visual inspection of the transmission and the many sensors that provide information to the E4OD's control module, the technician found faulty signals from the manual lever position (MLP) sensor. These signals explained the erratic shifting, because the computer uses the sensor to determine what gear the transmission should be in and how much modified line pressure to supply.

The technician used the required special tool to realign the sensor with the gear selector, but still found that the resistance readings were out of specifications. The sensor was replaced and the problem of erratic shifting solved.

Terms to Know

Aeration	Kickdown switch	Oxidation
Default mode	Limp-in	PM generator
DVOM	Lip seal	T.V. linkage
Grounding switches	MAP sensor	Throttle position sensor
Inhibitor switch	MLP	Transaxle identification number
Kickdown	Neutral safety switch	Vehicle speed sensor

ASE Style Review Questions

1. While diagnosing noises apparently coming from a transaxle assembly:
   Technician A says a knocking sound at low speeds is probably caused by worn CV joints.
   Technician B says a clicking noise heard when the vehicle is turning is probably caused by a worn or damaged outboard CV joint.
   Who is correct?
   A. A only   **C.** Both A and B
   B. B only   **D.** Neither A nor B

2. *Technician A* says if the shift for all forward gears is delayed, a slipping front or forward clutch is normally indicated.
   Technician B says a slipping rear clutch is indicated when there is a delay or slip when the transmission shifts into any forward gear.
   Who is correct?
   A. A only   **C.** Both A and B
   B. B only   **D.** Neither A nor B

3. *Technician A* says the only positive way to identify the exact design of the transmission is by the shape of its oil pan.
   Technician B says identification numbers only identify the manufacturer and assembly date of the transmission.
   Who is correct?
   A. A only   **C.** Both A and B
   B. B only   **D.** Neither A nor B

4. *Technician A* says low engine vacuum will cause a vacuum modulator to sense a load condition when it actually is not present, causing delayed and harsh shifts.
   Technician B says poor engine performance can cause delayed shifts through the action of the T.V. assembly.
   Who is correct?
   A. A only   **C.** Both A and B
   B. B only   **D.** Neither A nor B

5. *Technician A* says delayed shifting can be caused by worn planetary gear set members.
   Technician B says delayed shifts or slippage may be caused by leaking hydraulic circuits or sticking spool valves in the valve body.
   Who is correct?
   A. A only   C. Both A and B
   B. B only   D. Neither A nor B

6. While discussing proper band adjustment procedures:
   Technician A says on some vehicles the bands can be adjusted externally with a torque wrench.
   Technician B says a calibrated inch-pound torque wrench is normally used to tighten the band adjusting bolt to a specified torque.
   Who is correct?
   A. A only   C. Both A and B
   B. B only   D. Neither A nor B

7. While checking the condition of a car's ATF:
   Technician A says if the fluid has a dark brownish or blackish color and/or a burned odor, the fluid has been overheated.
   Technician B says if the fluid has a milky color, this indicates that engine coolant has been leaking into the transmission's cooler.
   Who is correct?
   A. A only   C. Both A and B
   B. B only   D. Neither A nor B

8. While discussing the proper way to diagnose a kickdown switch:
   Technician A says when the throttle pedal is fully depressed, a click should be heard just before the pedal reaches its travel stop. If the click is not heard, the switch should be replaced.
   Technician B says if the transmission cannot be forced to automatically downshift, the kickdown switch is open and should be replaced.
   Who is correct?
   A. A only   C. Both A and B
   B. B only   D. Neither A nor B

9. *Technician A* says on some transmissions, the throttle linkage may control both the downshift valve and the throttle valve.
   Technician B says some transmissions use a vacuum modulator to control the downshift valve.
   Who is correct?
   A. A only   C. Both A and B
   B. B only   D. Neither A nor B

10. While checking the engine and transmission mounts on a FWD car:
    Technician A says any engine movement may change the effective length of the shift and throttle cables and therefore may affect the engagement of the gears.
    Technician B says delayed or missed shifts are caused by hydraulic problems not linkage problems.
    Who is correct?
    A. A only   C. Both A and B
    B. B only   D. Neither A nor B

Table 3-1 ASE TASK

Diagnose the cause of transmission oil leaks.

Problem Area	Symptoms	Possible Causes	Classroom Manual	Shop Manual
LEAKS	Transmission leaks	1. Torque converter welds leak	39	38
		2. Torque converter drain plug leaks	39	38
		3. Oil pump seals are cut, worn, or deformed	52	39
		4. Oil pump leaks	52	39
		5. Manual or throttle valve lever seals leak	46	47
		6. Cooler line leaks	46	38
		7. Case is cracked or porous	42	38
		8. Oil pan or gasket leaks	42	38
		9. Extension housing is loose or damaged	43	38
		10. External bushings are worn	44	39
		11. Filler tube is damaged	52	33
		12. Defective vent tube	52	33
		13. External cover seals leak	43	39
		14. Excessive fluid level	51	33
	Oil is coming out of vent or dipstick tube	1. Improper fluid level	51	33
		2. Contaminated ATF	51	34
		3. Drain holes plugged	46	33
		4. Case thermo element does not close when hot	46	33
		5. Defective oil cooler	46	34
		6. Restricted vent	52	34

Table 3-2 ASE TASK

Diagnose the cause of transmission/transaxle overheating.

Problem Area	Symptoms	Possible Causes	Classroom Manual	Shop Manual
OVERHEATING	Transmission/transaxle overheats	1. Excessive loads	51	45
		2. Improper fluid levels	51	33
		3. Incorrect idle speed	53	33
		4. Poor engine performance	53	33
		5. Improper fluid pressures	52	35
		6. Restriction in cooler and/or cooler lines	46	38
		7. Seized one-way clutch	54	45
		8. Sticking valves or dirty valve body	51	46
		9. Incorrect throttle valve/linkage adjustment	53	47
		10. Defective pressure relief/regulator valve	52	46
		11. Defective oil pump	52	46
		12. Faulty band/clutch application	54	45
		13. Contaminated fluid	51	34
		14. Dirty or clogged oil filter	51	35
		15. Band out of alignment	54	55
		16. Defective flex-plate	52	46

Gear Train, Shaft, and Bearing Service

Upon completion and review of this chapter, you should be able to:

❏ Remove and install a transmission/transaxle assembly from a car or light truck.

❏ Diagnose noise and vibration problems and determine needed repairs.

❏ Disassemble, clean, and inspect a transmission/transaxle.

❏ Inspect, repair, and replace transmission case(s).

❏ Inspect and repair the bores, passages, bushings, vents, and mating surfaces of a transmission case.

❏ Inspect, repair, and replace extension housing.

❏ Inspect and replace speedometer drive gear, driven gear, and retainers.

❏ Inspect and replace parking pawl, shaft, spring, and retainer.

❏ Inspect, measure, and replace thrust washers and bearings.

❏ Inspect and replace shafts.

❏ Inspect and replace bushings.

❏ Inspect and measure a planetary gear assembly and replace parts as necessary.

❏ Inspect, repair, or replace transaxle drive chains, sprockets, gears, bearings, and bushings.

❏ Inspect, measure, repair, adjust, or replace transaxle final drive components.

❏ Assemble a transmission/transaxle after servicing it.

Basic Tools

Basic mechanic's tool set
Clean lintfree rags
Appropriate service manual

Diagnosis of Noise and Vibration Problems

Abnormal noises and vibrations can be caused by faulty bearings, damaged gears, worn or damaged clutches and bands, or a bad oil pump, as well as contaminated fluid or an improper fluid level. Torque converter and cracked flex plates problems can also be the cause of vibrations.

    A customer will often complain of a transmission noise which in reality is caused by something else in the driveline and not the transmission or torque converter. Bad CV or U-joints, wheel bearings, brake calipers, and dragging brake pads can generate noises that customers, and unfortunately some technicians, mistakenly blame on the transmission or torque converter. The entire driveline should be checked before assuming the noise is transmission related.

    Most vibration problems are caused by an unbalanced torque converter assembly, a poorly mounted torque converter, or a faulty output shaft. The key to determining the cause of the vibration is to pay particular attention to the vibration in relationship to engine and vehicle speed. If the vibration changes with a change in engine speed, the cause of the problem is most probably the torque converter. If the vibration changes with vehicle speed, the cause is probably the output shaft or the driveline connected to it. The latter type of problem can be a bad extension housing bushing or universal joint.

    Noise problems are also best diagnosed by paying a great deal of attention to the speed and the conditions at which the noise occurs. The conditions to pay most attention to are the operating gear and the load on the driveline. If the noise is engine speed-related and is present in all gears, including Park and Neutral, the most probable source of the noise is the oil pump because it rotates whenever the engine is running. However, if the noise is engine-related and is present in all gears except Park and Neutral, the most probable sources of the noise are those parts that rotate in all gears, such as the drive chain, the input shaft, and the torque converter.

    Noises that occur only when a particular gear is operating must be related to those components responsible for providing that gear, such as a band or clutch. If the noise is vehicle-related, the most probable causes are the output shaft and final drive assembly. Often, the exact cause of

This chapter covers the mechanical aspects of diagnosis, disassembly, inspection, service, and assembly of an automatic transmission. The hydraulic components are detailed in Chapter 5.

Refer to a good manual transmission textbook for procedures on diagnosing the entire driveline.

Shop Manual
Chapter 3, page 43

Classroom Manual
Chapter 4, page 74

noise and vibration problems can only be identified through a careful inspection of a disassembled transmission.

Prior to tearing down a transmission, it is a good idea to inspect the material trapped in the transmission's filter. Use a magnet to determine if the particles are steel or aluminum. Generally, steel deposits are caused by something in the transmission, whereas aluminum deposits are from the stator of the torque converter. Some transmissions use aluminum clutch drums and supports. If you are working on a transmission so equipped, the aluminum deposits may be from the transmission as well as the torque converter. Plastic deposits may also be from the torque converter. Some converters, such as most Ford converters, are fitted with stators made of phenolic plastic.

Regardless of the exact fault in the transmission that is causing a vibration or noise, the transmission will need to come out and the entire unit will have to be checked and carefully inspected. Proper diagnosis prior to disassembling the transmission will identify the specific areas that should be carefully looked at and can prevent unnecessary transmission removal and teardown.

Transmission/Transaxle Removal

Safe removal of transmissions requires the purchase or fabrication of tools to help support the engine in the chassis and to lift and carry the transmission away from the vehicle.

Special Tools

Drain pan

Transmission jack

Engine support and/or heavy-duty chain

Removing the transmission from a rear-wheel-drive car is generally more straightforward than removing one from a front-wheel-drive model, as there is typically one crossmember, one driveshaft, and easy access to cables, wiring, cooler lines, and bellhousing bolts. Transmissions in FWD cars, because of their limited space, can be more difficult to remove as you may need to disassemble or remove large assemblies such as engine cradles, suspension components, brake components, splash shields, or other pieces that would not usually affect RWD transmission removal.

Begin removal by placing the vehicle on a lift. However, before raising the vehicle, take a look around the engine bay to see if any interference will occur between the firewall and engine components, such as distributors, fans and fan shrouds, fuel lines, exhaust systems or electrical components, when the transmission is removed. If any causes for interference are found, these problems should be corrected before continuing. Also, at this time, any bellhousing bolts, wiring, or T.V. cables accessible from above should be removed. Before raising a FWD vehicle to begin transmission removal, a support fixture should be attached to the engine (Figure 4-1).

✓ **SERVICE TIP:** Some technicians are using instant cameras or video recording cameras to help recall the locations of underhood items by taking pictures before work is started. This technique can be quite valuable considering how complex the underhood systems of current cars have become.

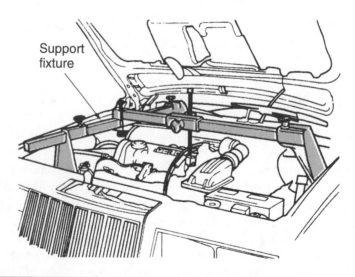

Figure 4-1 A typical engine support fixture for an FWD vehicle (Courtesy of Chrysler Corp.)

If the transmission is being removed to perform operations not related to the transmission, such as engine rear main oil seal replacement, the transmission need not be drained. However, suitable plugs must be used to seal the slip yoke opening and the dipstick tube opening to prevent fluid spills. Plastic plugs are available for sealing the slip yoke. An old slip yoke that you saved after doing driveshaft service is one of the best seals available, and the price is right! On FWD units, an old inner plunge joint tulip is a good plug; however, the plastic plugs used by the factory during shipping work the best as they allow more working space.

⚠ **WARNING:** Always begin the transmission removal procedure by disconnecting the battery ground cable. This is a safety-related precaution to help avoid any electrical surprises when removing starters or wiring harnesses. It is also possible to send voltage spikes, which may kill the ECM if wiring is disconnected when the battery is still connected.

On RWD vehicles, raise the vehicle to a comfortable height and remove all but the three or four corner bolts of the transmission pan, depending on the shape of the pan. Place a drain pan large enough to catch all the oil under the transmission. Carefully remove bolts from one side of the pan. Back off the bolt or bolts on the other side just enough to allow the pan to drop slightly as you pry it loose. Be careful, as some pans will come loose without being pried and if you loosened the pan too much, you may have a large mess to clean up! When the fluid stops draining, replace the pan with a minimum of bolts, taking care not to lose the remaining bolts.

On RWD vehicles, the driveshaft should be marked at the rear axle before disconnecting it to avoid runout-related vibrations. To remove FWD driveshafts, you must first loosen the large nut that retains the outer CV joint, which is splined shaft to the hub.

⚠ **WARNING:** It is recommended that this nut be loosened with the vehicle on the floor and the brakes applied, as this reduces possible damage to the CV joints and wheel bearings.

Now raise the FWD vehicle and remove the front wheels. Tap the splined CV joint shaft with a soft-faced hammer to see if it is loose. Most will come loose with a few taps. Many Ford FWD cars use an interference fit spline at the hub. You will need a special puller for this type CV joint. The tool pushes the shaft out and on installation pulls the shaft back into the hub.

The lower ball joint must now be separated from the steering knuckle. The ball joint (Figure 4-2) will either be bolted to the lower control arm or held into the knuckle with a pinch bolt. Once the ball joint is loose, the control arm can be pulled down and the knuckle can be pushed outward

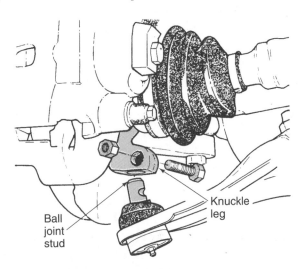

Figure 4-2 Typical ball joint-to-steering knuckle attachment (Courtesy of Chrysler Corp.)

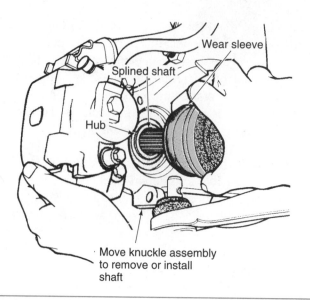

Figure 4-3 The knuckle assembly is moved out of the way to allow the stub shaft to be pulled out of the hub. (Courtesy of Chrysler Corp.)

to allow the splined CV joint shaft to slide out of the hub (Figure 4-3). The inboard joint can then be pried out or it will slide out.

The speedometer drive gears may need to be removed before pulling out the driveshafts, as they may be damaged when the shafts are removed. The inner CV joints, on some cars, have flange-type mountings. These must be unbolted for removal of the shafts. In some cases, the flange-mounted driveshafts may be left attached to the wheel and hub assembly and only unbolted at the transmission flange. Doing this will greatly decrease the amount of time needed to remove and install the transmission.

At this time, you should study the underbody layout to aid reinstallation. This is another time when use of an instant camera will help your memory when it comes time to install the transmission. In many shops, the technician repairing the transmission is not the same person that removes and installs it. If the installer does not remember the exact location of all the underbody and engine bay components, or the correct adjustments of control cables, the transmission may not operate properly when it is reinstalled. Improper installation can ruin a transmission or cause driveability problems.

At this point on both FWD and RWD vehicles, manual linkages, vacuum hoses, electrical connections, speedometer drives, and control cables should be disconnected (Figure 4-4). The exhaust system may also need to be lowered or partially removed. The inspection cover between the transmission and engine should be removed to allow access to the torque converter bolts. There will be three to six bolts or nuts securing the converter to the flex plate, depending on the application.

WARNING: To prevent accidental damage, pay attention to the proper location of all mounting bolts and hardware when the transmission is removed. For example, on Toyota torque converters, using longer than normal flywheel-to-converter bolts can dimple the torque converter clutch friction surface, causing the converter clutch to chatter.

Mark the position of the converter to the flexplate to help maintain balance or runout. It will be necessary to rotate the crankshaft to remove the converter bolts. This can be done by using a long ratchet and socket on the front crankshaft bolt or by using a flywheel turning tool, if space permits.

Carefully remove the cooler lines by holding the case fitting with one wrench and loosening the line nut with a line wrench. Doing this assures that you will not twist the steel lines, which will damage them and restrict their flow.

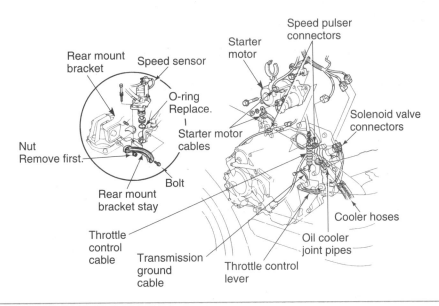

Figure 4-4 Location of the various switches, connectors, and levers on a typical transaxle (Courtesy of Honda Motor Co.)

Position a transmission jack before removing any crossmembers or bellhousing bolts. The use of a transmission jack also allows for easier access to parts hidden by crossmembers or hidden in the space between the transmission and the vehicle's floor pan.

⚠ **WARNING:** Never allow the transmission to hang by one bolt.

With the transmission jack supporting the transmission (Figure 4-5), remove the crossmember (RWD) or the transmission mounts (FWD). If the car is equipped with an engine cradle that will separate, remove the half of the cradle that allows for transmission removal.

Now remove the starter. The starter wiring may be left connected but you will need to hang the starter with heavy mechanic's wire to avoid damage to the cables. You can also completely remove the starter from the vehicle to get it totally out of the way.

⚠ **WARNING:** Never allow the starter to hang by the wires attached to it. The weight of the starter can damage the wires or, worse, break the wire and allow the motor to fall, possibly on you or someone else. Always securely support the starter and position it out of the way after you have unbolted it from the engine.

Shop Manual
Chapter 1, page 8

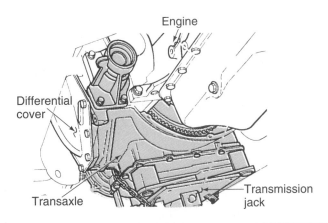

Figure 4-5 The transmission jack should be used to support the transmission while the final attaching bolts are loosened and removed. (Courtesy of Chrysler Corp.)

Now pull the transmission away from the engine. It may be necessary to use a pry bar between the transmission and engine block to separate the two units. Make sure the converter comes out with the transmission. This prevents bending the input shaft, damaging the oil pump, or distorting the drive hub. After separating the transmission from the engine, retain the torque converter in the bellhousing. This can be done simply by bolting a small combination wrench to a bellhousing bolt hole across the outer edge of the converter.

 WARNING: Never force the torque converter back into the oil pump if it has slipped out.

Installation

Transmission installation is generally a reverse of the removal procedure. Care must be taken to avoid destroying the new or rebuilt transmission during installation. A quick check of the following list will greatly simplify your installation and reduce the chances of destroying something during installation.

- Make sure the block alignment pins (dowels) are in good shape and that the alignment holes in the bellhousing are not damaged (Figure 4-6).
- Make sure the pilot hole in the crankshaft is smooth. This will allow the converter to move in and out on the flexplate.
- Make sure the pilot hub of the converter is smooth and cover it with a light coating of chassis lubricant to prevent chafing or rust.
- Make sure the converter's drive hub is smooth and coat it with ATF.
- Secure all wiring harnesses out of the way to prevent their being pinched between the bellhousing and engine block. If the wires get pinched, not only will there be a large electrical short, but you may also destroy the car's computer.
- Flush out the oil cooler lines and the cooler itself to remove any material from before that could damage the transmission. Likewise, the converter should be flushed. It is recommended that clutch-type converters be replaced, as it is not possible to tell how much the unit has been damaged.
- Always perform an endplay check and check the overall height before reinstalling a torque converter or installing a fresh unit out of the box.
- Pour one quart of the recommended fluid into the converter before mounting the converter to the transmission. This will assure that all parts in the converter have some lubrication before start-up.

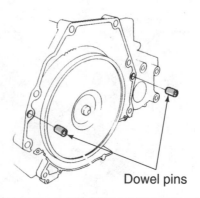

Dowel pins

Figure 4-6 The dowel pins may remain in the engine or the transmission case when the transmission is removed. In either case, they should be inspected. (Courtesy of Honda Motor Co.)

Alignment pins are commonly referred to as dowels.

Classroom Manual
Chapter 4, page 74

Slide the converter into the transmission, making sure that all drives are engaged. Double-check this by using the height dimension you measured during transmission removal. In older transmissions the torque converter must engage into the input shaft stator splines and the oil pump, whereas later models have those and possibly a direct drive shaft or an oil pump drive shaft.

Secure the converter in the transmission as you did during removal. Then transfer the transmission to the jack and move it under the car. Raise the transmission to get close alignment to the engine block. If the transmission has a full circle bellhousing, you will need to align the converter drive studs or bolt pads before you push the transmission up against the block, as this is difficult once the transmission is against the engine block. Make sure you set the torque converter onto the flexplate in alignment with the marks made during removal. These will provide proper balance for the unit. Once the converter is aligned, be sure the block dowel pins line up with the bellhousing, then push the transmission against the engine block. Check to be sure that nothing has been caught between the block and the bellhousing. Start two bellhousing bolts across from each other and use them to hold the transmission in place. Install the rest of the bolts and torque them to specifications.

On FWD cars, if the engine is held in place with a support bar, the transmission jack may be removed at this time. If the car has a split cradle-type subframe, it should be installed now.

On RWD vehicles, it may be necessary to leave the transmission jack in place while components such as exhaust crossovers or frame crossmembers are installed. Once the rear transmission mount is attached to the crossmember, the jack may be removed.

Install the converter drive bolts (or nuts). You should notice while installing these fasteners that the converter has some noticeable fore and aft movement. This amount varies with different transmissions, but it is generally between 1/8 to 1/4 inch. This is normal and necessary as it allows the converter to move on the flexplate and also allows for a noninterference fit at the oil pump drive gear. If there were no movement, premature pump failure would result.

WARNING: Never use an impact wrench on torque converter bolts. Impact wrenches can drive the bolts through the cover, which will warp the inside surface and prevent proper clutch apply or may damage the clutch pressure plate.

The cooler lines should be installed. Using the same technique as disconnecting the lines, tighten the line fittings, making sure you do not twist or distort the line. The remaining components—starter, throttle cables, wiring, and so on—can now be installed.

WARNING: After the transmission has been installed, make sure everything is properly realigned. Repositioning the transmission an inch or less can have a big effect on the manual shift linkage adjustment.

While connecting the manual shift linkage, pay attention to the condition of the plastic bushings. If these are worn or missing, hard or inaccurate linkage movement will result. Replace the bushings if needed.

Some other areas that require special attention during installation are the grounding straps and any rubber tubes. These are often overlooked during transmission removal and installation and can cause problems if not checked. Any rubber tube is suspect, as they are exposed to heat, which causes them to crack. New parts should be used if necessary. The ground straps must be in good condition and free of corrosion, as these provide an electrical ground path to the body of the car during operation. Failure to clean or attach these straps or cables can also result in poor signals to the ECM, voltage spikes that can damage the ECM, electrolysis through the fluid that welds internal transmission components, or even fused manual linkage cables as the current looks for a path to flow through.

On RWD vehicles, the driveshaft must be installed using the marks you made during removal. Be sure to coat the slip yoke with ATF before sliding it into the extension housing. The driveshafts of FWD cars are installed in the reverse manner in which they were removed and the related components (speedometer gears, strut lower ball joint bolts, and so on) should be installed just as they were before they were removed. Use a new nut on the outer CV joint stud, if the bolt is the self-locking type. The torque of this nut is critical. Torquing the nut should be done with the car's tires on the ground and/or with the brakes held. Air-type impact tools should not be used, as they can damage the wheel bearing or hub.

Connect the battery cables. Then add about one-half of the total quantity of the proper ATF to the transmission. This amount varies, but an average amount is four quarts. Some technicians will also connect a pressure gauge to the transmission and take pressure readings during the initial operation of the fresh transmission. Some imported transmission manufacturers do require this be done because line pressures must be set after start-up. Connecting the pressure gauges will also give you an indication of whether or not there is sufficient oil pressure to road test the car.

Start the engine and apply the brakes. Move the shift selector through all of the gear ranges. Then place the selector into Park. Check the fluid level and add ATF until you reach the ADD or COLD mark. Visually inspect the transmission for any signs of leakage. If a pressure gauge was attached and the pressure was fine, disconnect the gauge.

Road test the vehicle to check the operation of the transmission and anything else that may have been affected by your work. If any adjustments are necessary, make them. The road test should cover at least 20 miles in order to completely warm up the transmission. Recheck and adjust the fluid level. It should be between the ADD and FULL mark on the dipstick. DO NOT OVERFILL! After the road test, again visually inspect the transmission for signs of leakage. Also carefully look around the engine and transmission to see if any wires, hoses, or cables are disconnected or positioned in a undesirable spot.

Shop Manual
Chapter 3, page 33

SERVICE TIP: If the transmission is computer-controlled, check the service manual before taking it on its initial road test. Some computer-controlled automatic transmissions require that a "Learning Procedure" be followed, which includes various driving conditions. Since you need to road test the transmission and need to teach the transmission, you may as well do both at the same time.

Computer Relearn Procedures

Classroom Manual
Chapter 3, page 41

Vehicles equipped with engine or transmission computers may require a relearn procedure after the battery is disconnected. Many vehicle computers memorize and store vehicle operation patterns for optimum driveability and performance. When the vehicle battery is disconnected, this memory is lost. The computer will use default data until new data from each key start is stored. As the computer memorizes vehicle operation for each new key start, driveability is restored. A vehicle computer may memorize vehicle's operation patterns for 40 or more key starts.

Customers often complain of driveability problems during the relearn stage, because the vehicle acts differently than before it was serviced. Depending on the type and make of the vehicle and how it is equipped, the following complaints may exist: harsh or poor shift quality, rough or unstable idle, hesitation or stumble, rich or lean running, and poor fuel economy. These symptoms and complaints should disappear after a number of drive cycles have been memorized. To reduce the possibility of complaints, after any service that requires that the battery be disconnected, the vehicle should be road tested. If a specific relearn procedure is not available, the following procedure may be used on vehicles equipped with an automatic transmission:

1. Set the parking brake and start the engine in P or N. Allow the engine to warm up to normal operating temperature or until the electric cooling fan cycles on.
2. Allow the vehicle to idle for about a minute in the N position, then move the gear selector into the D position and allow it to idle in gear for one minute.
3. Road test the vehicle. Accelerate at normal throttle openings (20–50%) until the vehicle shifts into top gear.
4. Then maintain a cruising speed with a light-to-medium throttle.

5. Decelerate to a stop, make sure you allow the transmission to downshift, and use the brakes to bring the vehicle to a stop.
6. If a driveability problem still exists, repeat the sequence.

Some manufacturers recommend a specific relearn procedure, which is designed to establish good driveability during the relearn process. These procedures are especially important for all vehicles equipped with an electronically controlled converter and/or transmission. Always complete the procedure before returning the vehicle to the customer.

Chrysler recommends the following procedure for all vehicles equipped with 41TE and 42LE transaxles:

1. Warm the transaxle to normal operating temperatures by allowing the vehicle to idle.
2. Operate the vehicle and maintain a constant throttle opening during upshifts. This sets the transaxle into the upshift relearn process.
3. Accelerate the vehicle with a throttle opening of 10–50 degrees. Accelerating the vehicle from a stop to 45 mph with a moderate throttle is sufficient for this part of the procedure.
4. Operate the vehicle until the transaxle completes 1–2, 2–3, and 3–4 shifts at least 20 times.
5. Operate the vehicle with a vehicle speed of less than 25 mph and force downshifts with a wide-open throttle. Repeat this at least eight times.
6. Operate the vehicle at speeds above 25 mph and in fourth gear and force downshifts with a wide-open throttle. Repeat this at least eight times. The forced downshifts allow the computer to relearn kickdown operation.

Ford Motor Company also specifies a relearn procedure for some of their transmissions and transaxles. All of the specific procedures begin with a fluid check and warming the ATF to normal temperatures. The procedure has two segments: an idle relearn and a drive relearn. The idle relearn procedure begins with starting the vehicle in P, with all accessories off and the parking brake set. Then move the gear selector to N and allow the engine to idle for one minute. Move the gear selector to D and again allow the engine to idle for one minute. After the idle relearn procedure is complete, the drive relearn procedure can begin. The drive relearn procedures are specific for the different transmissions.

If the vehicle is equipped with an AOD-E, AXOD, or AXOD-E transmission, road test the vehicle. With the gear selector in overdrive, moderately accelerate the vehicle to 50 mph for a minimum of 15 seconds. The transmission should be in fourth gear at the end of that time. Then hold the speed with a steady throttle and lightly apply and release the brake for about 5 seconds. Then stop and park the vehicle with the gear selector in the D position for a minimum of 20 seconds. Repeat this procedure five times.

If the vehicle is equipped with an E4OD transmission, the drive procedure is as follows:

1. Put the gear selector in the D position and depress the O/D cancel button. (The LED should light.)
2. Moderately accelerate to 40 mph for a minimum of 15 seconds. The transmission should be in third gear at this time.
3. Hold the throttle steady and depress the O/D cancel button. (LED should go out.) Accelerate to 50 mph. The transmission should be in fourth gear at this time.
4. Hold that speed for 15 seconds, then lightly apply and release the brake. Maintain 50 mph while applying the brake.
5. Then stop and park the vehicle with the gear selector in the D position for a minimum of 20 seconds.
6. Repeat this procedure five times.

The disassembly of both a transmission and transaxle are similar; therefore, disassembly, inspection, and reassembly guidelines for both transaxles and transmissions can be safely grouped together.

This sequence covers the general disassembly of an automatic transmission. Complete overhaul of common transmissions is covered in Chapter 6 of this manual.

A transmission overhaul normally includes replacing all rubber and paper gaskets, friction materials, and worn bushings, cleaning the planetary gear sets and the valve body, and making correct endplay adjustments.

Classroom Manual
Chapter 4, page 76

Disassembly

Once the transmission has been removed from the vehicle, certain things should be checked before beginning to disassemble the transmission. Measure the depth that the converter fits into the transmission case (Figure 4-7). To do this, hold a straightedge across the bellhousing and measure into the pilot hub or a drive pad on the converter. Record this dimension and save it for use when installing the converter.

Also inspect the flexplate for warpage and cracks (Figure 4-8). Pay attention to the condition of the teeth on the starter ring gear. Also check the flexplate for excessive runout and elongated bolt holes. Replace the flexplate if there is evidence of damage. Also inspect the converter attaching bolts. Replace any worn bolts with suitable equivalents. Check the drive hub of the torque converter. It should be smooth and not worn. Pay particular attention to the area on which the seal rides. If the hub is worn, the torque converter should be replaced and the oil pump drive should be carefully inspected for scoring or other damage.

Before disassembling the transmission, check out the causes of any leakage. If there is evidence of leakage or leakage is the reason for the teardown, determine the path of the leakage before cleaning the area around the seals. At times, the leakage may be from sources other than the seal. Leakage could be from worn gaskets, loose bolts, cracked housings, or loose line connections.

Inspect the outside sealing area of the seal to see if it is wet or dry. If it is wet, see whether the oil is running out or if it is merely a lubricating film. Check both the inner and outer parts of the seals for wet oil, which means leakage.

While removing a seal, inspect the sealing surface, or lips, before cleaning it. Look for signs of unusual wear, warping, cuts and gouges, or particles embedded in the seal. On spring-loaded lip seals, make sure the spring is seated around the lip, and that the lip was not damaged when first installed. If the seal's lip is hardened, this was probably caused by heat from either the shaft or the fluid.

If the seal is damaged, check all shafts for roughness, especially at seal contact areas. Look for deep scratches or nicks that could have damaged the seal. Determine if a shaft spline, keyway, or burred end could have caused a nick or cut in the seal lip during installation. Inspect the bore into which the seal was fitted. Look for nicks and gouges that could create a path of oil leakage. A coarsely machined bore can allow oil to seep out through a spiral path. Sharp corners at the bore edges could have scored the metal case of the seal when it was installed. These scores can make a path for oil leakage.

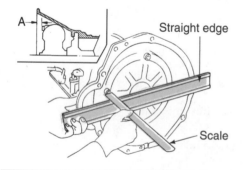

Figure 4-7 After the transmission and engine have been separated, measure and record the depth of the converter into the housing. This measurement will be used during reassembly to ensure the torque converter and oil pump are properly assembled. (Courtesy of Nissan Motors Co., Ltd.)

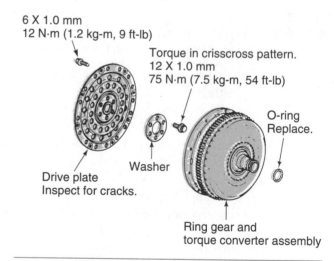

Figure 4-8 Carefully inspect the flexplate and torque converter mounting hardware. (Courtesy of Honda Motor Co.)

Cleaning and Inspection

✓ SERVICE TIP: Determine the model and serial number of the unit you are working on and record these numbers for reference. You will need these to purchase the correct parts during rebuild.

Before disassembling the automatic transmission, care should be taken to clean away any dirt, undercoating, grease, or road grime on the outside of the case. This ensures that dirt will not enter the transmission during disassembly. Once the transmission is clean outside, you may begin the disassembly.

When cleaning automatic transmission parts, avoid the use of solvents, degreasers, and detergents that can decompose the friction composites used in a transmission. Use compressed air to dry components; don't wipe down parts with a rag. The lint from a rag can easily destroy a transmission after it has been rebuilt.

There are many different methods used to clean automatic transmission parts. Some rebuilding shops use a parts washing machine, which takes little time to thoroughly clean a transmission case and associated parts. These parts washers use hot water and a special detergent that are sprayed onto the parts as they rotate inside the cleaner. Many rebuilders simply clean the parts in a mineral spirits tank, where the parts are brushed and hand-cleaned. No matter what type of cleaning procedure is followed, the transmission and parts should be rinsed with water and then air dried before reassembly.

CAUTION: Always wear safety goggles when using compressed air to dry something. The air pressure can easily move dirt, metal, or other debris from around the work area. If any of these gets into your eye, it can cause permanent damage to your eye.

After the case is clean, remove the torque converter and carefully inspect it for damage. Check the converter hub for grooves caused by hardened seals. Also check the bushing contact area. To remove the converter, slowly rotate it as you pull it from the transmission, have a drain pan handy to catch the fluid. It should come right out without binding. This is a good time to check the input shaft splines, stator support splines, and the converter's pump drive hub for any wear or damage. Converters with direct drive shafts should be checked to be sure that no excessive play is present at the drive splines of the shaft or the converter. If any play is found in the converter, the converter or the shaft must be replaced.

✓ SERVICE TIP: It is best to mount the transmission on a bench mount made especially for working on transmissions, like the one shown (Figure 4-9). These mounts are available from many tool suppliers.

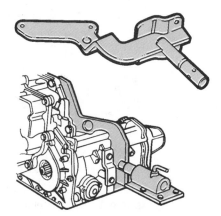

Figure 4-9 A typical transaxle holding fixture (Courtesy of Kent-Moore Tools)

Special Tools

Solvent

Scrub brush

Compressed air nozzle

The term *teardown* is often used to describe the process of disassembling a transmission.

Classroom Manual
Chapter 4, page 74

The AOD and 4T60 are examples of transmissions that use a direct drive shaft.

Position the transmission to perform an endplay check. The transmission endplay checks can provide the technician with information about the condition of internal bearings and seals, as well as clues to possible causes of improper operation found during the road test. These measurements will also determine the thickness of the thrust washer(s) during reassembly. Thrust-washer thickness sets the endplay of various components. Excessive endplay allows clutch drums to move back and forth too much, causing the transmission case to wear. Assembled endplay measurements should be between minimum and maximum specifications, but preferably at the low end of the specifications.

 SERVICE TIP: Record endplay measurements before and during disassembly.

Position the transmission so that the shaft centerline is vertical. This allows the weight of the internal components to load the shafts toward the rear of the transmission. Most GM transmissions can now be checked by mounting a dial indicator to read the input shaft movement. Zero the dial indicator and lift upwards on the input shaft. Most Chrysler and some Ford transmissions require removing the oil sump pan and valve body and prying the input shell upwards. Chrysler and Ford transmissions may be measured horizontally (Figure 4-10), but you may want to center the output shaft with a slip yoke for more accurate readings.

Input shaft endplay is measured with a dial indicator. The dial indicator should be solidly clamped to the bellhousing and the plunger positioned so that it is centered on the end of the input shaft. Move the input shaft in and out of the case and note the reading on the indicator. Compare this reading with the specifications for that transmission. If the endplay is incorrect, it should be corrected during assembly by installing a thrust washer, selective snap ring, or spacer of the correct thickness between the input shaft and the output shaft. Some transmissions require additional endplay checks during disassembly. The manufacturer's recommended procedure for checking endplay should always be followed.

Some transaxles may require more than one check. As most transaxles have their shaft centerlines to one side, the transaxle may need to be partially disassembled in order to gain access to the shafts.

It is possible that an output or final drive endplay check should be made on either a transmission or transaxle (Figure 4-11). The purpose of these settings is to provide long gear life and to eliminate engagement "clunk." All measurements taken during this phase of teardown should be kept during overhaul and used as a guide during the rebuild. Having this information gives the

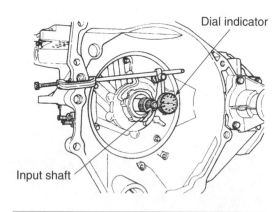

Figure 4-10 The endplay of the input shaft should be checked during disassembly and reassembly. (Courtesy of Chrysler Corp.)

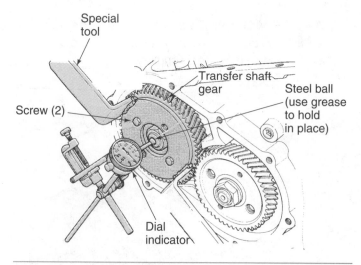

Figure 4-11 Some transaxles require additional endplay checks during assembly and disassembly. (Courtesy of Chrysler Corp.)

technician places to look for worn parts during the teardown and helps obtain the correct shims necessary to correct undesired endplay. If the endplay is excessive, this may indicate that the thrust washers or bearings are worn or the sealing rings and grooves are worn. During rebuild, endplay settings should be kept to the minimum allowable amount. Once the endplay checks have been made and recorded, disassembly can begin.

☑ **SERVICE TIP:** When working on automatic transmissions, there is no such thing as being too clean. All dirt, grease, and other materials should be cleaned off any parts that are going to be reused.

A clean work area, clean tools, and an orderly sequence during teardown will help you to avoid lost time and frustration during reassembly. As you remove the various assemblies from the transmission, place them on the workbench in the order that they were removed. The correct position of thrust washers, snap rings, and even bolts and nuts is of great importance and should be noted. Unless the transmission suffered a major breakdown or was overheated, the internal components of the transmission will be fairly clean due to the nature of the fluid. A light cleaning with mineral spirits or a simple blowing off with low pressure air is all that is necessary for cleaning most parts for inspection.

Each subassembly should be completely disassembled and the parts thoroughly cleaned and checked for signs of excessive wear, scoring, overheating, chipping, or cracking. If there is any question as to the condition of a part, replace it.

Basic Disassembly

Following is a general procedure for disassembling a transmission. Always refer to the manufacturer's recommended procedure for the specific transmission.

With the transmission mounted on a fixture, remove the oil pan. Carefully inspect the oil pan for types of foreign matter. An analysis of the type of matter may help you determine what parts need to be carefully examined.

Remove the bellhousing from the transmission case, if it is not part of the case. Unscrew and remove all externally mounted solenoids and switches, such as the electrical kickdown solenoid. Remove the O-rings and seals for each after the part is removed.

If the transmission has a vacuum modulator, unscrew it and remove it. Remove the speedometer drive assembly, with its gear and O-ring.

If the transmission has a solenoid assembly, unbolt it and remove it from the valve body or case. Pay attention to the wiring harness and connectors. Check them for fraying, corrosion, or other damage.

☑ **SERVICE TIP:** The magnets inside electronic shift control solenoids will attract any ferrous metal that is floating around the inside of the transmission. Thoroughly clean or replace these solenoids during a transmission rebuild.

Unbolt and remove the valve body from the case. Then remove the manual valve from the valve body to prevent the valve from dropping out. Back off the servo piston stem locknut and snugly tighten the piston stem to prevent the front clutch drum from dropping out when removing the front pump.

Using the correct puller, remove the front pump from the case (Figure 4-12). Lay this assembly to the side for further inspection.

Remove the front clutch thrust washer and bearing race. Now back off the front brake band servo piston stem to release the band. Remove the brake band strut and front brake band. The drum and band may be removed together.

Remove the front and rear clutch assemblies (Figure 4-13). Note the positions of the front pump thrust washers and rear clutch thrust washer, if the transmission has them.

Special Tools

Transmission mount
Mineral spirits
Snap-ring pliers
Dial indicator
Assortment of special tools as needed for particular transmission

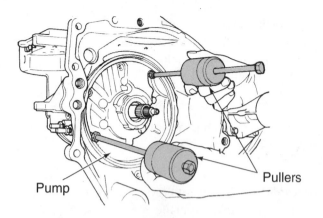

Figure 4-12 The proper tool should be used to pull the oil pump from the transmission housing. (Courtesy of Chrysler Corp.)

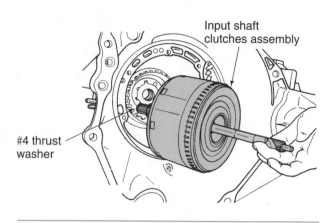

Figure 4-13 When removing clutch assemblies, be careful not to lose track of the placement of the thrust washers. (Courtesy of Chrysler Corp.)

Remove the rear clutch hub, front planetary carrier, and connecting shell. Note the positions of the thrust bearings and the front planetary carrier and thrust washer.

Remove the output shaft snap ring. It will often be easier to remove the snap ring if the carrier is removed first. Remove the carrier snap ring and remove the carrier. Now remove the output shaft snap ring.

Remove the rear connecting drum from the housing. Then using a screwdriver, remove the large retaining snap ring of the rear brake assembly. Tilt the extension housing upward and remove the rear brake assembly.

Then unbolt and remove the extension housing. Be careful not to lose the parking pawl, spring, and retaining washer. Pull out the output shaft, without the governor. Now remove the governor with its attachments (such as the oil distributor, thrust washer, and needle bearing assembly).

Remove the inner race of the one-way clutch, the thrust washer, piston return spring, and thrust ring. Using compressed air, remove the rear brake piston and front servo.

CAUTION: Cover the servo with a rag to prevent ATF from blowing into your face and to prevent the servo piston from popping into your face.

Band servos and accumulators are pistons with seals in a bore held in position by springs and retaining rings. Remove the retaining rings and pull the assembly from its bore for cleaning (Figure 4-14). Check the condition of the piston and springs. Cast iron seal rings may not need replacement, but elastomer seals should always be replaced.

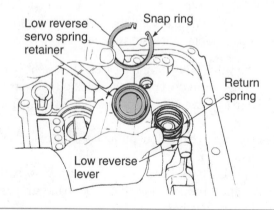

Figure 4-14 All servos and accumulators should be removed from their bores in the case and the bores carefully inspected. (Courtesy of Chrysler Corp.)

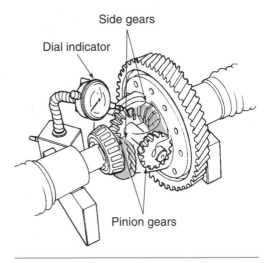

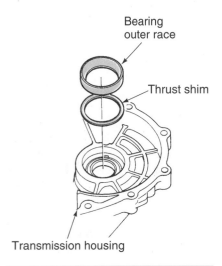

Figure 4-15 The backlash of the side gears in a final drive unit should be checked prior to assembling the transaxle. (Courtesy of Honda Motor Co.)

Figure 4-16 The side bearings of some final drive units are preloaded with selective shims. (Courtesy of Honda Motor Co.)

Once the transmission has been disassembled into its various subassemblies, each subassembly should be disassembled, cleaned, inspected, and reassembled.

Final Drive Components

Transaxle final drive units should be carefully inspected. Examine each gear, thrust washer, and shaft for signs of damage. If the gears are chipped or broken, they should be replaced. Also inspect the gears for signs of overheating or scoring on the bearing surface of the gears.

Final drive units may be helical gear or planetary gear units. The helical type should be checked for worn or chipped teeth, overloaded tapered roller bearings, and excessive differential side gear and spider gear wear. Excessive play in the differential is a cause of engagement clunk (Figure 4-15). Be sure to measure the clearance between the side gears and the differential case and to check the fit of the spider gears on the spider gear shaft. Proper clearances can be found in the appropriate shop manual. It is possible that the side bearings of some final drive units are preloaded with shims (Figure 4-16). Select the correct-size shim to bring the unit into specifications. With a torque wrench, measure the amount of rotating torque. Compare your readings against specifications (Figure 4-17).

The inspection, service, and testing of all hydraulic components are detailed in Chapter 5 of this manual.

Classroom Manual
Chapter 4, page 81

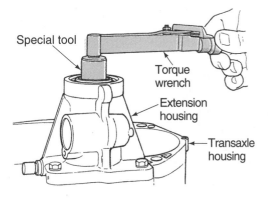

Figure 4-17 Using an inch-pound torque wrench, the turning torque of the transaxle assembly should be checked after assembly. (Courtesy of Chrysler Corp.)

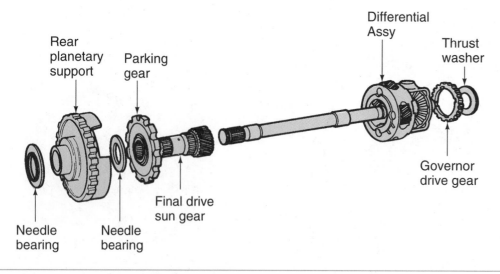

Figure 4-18 Some final drive units, such as this one from an AXOD, are planetary gear sets. (Reprinted with the permission of Ford Motor Co.)

The final drive units in Chrysler and some Ford transaxles typically have helical gears, whereas other Ford and GM transaxles use planetary gear sets.

If the bearing preload and endplay are fine and the bearings are in good condition, the parts can be reused. However, always install new seals during assembly. It should be noted that these bearings function the same as RWD rear axle side bearings and you should not set the preload to the specifications for a new bearing. Used bearings should be set to the amount found during teardown or about one-half the preload of a new bearing.

Planetary-type final drives (Figure 4-18) are also checked for the same differential case problems that the helical-type would encounter. The planetary pinion gears need to be checked for looseness or roughness on their shafts and for endplay. Any problems found normally result in the replacement of the carrier as a unit, since most pinion bearings and shafts are not sold as separate parts. Again, specifications for these parts are found in the shop manual.

Planetary-type final drives, like helical final drives, are available in more than one possible ratio for a given type of transaxle, so care should be taken to assure that the same gear ratios are used during assembly. This is not normally a problem when overhauling a single unit; however, in a shop where many transmissions are being repaired, it is possible to mix up parts, causing problems during the rebuild.

Speedometer Drive Service

Classroom Manual
Chapter 3, page 45

Speedometer gear problems normally result in an inoperative speedometer. However, this problem can also be caused by a faulty cable, drive gear, driven gear, or speedometer. A damaged drive gear can cause the driven gear to fail; therefore, both should be carefully inspected during a transmission overhaul (Figure 4-19). On some transmissions, the speedometer drive gear is a set of gear teeth machined into the output shaft. Inspect this gear. If the teeth are slightly rough, they can be cleaned up and smoothened with a file. If the gear is severely damaged, the entire output shaft must be replaced. Other transmissions have a drive gear that is splined to the output shaft, held in place by a clip (Figure 4-20), or driven and retained by a ball that fits into a depression in the shaft. If a clip is used, it should be carefully inspected for cracks or other damage.

The drive gear can be removed and replaced, if necessary. The driven gear is normally attached to the transmission end of the speedometer cable (Figure 4-21). If the drive gear is damaged, it is very likely that the driven gear will also be damaged. Most driven gears are made of plastic to reduce speedometer noise; therefore, they are weaker than most drive gears. Also check the retainer of the driven gear on the speedometer cable.

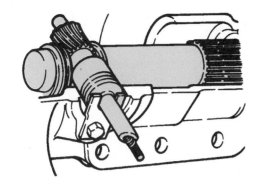

Figure 4-19 The relationship of the speedometer drive and driven gears (Reprinted with the permission of Ford Motor Co.)

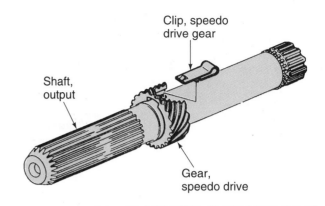

Figure 4-20 The speedometer drive gear is often retained to the output shaft by a clip. (Courtesy of the Hydra-Matic Division of General Motors Corp.)

Parking Pawl

The parking pawl can be inspected after the transmission is disassembled, or on some transmissions, while the transmission is still in the vehicle. Examine the engagement lug on the pawl, making sure it is not rounded off. If the lug is worn, it will allow the pawl to possibly slip out or not fully engage in the parking gear. Most parking pawls pivot on a pin. This also needs to be checked to make sure there is no excessive looseness at this point. The spring that pulls the pawl away from the parking gear must also be checked to make sure it can hold the pawl firmly in place. Also check the position and seating of the spring to make sure it will remain in that position during operation.

The pushrod or operating shaft (Figure 4-22) must provide the correct amount of travel to engage the pawl to the gear. Make sure the shaft is not bent or that the pivot holes in the internal shift linkage are not worn oblong.

Any components found unsuitable should be replaced. It should be noted that the components that make up the parking lock system are the only parts holding the vehicle in place when parked. If they do not function correctly, the car may roll or even drop into reverse when the engine is running, causing an accident or injury. Replace any questionable or damaged parts.

Classroom Manual
Chapter 3, page 38

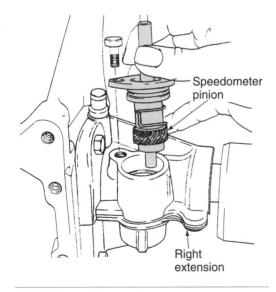

Figure 4-21 The speedometer cable and gear normally fit into a bore in the transmission case or extension housing. (Courtesy of Chrysler Corp.)

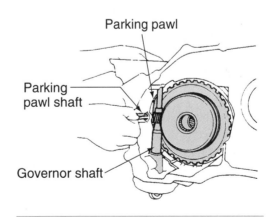

Figure 4-22 Make sure the parking pawl shaft is fully seated and capable of fully applying the pawl. (Courtesy of Nissan Motors Co., Ltd.)

▪ **CAUTION:** A careful inspection of the parking pawl assembly is essential to avoiding possible injury, death, and/or lawsuits.

Before installing the rear extension housing, assemble the parking pin, washer, spring, and pawl. Be sure they are assembled properly. Then install the housing and tighten the bolts to specifications.

Drive Chains

The drive chains used in some transaxles should be inspected for side play and stretch. These checks are made during disassembly and should be repeated as a double-check during reassembly. Chain deflection is measured between the centers of the two sprockets. Typically, very little deflection is allowed.

Classroom Manual
Chapter 4, page 85

Chain deflection is commonly referred to as chain slack.

Deflect the chain inward on one side until it is tight (Figure 4-23). Mark the housing at the point of maximum deflection. Then deflect the chain outward on the same side until it is tight (Figure 4-24). Again mark the housing in line with the outside edge of the chain at the point of maximum deflection. Measure the distance between the two marks. If this distance exceeds specifications, replace the drive chain.

Be sure to check for an identification mark on the chain during disassembly. These can be painted or dark-colored links, and may indicate either the top or the bottom of the chain, so be sure you remember which side was up.

The sprockets should be inspected for tooth wear and for wear at the point where they ride. If the chain was found to be too slack, it may have worn the sprockets in the same manner that engine timing gears wear when the timing chain stretches. A slightly polished appearance on the face of the gears is normal.

The bearings and bushings used on the sprockets need to be checked for damage. The radial needle thrust bearings must be checked for any deterioration of the needles and cage. The running surface in the sprocket must also be checked, as the needles may pound into the gear's surface during abusive operation. The bushings should be checked for any signs of scoring, flaking, or wear. Replace any defective parts. The removal and installation of the chain drive assembly of some transaxles requires that the sprockets be spread slightly apart (Figure 4-25). The key to doing this is to spread the sprockets just the right amount. If they are spread too far, they will not be easy to install or remove.

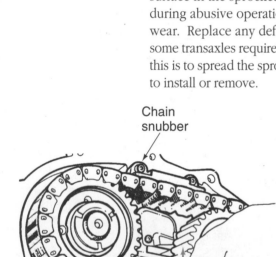

Chain snubber

Transfer chain   Screwdriver

Figure 4-23 While measuring the slack of the drive chain, outwardly deflect the chain and make a mark to that point of deflection. (Courtesy of Chrysler Corp.)

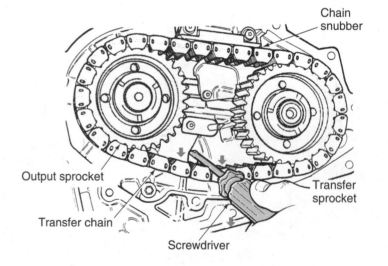

Chain snubber

Output sprocket   Transfer sprocket

Transfer chain   Screwdriver

Figure 4-24 Continue measuring the slack of the drive chain by deflecting the chain inwardly. Mark the point of deflection. The distance between the outward and inward marks is the amount of chain slack. (Courtesy of Chrysler Corp.)

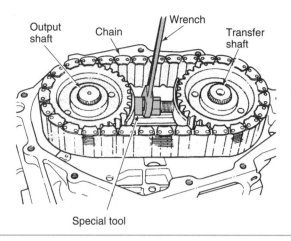

Figure 4-25 To remove and install some drive chains, they must be slightly spread apart with a special tool. (Courtesy of Chrysler Corp.)

Thrust Washers, Bushings, and Bearings

✓ **SERVICE TIP:** All bushings and thrust washers (Figure 4-26) should be prelubed during transmission assembly.

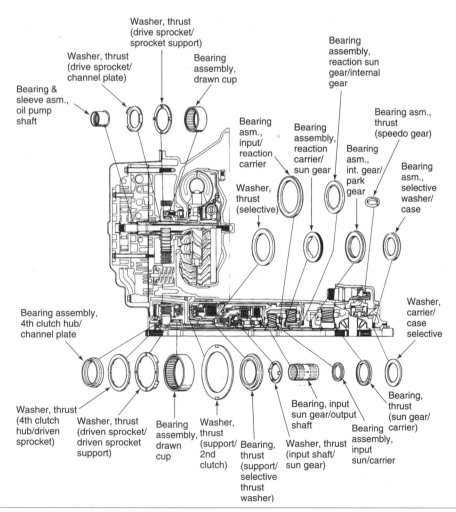

Figure 4-26 Location of thrust washers and bearings in a typical transaxle (Courtesy of the Buick Motor Division of General Motors Corp.)

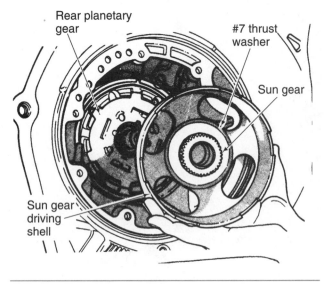

Figure 4-27 Placement of a thrust washer in a planetary gear set (Courtesy of Chrysler Corp.)

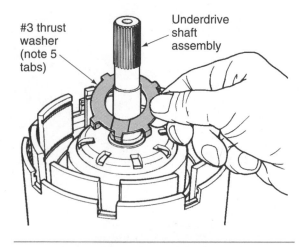

Figure 4-28 A thrust washer with tabs. These tabs must be carefully inspected for cracks or other damage. (Courtesy of Chrysler Corp.)

Thrust washers are often referred to as thrust plates.

Classroom Manual
Chapter 4, page 74

Thrust load is a load placed in parallel with the center of an axis.

Thrust washers are not included in an overhaul kit because they are select fit. Even if the existing thrust washers are the correct size, they should be replaced with new ones if they are pitted or scored.

Other transmissions have similar procedures. Always follow the recommended procedure given in the appropriate service manual.

The purpose of a thrust washer is to support a thrust load and keep parts from rubbing together, preventing premature wear on parts, such as planetary gear sets (Figure 4-27). Selective thrust washers come in various thicknesses to take up clearances and adjust shaft endplay.

Flat thrust washers and bearings should be inspected for scoring, flaking, and wear through to the base material. Flat thrust washers should also be checked for broken or weak tabs (Figure 4-28). These tabs are critical for holding the washer in place. On metal flat thrust washers, the tabs may appear cracked at the bend of the tab. This is a normal appearance due to the characteristics of the materials used to manufacture them. Plastic thrust washers will not show wear unless they are damaged. The only way to check their wear is to measure the thickness with a micrometer and compare it to a new part. All damaged and worn thrust washers and bearings should be replaced.

Proper thrust washer thicknesses are important to the operation of an automatic transmission. Photo Sequence 4 displays the procedure for selecting the proper thrust plate on a common transmission, the Chrysler A500. Incorrect thrust plate selection in a Chrysler A500 transmission can cause third to fourth gear bind-up, slipping in reverse, and/or no engine braking. After following the recommended procedures, refer to the manufacturer's chart for the proper thrust-plate thickness (see Table 4-1).

Table 4-1 CHRYSLER A500 THRUST PLATE SELECTION

Measurement in Inches	Spacer Thickness in Inches
1.250–1.264	0.108–0.110
1.265–1.279	0.123–0.125
1.280–1.294	0.138–0.140
1.295–1.309	0.153–0.155
1.310–1.324	0.168–0.170
1.325–1.339	0.183–0.185
1.340–1.354	0.198–0.200
1.355–1.369	0.213–0.215
1.370–1.384	0.228–0.230
1.385–1.399	0.243–0.245

Photo Sequence 4
Set-up and Procedure for Selecting the Proper Thrust Plate on a Common Transmission

P4-1 Assemble the entire overdrive section and sit it on a suitable workbench.

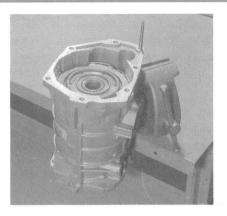

P4-2 Support the extension housing upright in a vise or holding fixture.

P4-3 Lay a straightedge across the extension housing gasket surface.

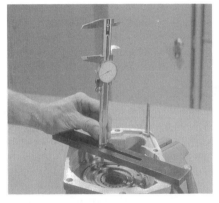

P4-4 Using a depth micrometer or vernier caliper, measure the distance to the sliding hub bearing seat, with the bearing removed.

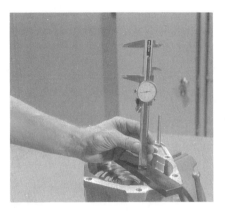

P4-5 Measure at four places, each approximately 90 degrees apart.

P4-6 Add all measurements together, then divide that number by four. This is the average depth measurement.

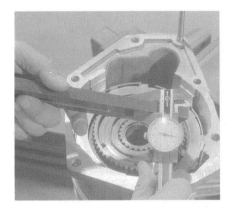

P4-7 Measure the thickness of the straightedge and subtract that amount from the average depth measurement.

P4-8 Compare your readings to the manufacturer's recommendations for the proper thrust plate thickness.

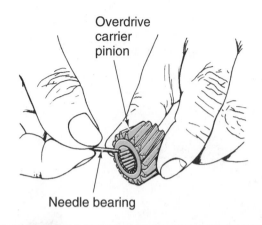

Overdrive carrier pinion

Needle bearing

Figure 4-29 Needle bearings are often located inside the small pinion gears. Coating the inside of the gear with petroleum jelly before the needle bearings are positioned will help to keep them in place. (Courtesy of the Hydra-Matic Division of General Motors Corp.)

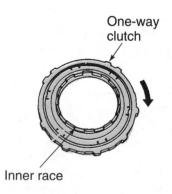

One-way clutch

Inner race

Figure 4-30 To check the action of a one-way clutch, hold the inner race and rotate the clutch in both directions. The clutch should rotate smoothly in one direction and lock in the other direction. (Courtesy of Mazda Motors)

Special Tools

Soft-jawed vise or holding fixture

Straightedge

Depth micrometer or vernier caliper

Etching is a discoloration or removal of some material caused by corrosion or some other chemical reaction.

Classroom Manual
Chapter 4, page 74

Classroom Manual
Chapter 4, page 75

Use a petroleum jelly-type lubricant to hold thrust washers in place during assembly. This will keep them from falling out of place, which will affect endplay. Besides petroleum jelly, there are special greases designed just for automatic transmission assembly that will work fine.

 WARNING: Under no circumstances should you use white lube or chassis lube. These greases will not mix in with the fluid and can plug up orifices and passages, and even hold check balls off of their seats.

All bearings should be checked for roughness before and after cleaning. Carefully examine the inner and outer races, and the rollers, needles (Figure 4-29), or balls for cracks, pitting, etching, or signs of overheating.

Sprag and roller clutches should be inspected in the same way as bearings. Check their operation by attempting to rotate them in both directions (Figure 4-30). If they are working properly, they will allow rotation in one direction only. Also, visually inspect each spring of the clutch unit.

Bushings should be inspected for pitting and scoring. Always check the depth to which bushings are installed and the direction of the oil groove, if so equipped, before you remove them. Many bushings that are used in the planetary gearing and output shaft areas have oiling holes in them. Be sure to line these up correctly during installation or you may block off oil delivery and destroy the geartrain. If any damage is evident on the bushing, it should be replaced.

Bushing wear can be visually checked, as well as checked by observing the lateral movement of the shaft that fits into the bushing. Any noticeable lateral movement indicates wear and the bushing should be replaced. The amount of clearance between the shaft and the bushing can be checked with a wire-type feeler gauge. Insert the wire between the shaft and the bushing. If the gap is greater than the maximum allowable gap, the bushing should be replaced. Bushings must normally fit the shafts they ride on with about a 0.0005–0.0015 inch clearance. You can check this fit by measuring the inside diameter of the bushing and the outside diameter of the shaft with a vernier caliper or micrometer (Figure 4-31). This is a critical fit throughout the transmission and especially at the converter drive hub where 0.0005 inch is the desired fit.

SERVICE TIP: If the converter's drive-hub bushing is excessively worn, check the vehicle's chassis ground.

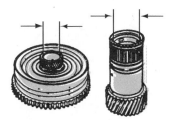

Figure 4-31 The inside diameter of bushings should be measured for wear with an inside micrometer or vernier caliper. (Courtesy of Mazda Motors)

Most bushings are press-fit into a bore. To remove them, they are driven out of the bore with a properly sized bushing tool (Figure 4-32). Some bushings can be removed with a slide hammer fitted with an expanding or threaded fixture that grips to the inside of the bushing. Another way to remove bushings is to carefully cut one side of the bushing and collapse it. Once collapsed, the bushing can be easily removed with a pair of pliers. Small-bore bushings that are located in areas

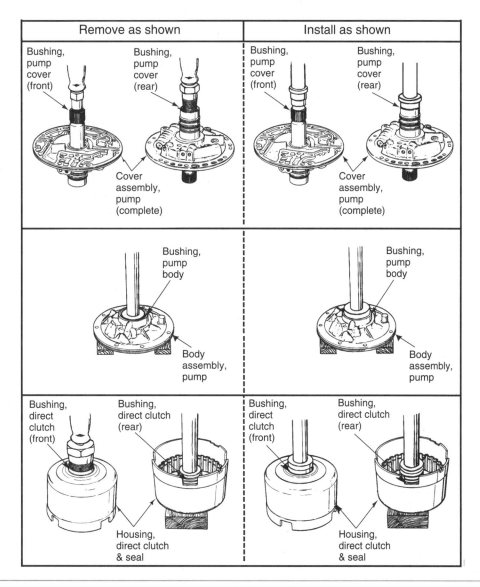

Figure 4-32 The proper procedure and tools for installing various bushings (Courtesy of the Hydra-Matic Division of General Motors Corp.)

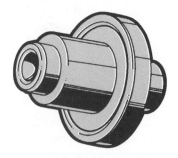

Figure 4-33   A typical bushing removal and installation tool (Courtesy of Kent-Moore Tools)

where it is difficult to use a bushing tool can be removed by tapping the inside bore of the bushing with threads that match a selected bolt, which fits into the bushing. After the bushing has been tapped, insert the bolt and use a slide hammer to pull the bolt and bushing out of its bore.

Whenever possible, all new bushings should be installed with the proper bushing driver. The use of these tools prevents damage to the bushing and allows for proper seating of the bushing into its bore (Figure 4-33).

Planetary Gear Assemblies

Classroom Manual
Chapter 4, page 61

Inspection at this point is to try and eliminate the possibility of putting noise into a newly rebuilt unit. This would result in a costly comeback, so close inspection of the planetary gear set is a must. All planetary gear teeth should be inspected for chips or stripped teeth. Any gear that is mounted to a splined shaft needs the splines checked for mutilation or shifted splines. The planetary gears used in automatic transmissions are the helical-type gear, just like the ones used in most manual transmissions. This type of gear provides low noise in operation, but makes it necessary to check endplay of individual gears during your inspection. The helical cut of the gears makes them thrust to one side during operation. This can put a lot of load on the thrust washers and may wear them beyond specification. Checking these was discussed already, but particular attention should be given to the planetary carriers.

Look first for obvious problems like blackened gears or pinion shafts. These conditions indicate severe overloading and require that the carrier be replaced. Occasionally, the pinion gear and shaft assembly can be replaced individually. When looking at the gears themselves, a bluish condition can be a normal condition, as this is part of a heat-treating process used during manufacture. Check the planetary pinion gears for loose bearings. Check each gear individually by rolling it on its' shaft to feel for roughness or binding of the needle bearings. Wiggle the gear to be sure it is not loose on the shaft. Looseness will cause the gear to whine when it is loaded. Also, inspect the gears' teeth for chips or imperfections, as these will also cause whine.

Check the gear teeth around the inside of the front planetary ring gear. Check the fit between the front planetary carrier to the output shaft splines. Remove the snap ring and thrust washer from the front planetary ring gear (Figure 4-34). Examine the thrust washer and the outer splines of the front drum for burrs and distortion. The rear clutch friction discs must be able to slide on these splines during engagement and disengagement. With the snap ring removed, the front planetary carrier can be removed from the ring gear. Check the planetary carrier gears for endplay by placing a feeler gauge between the planetary carrier and the planetary pinion gear (Figure 4-35). Compare the endplay to specifications. On some Ravigneaux units, the clearance at both ends of the long pinion gears must also be checked and compared to specifications.

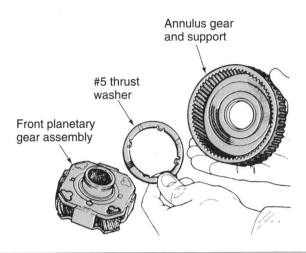

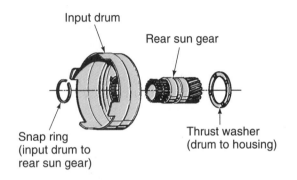

Figure 4-34 The planetary assembly, its thrust washer, and annulus gear should be inspected for signs of abnormal wear. (Courtesy of Chrysler Corp.)

Figure 4-35 The clearance between the pinion gears and the planetary carrier should be checked and compared to specifications. (Courtesy of Mazda Motors)

Figure 4-36 The fit of all drums onto the splines of their mating shafts should be checked. (Courtesy of the Oldsmobile Division of General Motors Corp.)

Check the splines of the sun gear. Sun gears should have their inner bushings inspected for looseness on their respective shafts. Also check the fit of the sun shell to the sun gear (Figure 4-36). The shell can crack where the gears mate with the shell. The sun shell should also be checked for a bell-mouthed condition where it is tabbed to the clutch drum. Any variation from a true round should be considered junk and should not be used. Look at the tabs and check for the best fit into the clutch drum slots. This involves trial fitting the shell and drum at all the possible combinations and marking the point where they fit the tightest. A snug fit here will eliminate bellmouthing due to excess play at the tabs. It can also reduce engagement noise in reverse, second, and fourth gears. This excess play allows the sun shell tabs to strike the clutch drum tabs as the transmission shifts from first to second or when the transmission is shifted into reverse.

The gear carrier should have no cracks or other defects. Replace any abnormal or worn parts. Check the thrust bearings for excessive wear and, if required, correct the input shaft thrust clearance by using a washer with the correct thickness. Determine the correct thickness by measuring the thickness of the existing thrust washer and comparing it to the measured endplay. Now move the gear back and forth to check its endplay. Some shop manuals will give a range for this check, but if none can be found, you can figure about 0.007 to 0.025 inch as an average amount. All the pinions should have about the same endplay.

Shafts

Classroom Manual
Chapter 4, page 74

Carefully examine the area on all shafts that rides in a bushing, bearing, or seal. Also inspect the splines for wear, cracks, or other damage (Figure 4-37). A quick way to determine spline wear is to fit the mating splines and check for lateral movement.

Shafts are checked for scoring in the areas where they ride in bushings. Since the shaft is a much harder material than the bushings, any scoring on the shaft indicates a lack of lubrication at that point. The affected bushing should appear worn into the backing metal. Because shaft-to-bearing fit is critical to correct oil travel throughout the transmission, a scored shaft should be replaced. Lubricating oil is carried through most shafts; therefore, an internal inspection for debris is necessary. A blocked oil delivery hole can starve a bushing, resulting in a scored shaft. The internal oil passage of a shaft may not be able to be visually inspected and only observation during cleaning will give an indication of the openness of the passage. Washing the shaft passage out with a solvent and possibly running a piece of small diameter wire through the passage will dislodge most particles. Be sure to check that the ball that closes off the end of the shaft, if the shaft is so equipped, is securely in place. A missing ball could be the cause of burned planetary gears and scored shafts due to a loss of oil pressure. Any shaft that has an internal bushing, such as the GM THM-350, should be inspected as described earlier. Replace all defective parts as necessary.

Input and output shafts can be either solid, drilled, or tubular. The solid and drilled shafts are supported by bushings, so the bushing journals of the shafts should be free of noticeable wear at these points. Small scratches can be removed with 320-grit emery paper. Grooved or scored shafts require replacement. The splines should not show any sign of waviness along their length. Check drilled shafts to be sure the drilled portion is open and free of any foreign material. Wash out the shaft with solvent and run a small diameter wire through the shaft to dislodge any particles. After running the wire through the opening, wash out the shaft once more and blow it out with compressed air.

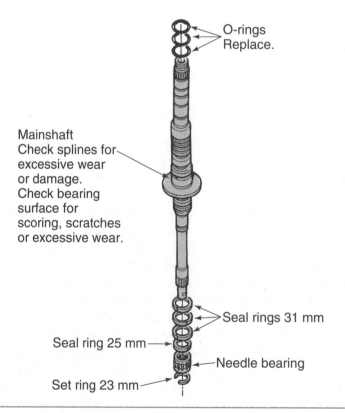

Figure 4-37 All shafts, including their splines and ring grooves, should be carefully inspected for wear or other damage. (Courtesy of Honda Motor Co.)

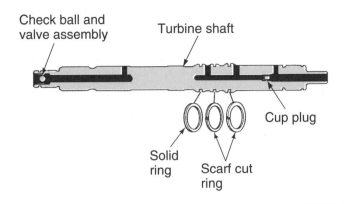

Figure 4-38 If the turbine shaft is fitted with a check ball, make sure it is able to seat and unseat. (Courtesy of the Oldsmobile Division of General Motors Corp.)

If the shaft has a check ball, such as the 4L60 turbine shaft (Figure 4-38), be certain the ball seats in the correct direction. This particular check ball controls oil flow direction to the converter. Some shafts have a ball pressed into one end to block off one end of the shaft. This is used to hold oil in the shaft so the oil is diverted through holes in the side of the shaft. These holes supply oil to bushings, one-way clutches, and planetary gears. If the ball does not fully block the end of the shaft, oil pressure can be lost, causing failure of these components. Some shafts may be used to support another shaft (Figure 4-39), as in the GM THM-350. The output shaft uses the rear of the input shaft to center and support itself. The small bushing found in the front end of the output shaft should always be replaced on this transmission during rebuild. If the GM THM-350 input shaft pilot is worn or scored, a replacement shaft will be necessary.

All hubs, drums, and shells should be carefully examined for wear and damage. Especially look for: nicked or scored band application surfaces on drums, worn or damaged lug grooves in clutch drums, worn splines, and burned or scored thrust surfaces. Minor scoring or burrs on band application surfaces can be removed by lightly polishing the surface with a 600-grit crocus cloth. Any part that is heavily scored or scratched should be replaced.

Crocus cloth is a very fine polishing paper. It is designed to remove very little metal, so it is safe to use on critical surfaces. Never use regular sandpaper, as it will remove too much metal.

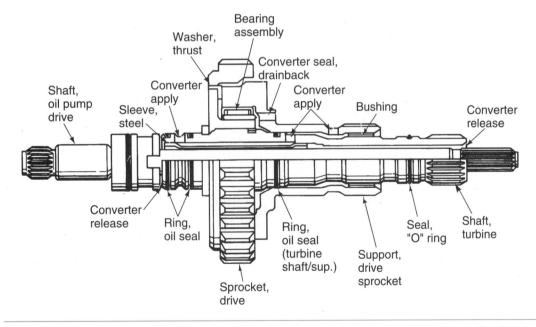

Figure 4-39 Some transmission shafts support another shaft through the bushings fitted to the inside diameter of one shaft. These bushings should be carefully inspected. (Courtesy of the Hydra-Matic Division of General Motors Corp.)

Transmission Case Service

Classroom Manual
Chapter 4, page 76

The transmission case should be thoroughly cleaned and all passages blown out. After the case has been cleaned, all bushings, fluid passages, bolt threads, clutch plate splines, and the governor bore should be checked. The passages can be checked for restrictions and leaks by applying compressed air to each one. If the air comes out the other end, there is no restriction. To check for leaks, plug off one end of the passage and apply air to the other. If pressure builds in that passage, there are probably no leaks in it.

Modern transmission cases are made of aluminum, primarily to save weight. Aluminum is a soft material that can be deformed, scratched, cracked, or scored much more easily than cast iron. Special attention should be given to the following areas: clutch, oil pump, servo, and accumulator bores. All bores should be smooth to avoid scratching or tearing the seals. The servo piston could also hang up in a bore that is deeply scored. Check the fit of the servo piston in the bore without the seal, if possible, to be sure it has free travel. There should be no tight spots or binding over the whole range of travel. Any deep scratches or gouges that cause binding of the piston will require case replacement.

Case-mounted accumulator bores are checked the same as servo bores. The oil pump bore at the front of the case should be free of any scratches that would keep the O-ring from sealing the outer diameter of the pump to the front of the case. Case-mounted hydraulic clutch bores are prone to the same problems as servo bores. Look for any scratches or gouges in the sealing area that would affect the rubber seals. It is possible to damage these areas during disassembly, so be careful with tools used during overhaul.

Sealing surfaces of the case should be inspected for surface roughness, nicks, or scratches where the seals ride (Figure 4-40). Any problems found in servo bores, clutch drum bores, or governor support bores can cause pressure leakage in the affected circuit. Imperfections in steel or cast iron parts can usually be polished out with crocus cloth. Care should be taken so as to not disturb the original shape of the bore. Under no circumstances should sandpaper be used. Sandpaper will leave too deep a scratch in the surface. Use the crocus cloth inside clutch drums to remove the polished marks left by the cast iron sealing rings. This will help the new rings to rotate with the

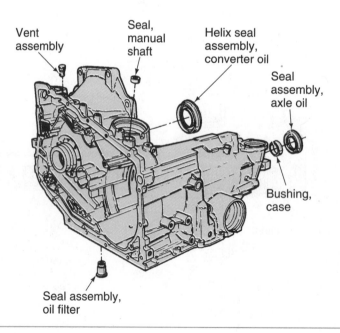

Figure 4-40 All sealing surfaces and bores of the transmission case should be carefully inspected for cracks, grooves, and scratches. (Courtesy of the Buick Motor Division of General Motors Corp.)

drum as designed. As a rule, all sealing rings, either cast iron or Teflon, are replaced during overhaul, as this gives the desired sealing surface required for proper operation.

Passages in the case guide the flow of fluid through the case. Although not that common, porosity in this area can cause cross-tracking of one circuit to another. This can cause bind up (two gears at once) or a slow bleed off of pressure in the affected circuit, which can lead to slow burnout of a clutch or band. If this is suspected, try filling the circuit with solvent and watching to see if the solvent disappears or leaks away. If the solvent goes down, you will have to check each part of the circuit to find where the leak is. Be sure to check that all necessary check balls were in position during disassembly.

Check the valve body mounting area for warpage with a straightedge and feeler gauge. This should be done in several locations so that any crossover from one worm track to another is evident. If there is a slight burr or high spot, it can be removed by flat filing the surface.

A long straightedge should be laid across the lower flange of the case to check for distortion. Any warpage found here may result in circuit leakage, causing any number of hydraulically related problems. Case warpage should be less than 0.010 inch. Cases with bolted-in center supports, such as 3L80, 4L80, 200R4, and E4OD models, should be checked with the support bolted in place. Cases are being made lighter and often will distort during service. This causes a pressure loss to some circuits, causing band or clutch burnouts.

> ✓ **SERVICE TIP:** The Ford E4OD has shown this to be a problem with as much as 1/16-inch warpage across the lower surface. Since the E4OD case costs approximately $625.00, replacing the case is rather expensive. It is possible to push the case back into the correct shape and then, with the center support bolted in, flat file the case to obtain the desired flatness. This type of repair is used in the aftermarket to save an otherwise good case and to reduce costs. This is not a Ford authorized repair.

> ✓ **SERVICE TIP:** The oil pump gears may seize to the pump plate on Ford A4LD transmissions if the bellhousing is warped around the bushing bore. Be extra careful when inspecting this bellhousing as this fault can destroy a transmission rebuild.

Be sure to check all bellhousing bolt holes and dowel pins. Cracks around the bolt holes indicate that the case bolts were tightened with the case out of alignment with the engine block. The case should be replaced if the following problems are present: broken worm tracking, cracked case at the oil pump to case flange, case cracked at clutch housing pressure cavity, ears broken off the bellhousing, or oil pan flange broken off the case. Although it is possible to weld the aluminum case, it is not possible to determine if the repair will hold. A transmission case is very thin and welding may distort the case.

If any of the bolts that were removed during disassembly have aluminum on the threads, the thread bore is damaged and should be repaired. Thread repair entails installing a thread insert, which serves as new threads for the bolt, or retapping the bore. After the threads have been repaired, make sure you thoroughly clean the case.

The small screens found during teardown should be inspected for foreign material. These screens are used to prevent valve hang-up at the pressure regulator and governor and they must be in place. Most screens can be removed easily. Care should be taken when cleaning, because some cleaning solvents will destroy the plastic screens. Low air pressure (approximately 30 psi) can be used to blow the screens out in a reverse direction.

Bushings in a transmission case are normally found in the rear of the case and require the same inspection and replacement techniques as other bushings in the transmission. Always be sure that the oil passage to a pressure-fed bushing or bearing is open and free of dirt and foreign material. It does no good to replace a bushing without checking to be sure it has good oil flow.

Vents are located in the pump body or transmission case and provide for equalization of pressures in the transmission. These vents can be checked by blowing low-pressure air through them,

The oil passages in the case are commonly called worm tracks.

Classroom Manual
Chapter 4, page 76

Shop Manual
Chapter 5, page 105

squirting solvent or brake cleaning spray through them, or by pushing a small-diameter wire through the vent passage. A clean, open passage is all you need to verify proper operation.

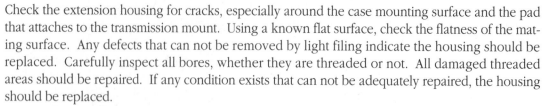

Extension Housing

Classroom Manual
Chapter 4, page 76

Check the extension housing for cracks, especially around the case mounting surface and the pad that attaches to the transmission mount. Using a known flat surface, check the flatness of the mating surface. Any defects that can not be removed by light filing indicate the housing should be replaced. Carefully inspect all bores, whether they are threaded or not. All damaged threaded areas should be repaired. If any condition exists that can not be adequately repaired, the housing should be replaced.

At the rear of the extension housing the slip-yoke bushing, this bushing will normally wear to one side due to loads imparted on it during operation. Oil-feed holes at this bushing must be checked to make sure oil can get to this bushing. The speedometer drive gear is often responsible for throwing oil back to the rear bushing. A sheared or otherwise inoperative speedometer gear could cause the extension housing bushing to fail. Always make sure this bushing is aligned correctly during replacement or premature failure can result.

Before installing the rear extension housing, assemble the parking pawl pin, washer, spring, and pawl, and any other assembly that is enclosed by the extension housing. Be sure they are assembled properly. Install the housing and tighten the bolts to specifications.

Automatic transmission parts distributors package part kits to fit the level of service to be performed. Two of the most common kits are the resealing and reconditioning kits. The resealing kit includes all of the necessary parts, such as gaskets and cast iron, Teflon, rubber, and metal rings and seals to reseal a transaxle or transmission. A reconditioning kit includes all of the necessary seals, gaskets, filters, bushings, clutch friction and steel discs, and brake bands.

Reassembly and Testing

Before proceeding with the final assembly of all components, it is important to verify that the case, housing, and parts are clean and free from dust, dirt, and foreign matter (use an air gun). Have a tray available with clean ATF for lubricating parts. Also have a jar or tube of vaseline for securing washers during installation.

SERVICE TIP: When tightening any fastener that directly or indirectly involves a rotating shaft or other part, rotate that part during and after tightening to ensure that the part does not bind.

Coat all parts with the proper type of ATF. Soak bands and clutches in the fluid for at least 15 minutes before installing them. All new seals and rings should have been installed before beginning final assembly.

Special Tools

Appropriate set of
seal removers and
installers
Fresh ATF
Feeler gauge
Torque wrench

Seal Installation

All seals should be checked in their own bores (Figure 4-41). They should be slightly smaller or larger (+ or - 3%) than their groove or bore. If a seal is not the proper size, find one that is. Do not assume that because a particular seal came with the overhaul kit, it is the correct one.

Never install a seal when it is dry. The seal should slide into position and allow the part it seals to slide into it. A dry seal is easily damaged during installation. Following are some guidelines to follow while installing a seal in an automatic transmission:

1. Install only genuine seals recommended by the manufacturer of the transmission.
2. Use only the proper fluids as stated in the appropriate service manual.
3. Keep the seals and fluids clean and free of dirt.
4. Before installing seals, clean the shaft and/or bore area. Carefully inspect these areas for damage. File or stone away any burrs or bad nicks and polish the surfaces with a fine crocus cloth. Then clean the area to remove the metal particles. In dynamic applications, the sliding surface for the seal should have a mirror finish for best operation.

Classroom Manual
Chapter 4, page 76

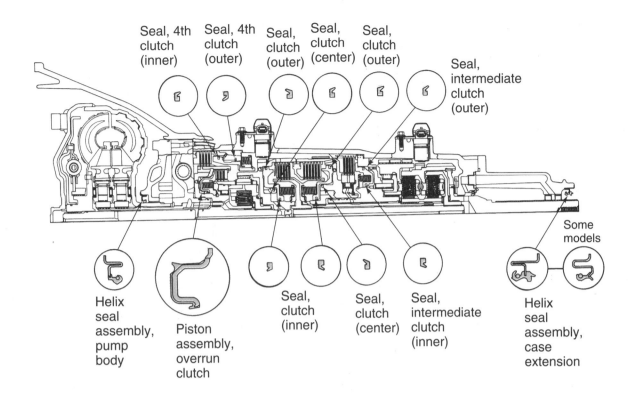

Figure 4-41 Location and position of the various seals in a typical transmission (Courtesy of the Hydra-Matic Division of General Motors Corp.)

5. Lubricate the seal, especially any lips, to ease installation.
6. All metal sealing rings should also be checked for proper fit. Since these rings seal on their outer diameter, the seal should be inserted in its bore and should feel tight there. If the seal has some form of locking ends, these should be interlocked prior to trying the seal in its bore (Figures 4-42 and 4-43). The fit of the sealing rings in their shaft groove should also be checked. If the ring can move laterally in the groove, the groove is worn and this will cause internal fluid leaks. To check the side clearance of the ring, place the ring into its groove and measure the clearance between the ring and the groove with a feeler gauge. Typically, the clearance should not exceed 0.003 inch.

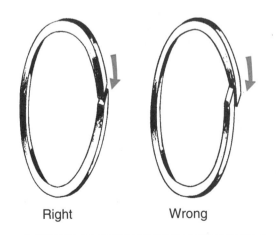

Right   Wrong

Figure 4-42 Installation of Teflon oil seal rings (Courtesy of the Oldsmobile Division of General Motors Corp.)

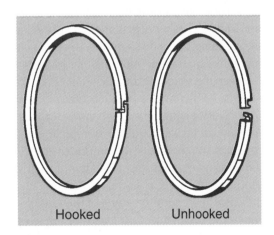

Hooked   Unhooked

Figure 4-43 Installation of cast iron oil seal rings (Courtesy of the Oldsmobile Division of General Motors Corp.)

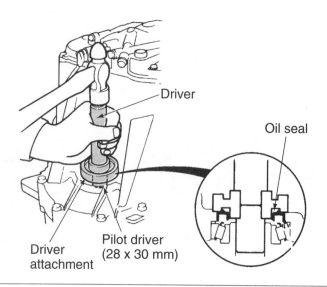

Figure 4-44 Using a special driver to install an oil seal. Note how the tool prevents damage to the seal. (Courtesy of Honda Motor Co.)

7. While checking the clearance, look for nicks in the grooves and for evidence of groove taper or stepping. If the grooves are tapered or stepped, the shaft will need to be replaced. If there are burrs or nicks in the grooves, they can be filed away.
8. Any distorted or undersized sealing ring should be replaced.
9. Always use the correct driver when installing a seal (Figure 4-44) and be careful not to damage the seal during installation.

Transmission Reassembly

Carefully examine all thrust washers and coat them with petroleum jelly before placing them in the housing. Install the thrust ring, piston return spring, thrust washer, and one-way clutch inner race into the case. Align and start the bolts into the inner race from the rear of the case. Torque the bolts to specifications.

 WARNING: Check and verify that the return spring is centered onto the race before tightening.

Lubricate and install the rear piston into the case. After determining the correct number of friction and steel plates, install the steel dished plate first, then the steel and friction plates, and finally the retaining plate and snap-ring. Using a suitable blow gun with a rubber tapered tip, air check the rear brake operation. After the rear brake has been completely assembled, measure the clearance between the snap-ring and the retainer plate. Select the proper thickness of retaining plate that will give the correct ring-to-plate clearance if the measurement does not meet the specified limits.

Slide the governor distributor assembly onto the output shaft from the front of the shaft. Install the shaft and governor distributor into the case, using care not to damage the distributor rings.

On some models, the output shaft, bearing, and appropriate gauging shims are placed into the transmission housing. The output shaft washer and bolt are then installed. While holding the output shaft and gear assembly, torque the output shaft nut to specifications. Then install a dial indicator and check the travel of the output shaft as it is pushed and pulled. Remove the gauging shims and install the correct sizes of service shims, output shaft gear, washer, and nut. Torque the output shaft nut to specifications. Using an inch-pound torque wrench, check the turning torque of the output shaft and compare this reading to specifications.

Place the small thrust washer on the pilot end of the transaxle output shaft. Then place the rear clutch assembly, front clutch drum, turbine shaft, and thrust washer into the housing. Locate and align the rear clutch over its hub. Gently move the rear clutch and turbine shaft around, rotating the assembly to engage the teeth of the friction discs with the rear clutch hub. Align the direct clutch assembly over the front clutch hub. Move the input shaft back and forth, rotating it so the front clutch friction disks engage with the front clutch hub.

Position the thrust washer to the back of the rear planetary carrier. Install the rear planetary carrier and thrust washer into the housing to engage the rear planetary ring gear. Install the front thrust washer and the drive shell assembly, engaging the common sun gear with the planetary pinions in the rear planetary carrier. Assemble the front planetary gear assembly into the front planetary ring gear. Make sure the planetary pinion gear shafts are securely locked to the planetary carrier.

Install the one-way sprag into the one-way clutch outer race with the arrow on the sprag facing the front of the transmission. Install the connecting drum with sprag by rotating the drum clockwise, using a slight pressure and wobbling to align the plates with hub and sprag assembly. The connecting drum should now be free to rotate clockwise only. This check will verify that the sprag is correctly installed and operative. Now install the rear internal gear and the shaft's snapring.

Secure the thrust bearing with petroleum jelly and install the rear planet carrier and the snap ring.

SERVICE TIP: This snap-ring may be thinner than the clutch drum snap-ring, so be sure you are using the correct size. Should you encounter trouble in having sufficient space to install the snap ring into the drum groove, pull the connecting drum forward as far as possible. This will give you sufficient groove clearance to install the drum snap ring.

Assemble the front and rear clutch drum assemblies together and lay them flat on the bench. Be sure the rear hub thrust bearing is properly seated, then measure from the face of the front clutch drum to the top of the thrust bearing. Install the thrust washer and pump front bearing race to the pump. Then measure from the pump shaft (bearing race included) to the race of the thrust washer. Normally, the difference in measurements should be about 0.020 inch. If the thrust washer is not within the limits, replace it with one of the correct thickness.

Total endplay should now be checked. Set the transmission case on end, front end up. Be sure the thrust bearings are secure with petroleum jelly. Pick up the complete front clutch assembly and install it into the case. Be sure all parts are seated before proceeding with the measurement. Using a dial indicator or caliper, measure the distance from the rear hub thrust bearing to the case. Next measure the pump with the front bearing race and gasket installed. Tolerance should fall within specifications. If the difference between the measurements is not within tolerance, select the proper size front bearing race. If it is necessary to change the front bearing race, be sure to change the front clutch thrust washers the same amount.

Install the brake band servo. Use extreme care not to damage the O-rings. Lubricate around the seals. Then install and torque the retainer bolts to specifications. Loosen the piston stem. Install the brake band strut and finger tighten the band servo piston stem just enough to keep the band and strut snug or from falling out. Do not adjust the band at this time. Air check for proper performance.

Place some petroleum jelly in two or three spots around the oil pump gasket and position it on the transaxle housing. Next align the pump and install it with care. Tighten the pump-attaching bolts to specifications in the specified order. After the bolts are tight, check the rotation of the input shaft. If the shaft does not rotate, disassemble the transmission to locate the misplaced thrust washer.

Before installing the bellhousing, check the bolt hold alignment. Install the bellhousing and torque the retaining bolts to specifications.

Now adjust the band after you check to make sure that the brake band strut is correctly installed. Torque the piston stem to specifications. Back off two (or the number specified by the manufacturer) full turns and secure with a locknut. Tighten the locknut to specifications.

Before proceeding with the installation of the valve body assembly, it is good practice to perform a final air check of all assembled components. This will ensure that you have not overlooked the tightening of any bolts or damaged any seals during assembly.

Assemble the parking pawl assembly. Place the assembly into its position and install the extension housing with a new gasket, then tighten the attaching bolts to the proper specifications.

On transaxles, the differential assembly should be disassembled, cleaned, inspected, and reassembled. After it has been reassembled, measure its endplay with gauging shims. Then select a shim thick enough to correct the endplay. After installing the proper shims, measure the differential turning torque with an inch-pound torque wrench. Increase or decrease the shim thickness to provide the correct turning torque.

Install the valve body. Be sure the manual valve is in alignment with the selector pin. Tighten the valve body attaching bolts to the specified torque.

 WARNING: Be certain that the proper length bolts are installed in the related depth holes. The length of the bolts varies.

Before installing the vacuum modulator valve, it is good practice to measure the depth of the hole in which it is inserted. This measurement determines the correct rod length to ensure proper performance. Refer to the service manual to determine the correct rod length based on your measurements. You should note that the actual rod size is slightly longer than the measurement taken. If you do not have the correct chart, it is fairly safe to simply add 0.070 inch to the measurement taken.

Before installing the kickdown solenoid or other solenoids, check to verify that they are operating properly. Connect the solenoid to a 12-volt source and ground the other terminal. You should hear the solenoid click on. If the solenoid is good, install it. If the solenoid does not check out good, replace it.

Install the kickdown switch. Again check the operation of the switch. This is best done by connecting an ohmmeter across the switch and moving the switch through its different settings.

Before installing the oil pan, check the alignment and operation of the control lever and parking pawl engagement. Make a final check to be sure all bolts are installed in the valve body. Install the oil pan with a new gasket. Torque the bolts to specifications.

Lubricate the oil pump's lip seal and the converter neck before installing the converter. Install the converter, making sure that the converter is properly meshed with the oil pump drive gear.

 SERVICE TIP: It is a good idea to partially prefill all torque converters before installing them in the vehicle.

The transmission is now ready for installation into the vehicle. Use the reverse of the removal procedures. Remember to follow proper fluid-filling procedures.

CASE STUDY

A customer brought a late model GM vehicle with a 2.8 L engine into a transmission rebuilding shop to have the transmission rebuilt. This is the second time in the last three years that the transmission went out.

The technician diagnosed the problem as low pump pressure in all gears. A faulty pump was suspected. While pulling out the transmission, she noticed that the flexplate was cracked. This somewhat verified the diagnosis, since a faulty or excessively worn oil pump body bushing would cause this problem. She assumed that by replacing the oil pump and the flexplate, the customer would be happy and have many years of troublefree transmission service.

After completing the repairs and during the installation of the transmission, the technician noticed the dowel pins. There was nothing wrong with them, but they made her remember something she had read in a technical bulletin. She researched the TSBs, and found the one she was looking for. The two 15.0-mm long dowels pins at the rear of the engine should be replaced with two 19.0-mm long dowels. This extra length gives more rigidity to the mating of the transmission and the engine.

The slight shifting of the transmission against the engine caused both the oil pump bushings and the flexplate to go bad. Had the technician not read the TSB before, she would have replaced the flexplate and the oil pump only to have the customer come back again for the same problem.

Terms to Know

Chain slack	Dowels	Teardown
Crocus cloth	Etching	Thrust loads

ASE Style Review Questions

1. *Technician A* says before tearing down a transmission, you should inspect the material trapped in the fluid filter.
   Technician B says installing an overhaul kit will take care of nearly all the problems with a transmission.
   Who is correct?
   A. A only   **C.** Both A and B
   B. B only   **D.** Neither A nor B

2. While discussing endplay checks:
   Technician A says these checks should be taken before the transmission is disassembled.
   Technician B says these checks should be taken after the transmission is reassembled.
   Who is correct?
   A. A only   **C.** Both A and B
   B. B only   **D.** Neither A nor B

3. *Technician A* says abnormal noises from a transmission will never be caused by faulty clutches or bands.
   Technician B says abnormal noises from a transmission are typically caused by a faulty torque converter.
   Who is correct?
   A. A only   **C.** Both A and B
   B. B only   **D.** Neither A nor B

4. *Technician A* says bushings should be heated with a torch to easily remove them.
   Technician B says bushings should be removed with a slide hammer and the correct attachment.
   Who is correct?
   A. A only   **C.** Both A and B
   B. B only   **D.** Neither A nor B

5. While checking a planetary gear set:
   Technician A says the end clearance of the pinion gears should be checked with a feeler gauge.
   Technician B says the end clearance of the long pinions in a Ravigneaux gear set should be at both ends.
   Who is correct?
   A. A only   **C.** Both A and B
   B. B only   **D.** Neither A nor B

6. *Technician A* says most vibration problems are caused by an unbalanced torque converter.
   Technician B says vibration problems can be caused by a faulty output shaft.
   Who is correct?
   A. A only   **C.** Both A and B
   B. B only   **D.** Neither A nor B

7. *Technician A* says a blocked oil delivery passage will cause the shaft to score.
Technician B says if the shaft is fitted with a check ball and the check ball does not seat properly, low oil pressure will result.
Who is correct?
A. A only   **C.** Both A and B
B. B only   **D.** Neither A nor B

8. While checking a transmission's vent:
Technician A applies a vacuum to it and says that if the vent leaks, it should be replaced.
Technician B runs ATF through the vent and says that if the vent cannot hold fluid, it must be replaced.
Who is correct?
A. A only   **C.** Both A and B
B. B only   **D.** Neither A nor B

9. *Technician A* says the seals should be kept clean and free of dirt, before and during installation.
Technician B says most seals should be installed dry.
Who is correct?
A. A only   **C.** Both A and B
B. B only   **D.** Neither A nor B

10. While discussing computer relearn procedures:
Technician A says this type of computer strategy is designed to optimize driveability.
Technician B says the computer on some cars needs to learn about the vehicle before it can control the systems properly.
Who is correct?
A. A only   **C.** Both A and B
B. B only   **D.** Neither A nor B

Table 4-2 ASE TASK

Diagnose noise and vibration problems and determine needed repairs.

Problem Area	Symptoms	Possible Causes	Classroom Manual	Shop Manual
NOISE	Buzzing noise	1. Improper fluid level or condition	50	33
		2. Defective governor	54	111
		3. Damaged one-way clutch	51	82
		4. Damaged or defective torque converter	39	113
		5. Defective oil pump	50	113
		6. Defective valve body and/or accumulator	51	105
		7. Water in ATF	50	34
		8. Damaged planetary gear set	61	84
	General operation is noisy	1. Improper fluid level	50	33
		2. Linkage out of adjustment	46	47
		3. Damaged planetary gear set	61	84
		4. Internal leakage or pump cavitation	50	113
		5. Sticking valves or dirty valve body	51	105
		6. Faulty band or clutch application	51	119
		7. Transmission cooler lines hitting against the frame or engine	50	142
		8. Damaged roller bearings	74	79
	Growling, grating, or scraping noises	1. Improper fluid level or condition	50	33
		2. Worn bands and/or clutches	51	119
		3. Defective torque converter	39	113
		4. Defective one-way/overrun clutch	51	141
		5. Damaged or worn oil pump	50	113
		6. Clogged oil filter	50	34
		7. Defective flexplate	50	113
		8. Damaged planetary gear set	61	84
		9. Low pressures	50	113

97

Hydraulic System and Reaction Unit Service

Upon completion and review of this chapter, you should be able to:

❏ Diagnose mechanical and vacuum control systems; determine needed repairs.

❏ Inspect, adjust, and replace the vacuum modulator, lines, and hoses.

❏ Air test the operation of the clutch pack and servo assemblies.

❏ Perform oil pressure tests; determine needed repairs.

❏ Inspect and measure valve body mating surfaces, bores, springs, sleeves, retainers, brackets, check balls, screens, spacers, and gaskets while the transmission is in or out of the vehicle.

❏ Check and/or adjust valve body bolt torque while the transmission is in or out of the vehicle.

❏ Inspect, adjust, repair, and replace the governor cover, seals, sleeve, valve, weights, springs, retainers, and gear while the transmission is in or out of the vehicle.

❏ Inspect, measure, and replace oil pump housings and parts.

❏ Inspect and replace bands and drums.

❏ Adjust bands, internally and externally.

❏ Inspect and replace the servo, including the bore, piston, seals, pin, spring, and retainers and repair or replace as necessary.

❏ Inspect an accumulator bore, piston, seals, spring, and retainers while the transmission is in or out of the vehicle.

❏ Inspect a multiple-disc clutch assembly and replace as necessary.

❏ Measure and adjust clutch pack clearance.

❏ Air test the operation of the clutch pack and servo assemblies.

❏ Inspect and replace external seals and gaskets.

❏ Inspect oil delivery seal rings, ring grooves, and sealing surface areas.

❏ Inspect one-way clutch assemblies and replace parts as necessary.

❏ Inspect, test, flush, and replace the cooler, lines, and fittings.

Basic Tools

Basic mechanic's tool set

Appropriate service manual

Torque wrench

Lintfree rag

Classroom Manual
Chapter 5, page 121

A transmission's oil circuit chart is often referred to as the transmission's flow chart.

Diagnose Hydraulic and Vacuum Control Systems

The best way to identify the exact cause of transmission problems is to use the results of a road test, logic, and oil circuit charts for the transmission being worked on. Using these circuits, you can trace problems to specific valves, servos, clutches, and bands.

The basic oil flow is the same for all transmissions. The oil pump supplies the pressure that is used throughout the transmission. Pressure from the pump always goes to the pressure-regulating valve. From there, the pressurized fluid is directed to the manual shift valve. When the gear selector is moved, the manual valve directs the oil to other valves and to the apply devices. By following the flow of the oil on the oil circuit chart, you can identify which valves and apply devices should be operating in each particular selector position. Through a process of elimination, you can identify the most probable cause of the problem.

In most cases, the transmission or transaxle will need to be removed to repair or replace the items causing the problem. However, some transmissions allow for a limited amount of service to the apply devices and control valves.

✓ **SERVICE TIP:** Whenever there is a shifting problem that is not readily diagnosed, check the service manual and technical service bulletins before making assumptions about the problem or getting frustrated. Sometimes, the cause of a problem is something that you would least expect. For example: A single, and hard to diagnose, problem in a THM200-R4 transmission can cause it to have no movement, forward movement in neutral, no forward movement, or no reverse gear. All of these complaints may be caused by a defective center support. The center support is made of two pieces, the aluminum housing and the steel sleeve, which are pressed into the center of the support. Grooves machined into the sleeve provide channels for several oil paths through the support. If the sleeve rotates, the potential for several problems exists. The most common complaint is a bind-up in reverse. It is not recommended that the sleeve be removed and a new one installed, rather the entire center support assembly should be replaced.

Shop Manual
Chapter 3, page 47

Mechanical and/or vacuum controls can also contribute to shifting problems. The condition and adjustment of the various linkages and cables should be checked whenever there is a shifting problem. If upshifts do not occur at the correct speeds or do not occur at all, the problem may be a faulty vacuum modulator or the vacuum supply line is damaged. Diagnosing a vacuum modulator system begins with checking the vacuum at the line to the modulator. Normal vacuum readings indicate that there are no vacuum leaks in the line and it can be assumed that the engine's condition is satisfactory. If you find transmission fluid in the vacuum line to the modulator, the vacuum diaphragm in the modulator is leaking and you should replace the modulator. To verify this, apply a vacuum to the valve with a hand-held vacuum pump. The valve will not hold a vacuum if the diaphragm is leaking. If the problem seems to be modulator-related, but the vacuum source, vacuum lines, and vacuum modulator all check out fine, the modulator may need to be adjusted.

The modulator must be removed from the transmission in order to adjust it. Some modulators are screwed into the transmission case, while others are retained by a clamp and cap screw. While removing the modulator valve assembly, take care not to lose the actuating pin, which may drop out as the valve is removed. Use a hand-held vacuum pump and the required special gauge pins to bench test and adjust the modulator. Always follow the manufacturer's guidelines when making this adjustment.

If all checks indicate that the problem is either an apply device or in the valving, an air pressure test can help identify the exact problem. Air pressure tests are also performed during disassembly to locate leaking seals and during reassembly to check the operation of the clutches and servos.

An air pressure test is conducted by applying clean, moisturefree air, at approximately 40 psi, through a rubber-tipped air nozzle. With the valve body removed, direct air pressure to the case holes that lead to the servo and clutch apply passages. You should clearly hear the action of the holding devices. If a hissing noise is heard, a seal is probably leaking in that circuit (Figure 5-1). If you can not hear the action of the servo or clutch, apply air pressure again and watch the assembly to see if it is reacting to the air. A clutch or servo should react immediately to the introduction of the apply air. If it does not, something is making it stick. Repair or replace the apply devices if they do not operate normally.

Air pressure can normally be directed to the following circuits, through the appropriate hole for each: front clutch, rear clutch, kickdown servo, low servo, and reverse servo. Some manufacturers recommend the use of a specially drilled plate, which is bolted to the transmission case. This plate not only clearly identifies which passages to test but also seals off the other passages. Air is applied directly through the holes in the plate.

Pressure Tests

If you cannot identify the cause of a transmission problem from your inspection or road test, a pressure test should be conducted. This test measures the fluid pressure of the different transmission circuits during the various operating gears and gear selector positions. Refer to the illustrations shown (Figures 5-2, 5-3, 5-4, and 5-5) for a quick review of power flow in the various gears.

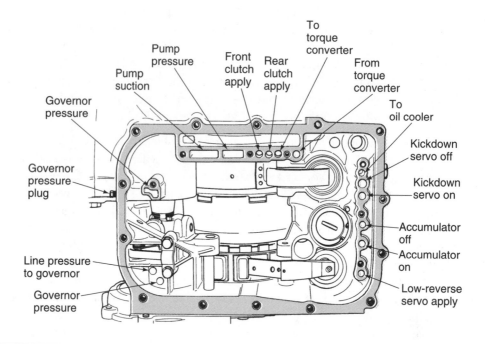

Figure 5-1 Air testing points on a typical transmission (Courtesy of Chrysler Corp.)

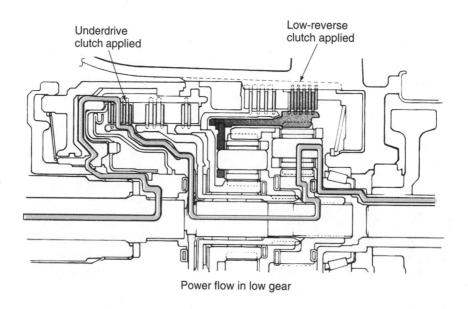

Power flow in low gear

Figure 5-2 Power flow in low gear (Courtesy of Chrysler Corp.)

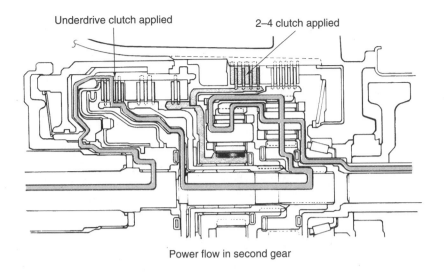

Power flow in second gear

Figure 5-3 Power flow in second gear (Courtesy of Chrysler Corp.)

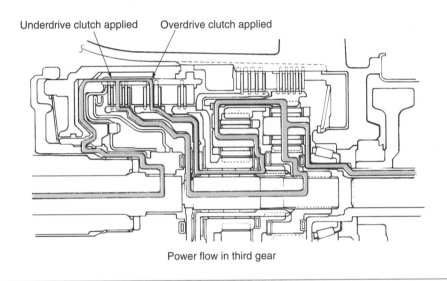

Power flow in third gear

Figure 5-4 Power flow in third gear (Courtesy of Chrysler Corp.)

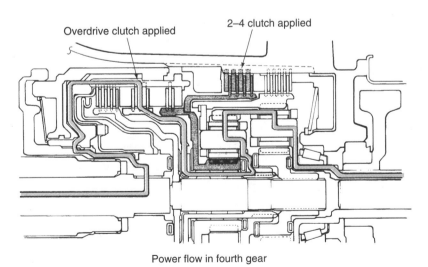

Power flow in fourth gear

Figure 5-5 Power flow in fourth gear (Courtesy of Chrysler Corp.)

VFM = variable force motor.

The number of hydraulic circuits that can be tested varies with the different makes and models of transmissions. However, all transmissions are equipped with pressure taps, which allow the pressure test equipment to be connected to the transmission's hydraulic circuits (Figures 5-6 and 5-7). Pressure testing checks the operation of the oil pump, pressure regulator valve, throttle valve, and vacuum modulator system (if the vehicle is equipped with one), plus the governor assembly. Pressure tests can be conducted on all transmissions whether their pressures are regulated by a VFM, vacuum modulator, or through conventional valving.

Most transmission problems can be identified without conducting a pressure test; therefore, a pressure test should never be heavily relied on or used to begin your diagnostics. Also, the results from the pressure tests may have little value when diagnosing some problems. For example, if the transmission does not operate in a particular gear but operates fine in other gears, a pressure test will not identify the problem. If there is enough oil pressure to operate the transmission in the other gears, then there is certainly enough pressure to operate in the malfunctioning one. When there is a specific failure or slippage in any gear, the cause of the problem is identified more easily by a complete visual inspection, a road test, and the use of logic.

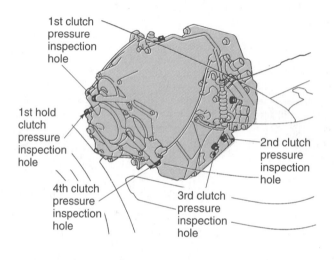

Figure 5-6 Pressure taps on the outside of a typical Honda transaxle case (Courtesy of Honda Motor Co.)

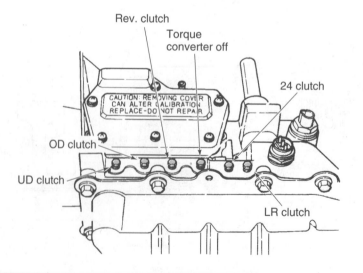

Figure 5-7 Pressure taps on the outside of a typical Chrysler transaxle (Courtesy of Chrysler Corp.)

A pressure test has its greatest value when the transmission shifts roughly or when the shift timing is wrong. Both of these problems may be caused by excessive line pressure, which can be verified by a pressure test.

Conducting a Pressure Test

To conduct a pressure test, use a tachometer, two pressure gauges, the correct manufacturer specifications (Figure 5-8), and (for vehicles equipped with a vacuum modulator) a vacuum gauge and

PRELIMINARY CHECK PROCEDURE

CHECK TRANSMISSION OIL LEVEL • CHECK AND ADJUST T.V. CABLE
CHECK OUTSIDE MANUAL LINKAGE AND CORRECT • CHECK ENGINE TUNE
INSTALL OIL PRESSURE GAGE* • CONNECT TACHOMETER TO ENGINE
CHECK OIL PRESSURES IN THE FOLLOWING MANNER:

Minimum T.V. Line Pressure Check
Set the T.V. cable to specification; and with the brakes applied, take the line pressure readings in the ranges and at the engine r.p.m.'s indicated in the chart below.

Full T.V. Line Pressure Check
Full T.V. line pressure readings are obtained by tying or holding the T.V. cable to the full extent of its travel; and with the brakes applied, take the line pressure readings in the ranges and at the engine r.p.m.'s indicated in the chart below.

*For line pressure tap location see Fig. A-10, Item 405.

CAUTION Brakes must be applied at all times.   **NOTICE** Total running time for this combination not to exceed 2 minutes.

RANGE	MODEL	MINIMUM T.V.		MAXIMUM T.V.	
		kPa	P.S.I.	kPa	P.S.I.
Park @ 1000 RPM	7BPC,7BJC,7BMC,7CAC,7CBC,7CDC,7CNC,7CPC,7CTC,7CUC,7CXC,7HRC, 7JPC,7JTC,7KDC,7LHC,7PDC,7PKC,7PMC,7PNC,7PSC,7PTC,7PWC	459-507	66-74	459-507	66-74
	7HLC,7LKC,7PHC	511-581	74-84	511-581	74-84
	7HZC,7JAC,7JDC,7JFC,7JKC,7JMC,7JNC,7JUC,7JWC,7RAC,7RCC,7TAC,7TBC	396-436	57-63	396-436	57-63
Reverse @ 1000 RPM	7BPC,7CTC,7CXC	804-887	117-129	1630-1847	236-268
	7BJC,7JPC,7PTC	804-887	117-129	1487-1653	216-240
	7BMC,7HRC,7JTC,7KDC,7PDC,7PSC	804-887	117-129	1748-1984	254-288
	7CAC,7CBC,7PKC,7PNC,7PWC	804-887	117-129	1512-1710	219-248
	7CDC,7CNC,7CPC,7CUC	804-887	117-129	1601-1781	232-258
	7HLC,7PHC	895-1018	130-148	1721-1978	250-287
	7HZC	760-837	110-121	1758-1956	255-284
	7JAC,7JUC,7RCC,7TAC	760-837	110-121	1633-1816	237-263
	7JDC,7JMC	694-764	101-111	1402-1587	203-230
	7JFC,7JKC	694-764	101-111	1605-1786	233-259
	7JNC	694-764	101-111	1185-1317	172-191
	7JWC	694-764	101-111	1456-1612	211-234
	7LHC	804-887	117-129	1866-2121	271-308
	7LKC	895-1018	130-148	1957-2252	284-327
	7PMC	780-861	113-125	1810-2057	263-298
	7RAC	694-764	101-111	1377-1530	200-222
	7TBC	760-837	110-121	1509-1676	219-243
Neutral/ Drive @ 1000 RPM	7BPC,7CTC,7CXC	459-507	67-74	931-1055	135-153
	7BJC,7JPC,7PTC	459-507	67-74	849-944	123-137
	7BMC,7HRC,7JTC,7KDC,7PDC,7PSC	459-507	67-74	998-1133	145-164
	7CAC,7CBC,7PKC,7PNC,7PWC	459-507	67-74	863-976	125-142
	7CDC,7CNC,7CPC,7CUC	459-507	67-74	914-1017	133-147
	7HLC,7PHC	511-581	74-84	983-1130	143-164
	7HZC,7JFC,7JKC	396-436	57-63	917-1020	133-148
	7JAC,7JUC	396-436	57-63	851-947	123-137
	7JDC,7JMC	396-436	57-63	801-906	116-131
	7JNC	396-436	57-63	677-752	98-109
	7JWC	396-436	57-63	831-921	121-134
	7LHC,7PMC	459-507	67-74	1066-1211	155-176
	7LKC	511-581	74-84	1118-1286	162-187
	7RAC,7TBC	396-436	57-63	786-874	114-127
	7RCC,7TAC	396-436	57-63	851-947	123-137
Intermediate/ Lo @ 1000 RPM	7BPC,7BJC,7BMC,7CAC,7CBC,7CDC,7CNC,7CPC,7CTC,7CUC,7CXC, 7HRC,7JPC,7JTC,7KDC,7LHC,7PDC,7PKC,7PNC,7PSC,7PTC,7PWC	788-869	114-126	788-869	114-126
	7HLC,7LKC,7PHC	877-998	127-145	877-998	127-145
	7HZC,7JAC,7JUC,7RCC,7TAC,7TBC	827-910	120-132	827-910	120-132
	7JDC,7JFC,7JKC,7JMC,7JNC,7JWC,7RAC	680-749	99-109	680-749	99-109
	7PMC	958-1057	139-153	958-1057	139-153

Line pressure is basically controlled by pump output and the pressure regulator valve. In addition, line pressure is boosted in Reverse, Intermediate and Lo by the reverse boost valve.

Also, in the Neutral, Drive and Reverse positions of the selector lever, the line pressure should increase with throttle opening because of the T.V. system. The T.V. system is controlled by the T.V. cable, the throttle lever and bracket assembly and the T.V. link, as well as the control valve pump assembly.

Figure 5-8 Typical manufacturer's line pressure chart. This type chart should be constantly referred to when checking oil pressure to diagnose a problem. (Courtesy of the Chevrolet Motor Division of General Motors Corp.)

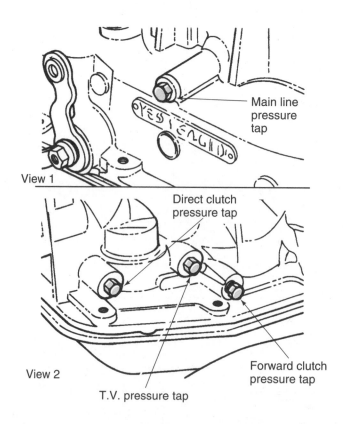

View 1

Main line pressure tap

Direct clutch pressure tap

View 2

T.V. pressure tap

Forward clutch pressure tap

Figure 5-9 Location of the pressure taps used to connect the pressure gauge while conducting a pressure test on a transaxle (Reprinted with the permission of Ford Motor Co.)

a hand-held vacuum pump. The pressure gauges are connected to the pressure taps (Figure 5-9) in the transmission housing and routed so that the gauges can be seen by the driver. The vehicle is then road tested and the gauge readings observed during the following operational modes: slow idle, fast idle, and wide-open throttle.

During the road test, observe the starting pressures and the steadiness of the increases that should occur with slight increases in load. The amount the pressure drops as the transmission shifts from one gear to another should also be noted. The pressure should not drop more than 15 psi between shifts.

Any pressure reading not within the specifications indicates a problem. Typically, when the fluid pressures are low, there is an internal leak, clogged filter, low oil pump output, or faulty pressure regulator valve. If the fluid pressure increased at the wrong time or the pressure was not high enough, sticking valves or leaking seals are indicated. If the pressure drop between shifts was greater than 15 psi, an internal leak at a servo or clutch seal is indicated.

To maximize the usefulness of a pressure test and to be better able to identify specific problems, begin the test by measuring line pressure. Mainline pressure should be checked in all gear ranges and at the three basic engine speeds. It is very helpful for you to make a quick chart such as the one shown (Figure 5-10) to record the pressures during the test.

If the pressure in all operating gears is within specifications at slow idle, the pump and pressure regulator are working fine. If all pressures are low at slow idle, it is likely that there is a problem in the pump, pressure regulator, filter, fluid level, or there is an internal pressure leak. To further identify the cause of the problem, check the pressure in the various gears while the engine is at a fast idle.

If the pressures at fast idle are within specifications, the cause of the problem is normally a worn oil pump; however, the problem may be an internal leak. Internal leaks typically are more evident in a particular gear range, because that is when ATF is being sent to a particular device

	Slow idle	Fast idle	WOT
P			///
R			
N			///
D			
3			
2			
1			

Figure 5-10 Pressure chart to aid in diagnostics

through a particular set of valves and passages. If any of these leak, the pressure will drop when that gear is selected or when the transmission is operating in that gear.

Further diagnostics can be made by observing the pressure change when the engine is operated at WOT in each gear range. A clogged oil filter will normally cause a gradual pressure drop at higher engine speeds because the fluid cannot pass through the filter fast enough to meet the needs of the transmission and the faster turning oil pump. If the fluid pressure did not change with the increase in engine speed, a stuck pressure regulator is the most probable cause of the problem. A stuck pressure regulator may still allow the pressure to build with increases in engine speed, but it will not provide the necessary boost pressures.

If the pressures are high at slow idle, a faulty pressure regulator or throttle valve problem is indicated. If all of the pressures are low at WOT, pull on the T.V. cable or disconnect the vacuum hose to the vacuum modulator. If this causes the pressures to be in the normal range, the low pressure is caused by a faulty cable or there is a problem in the vacuum modulator or vacuum lines. If the pressures stay below specifications, the most likely causes of the problem are the pump or the control system. If all pressures are high at WOT, compare the readings to those taken at slow idle. If they were high at slow idle and WOT, a faulty pressure regulator or throttle system is indicated. If the pressures were normal at slow idle and high at WOT, the throttle system is faulty. To verify that the low pressures are caused by a weak or worn oil pump, conduct a reverse stall test. If the pressures are low during this test but are normal during all other tests, a weak oil pump is indicated.

> **CAUTION:** Always wear safety goggles when connecting and disconnecting the pressure gauges at the transmission. The fluid is under pressure and can easily spray at you and get in your eyes.

WOT is a common acronym for wide-open throttle.

The reverse stall test is actually a way to test maximum pump output.

Valve Body Service

If the pressure test indicated that there is a problem associated with the valves in the valve body, a thorough disassembly, cleaning in fresh solvent, careful inspection, and the freeing up and polishing of the valves may correct the problem. Sticking valves and sluggish valve movements are caused by poor maintenance, the use of the wrong type of fluid, and/or overheating the transmission. The valve body of most transmissions can be serviced when the transmission is in the vehicle, but it is typically serviced when the transmission has been removed for other repairs.

Classroom Manual
Chapter 5, page 110

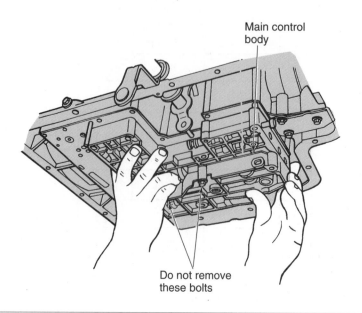

Main control body

Do not remove these bolts

Figure 5-11 When removing a valve body from the transmission, remove only the bolts that are necessary to lower the valve body. Also, be careful not to lose any springs and check balls. (Reprinted with the permission of Ford Motor Co.)

To remove a valve body, begin by draining the fluid and removing the oil pan. Then disconnect the manual and throttle lever assemblies. Carefully remove the detent spring and screw assemblies. Loosen and remove the valve body screws (Figure 5-11). Before lowering the valve body and separating the assembly, hold the assembly with the valve body on the bottom and the transfer and separator plates on top. Holding the assembly in this way will reduce the chances of dropping the steel balls that are located in the valve body (Figure 5-12). Lower the valve body and note where these steel balls are located in the valve body. Remove them and set them aside, along with the various screws.

✓ **SERVICE TIP:** To avoid having to spend hours crawling on the floor of the shop looking for lost parts, place your hand or fingers over spring-loaded valves or plugs when removing them from the valve body.

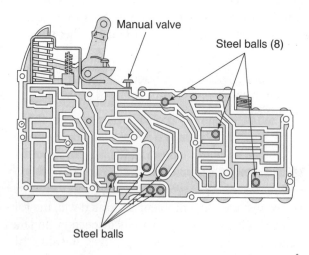

Manual valve

Steel balls (8)

Steel balls

Figure 5-12 Location of the manual shift valve and check balls in a typical valve body (Courtesy of Chrysler Corp.)

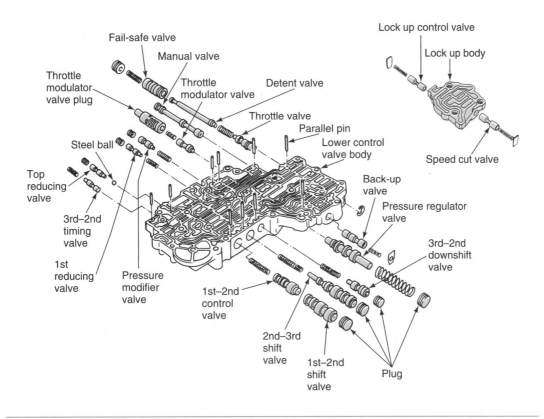

Figure 5-13 Before disassembling a valve body, note the location of the various springs and their relationship with the various shift valves. (Courtesy of Nissan Motor Co., Ltd.)

Disassembly

Begin disassembly by removing the manual shift valve from the valve body. Then remove the pressure regulator retaining screws while keeping one hand around the spring retainer and adjusting screw bracket. Remove the pressure regulator valve. Then remove all of the valves and springs from the valve body. Make sure that you keep all springs and other parts with their associated valves.

✓ **SERVICE TIP:** Lay out lintfree wiping towels on a bench before removing the valves and springs from the valve body. As each valve and spring is removed from the valve body, arrange them in order on the towel.

It is important that you somehow keep track of the position of the springs in relationship to the valves (Figure 5-13). You can draw diagrams on a piece of paper, which can serve as a quick reference during reassembly. You can also use a Polaroid or other type of instant camera to photograph the valve body with its channel plate exposed and the check balls in place (Figure 5-14). This will make an excellent reference for reassembly.

✓ **SERVICE TIP:** Another trick is to use the cardboard sheet included in every gasket set. Fold this sheet in one-inch pleats like an accordion and lay it on your bench. Take the valves and springs out of the valve body and lay them in the different grooves in the sequence they were removed. Reassembly after cleaning is easy because the parts won't roll around on the bench.

Cleaning and Inspection

After all of the valves and springs have been removed from the valve body, soak the valve body and separator and transfer plates in mineral spirits for a few minutes. Some rebuild shops soak the

Special Tools

Mineral spirits

Flat file

Crocus cloth

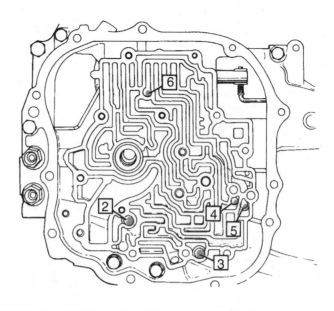

Figure 5-14 Always note the exact location of the check balls in the valve body. (Courtesy of the Chevrolet Motor Division of General Motors Corp.)

Mineral spirits can be purchased at most hardware stores.

Classroom Manual
Chapter 5, page 108

valve body and its associated parts in carburetor cleaner, then wash off the parts with water. Thoroughly clean all parts and make sure all passages within the valve body are clear and free of debris. Carefully blow-dry each part individually with dry compressed air. Never wipe the parts of a valve body with a rag or paper towel. Lint from either will collect in the valve body passages and cause shifting problems.

WARNING: Do not leave valve bodies or other parts submerged in carburetor cleaner longer than five minutes.

Check the separator plate for scratches or other damage (Figure 5-15). Scratches or score marks can cause oil to bypass correct oil passages and result in system malfunction. If the plate is defective in any way, it must be replaced. Check the oil passages in the upper and lower valve bodies for varnish deposits, scratches, or other damage that could restrict the movement of the valves. Check all of the threaded holes and related bolts and screws for damaged threads, and replace as needed.

Examine each valve for nicks, burrs, and scratches. Make sure that each valve fits properly into its respective bore. To do this, hold the valve body vertically and install an unlubricated valve into its bore. Let the valve fall of its own weight into the valve body until the valve stops. Then place your finger over the valve bore and turn the valve body over. The valve should again drop by its own weight. If the valve moves freely under these conditions, it will operate freely with fluid pressure. Repeat this test on all valves.

If a valve cannot move freely within its bore, it may have small burrs or nicks that can be removed. These flaws can and should be removed. To do this, never use sandpaper or a file, rather use products such as Arkansas stone or crocus cloth, which are designed to polish the surface without removing metal from the valve. Sandpaper and emery cloth will remove metal, as well as scratch and leave a rough surface. Normally valve bodies are replaced if they have damaged bores.

An alternative to using Arkansas stone or crocus cloth is the use of toothpaste or a mixture of ATF and a very fine abrasive cleaning paste. Place the paste in your hand, then wrap your hand around the valve and move it back and forth, letting the abrasives remove the burr or nick.

After using any of these polishing methods to remove a nick or burr, the valve must be thoroughly cleaned to remove all of the cleaning and abrasive materials. After the valve has been recleaned, it should be tested in its bore again.

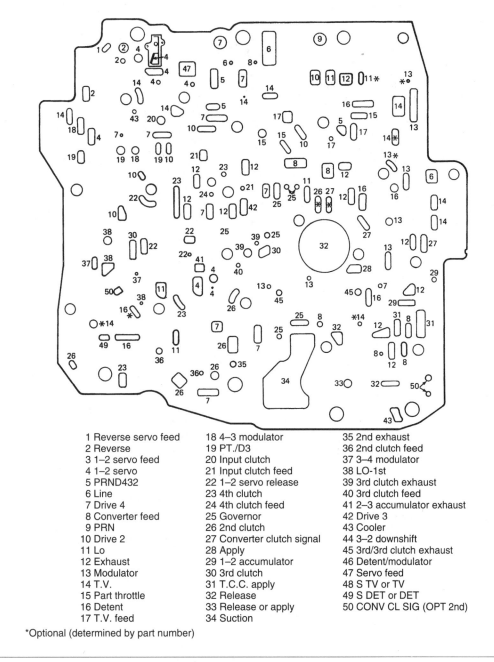

1 Reverse servo feed	18 4–3 modulator	35 2nd exhaust
2 Reverse	19 PT./D3	36 2nd clutch feed
3 1–2 servo feed	20 Input clutch	37 3–4 modulator
4 1–2 servo	21 Input clutch feed	38 LO-1st
5 PRND432	22 1–2 servo release	39 3rd clutch exhaust
6 Line	23 4th clutch	40 3rd clutch feed
7 Drive 4	24 4th clutch feed	41 2–3 accumulator exhaust
8 Converter feed	25 Governor	42 Drive 3
9 PRN	26 2nd clutch	43 Cooler
10 Drive 2	27 Converter clutch signal	44 3–2 downshift
11 Lo	28 Apply	45 3rd/3rd clutch exhaust
12 Exhaust	29 1–2 accumulator	46 Detent/modulator
13 Modulator	30 3rd clutch	47 Servo feed
14 T.V.	31 T.C.C. apply	48 S TV or TV
15 Part throttle	32 Release	49 S DET or DET
16 Detent	33 Release or apply	50 CONV CL SIG (OPT 2nd)
17 T.V. feed	34 Suction	

*Optional (determined by part number)

Figure 5-15 Typical bores of a separator plate (Courtesy of the Buick Motor Division of General Motors Corp.)

If the valve cannot be cleaned well enough to move freely in its bore, the valve body should be replaced. Individual valve body parts are not available. Individual valves are lapped to a particular valve body and therefore if any parts need to be replaced, the entire valve body must be replaced.

Although it is desirable to have the valves move freely in their bores, excessive wear is also a problem. There should never be more than 0.001-inch clearance between the valve and its bore. If the bore or piston is worn, the entire valve body needs to be replaced.

✓ **SERVICE TIP:** Make sure you include the manual valve while you are inspecting and cleaning the valves. It is easy to overlook this valve because it is removed so early during the disassembly of the valve body and it won't be inserted into the valve body until the valve body is installed into the transmission case.

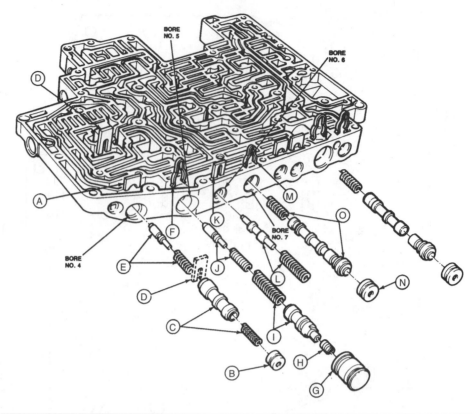

	Bore no. 4		Bore no. 5		Bore no. 6		Bore no. 7
(A)	Spring retainer plate	(F)	Clip	(K)	Retainer plate	(M)	Clip
(B)	Bore plug 1	(G)	Sleeve	(L)	TV limit valve and spring	(N)	Bore plug
(C)	Orifice control valve and spring	(H)	Plug			(O)	1–2 shift valve and spring
(D)	Spring retainer plate 1	(I)	3–4 shift valve and spring				
(E)	2–3 capacity modulator valve and spring	(J)	3–4 TV modulator valve and spring				

Figure 5-16 Typical location of some of the valves and springs in a valve body (Reprinted with the permission of Ford Motor Co.)

With a straightedge and feeler gauge, check the flatness of the valve body's sealing surface. If it is slightly warped, it can be flat filed.

Reassembly

SERVICE TIP: When assembling a valve body, always have your notes, photograph, and the correct service manual close for quick reference. Many valves and springs look similar; however, each has its own bore and purpose (Figure 5-16).

Before beginning to reassemble a valve body, check the new valve body gasket to make sure it is the correct one by laying it over the separator plate and holding it up to the light. No oil holes should be blocked (Figure 5-17). Then install the bolts to hold valve body sections together and

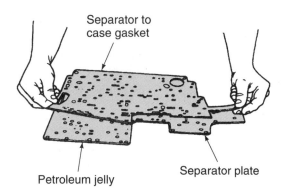

Separator to
case gasket

Petroleum jelly

Separator plate

Figure 5-17 Always check a new valve body gasket by laying it over the separator plate and making sure it does not block any holes. If the gasket is correct, use clean petroleum jelly to hold it in place during assembly. (Reprinted with the permission of Ford Motor Co.)

the valve body to the case. Tighten the bolts to the torque specifications to prevent valve body warpage and possible leakover. Over-torquing can also cause the bores to distort, which would not allow the valves to move freely once the valve body is tightened to the transmission case.

 WARNING: The primary cause of valve sticking is the overtightening of the valve body bolts. This is especially true of aluminum valve bodies. Always be careful when handling a valve body. They are very precise components.

WARNING: Be certain that the proper length bolts are installed in the related depth holes. The length of the bolts varies.

Governor Service

If the pressure tests suggested that there was a governor problem, it should be removed, disassembled, cleaned, and inspected. Some governors are mounted internally and the transmission must be removed to service the governor. Other governors can be serviced by removing the extension housing or oil pan, or by detaching an external retaining clamp and then removing the unit (Figure 5-18).

Classroom Manual
Chapter 5, page 115

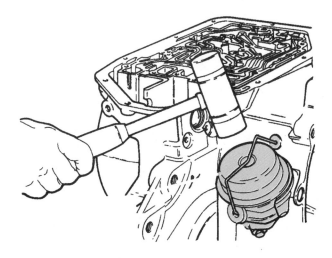

Figure 5-18 Some governor assemblies are contained in a separate housing and retained by a cover and external retaining clamp. (Reprinted with the permission of Ford Motor Co.)

Improper shift points are typically caused by a faulty governor or governor drive gear system. However, some electronically controlled transmissions do not rely on the hydraulic signals from a governor, rather they rely on the electrical signals from these sensors. Sensors, such as speed and load sensors, signal to the transmission's computer when gears should be shifted. Faulty electrical components and/or loose connections can also cause improper shift points.

Disassembly

To disassemble a typical governor, remove the primary governor valve from its bore in the governor housing. Then remove the secondary valve retaining pin, secondary valve spring, and valve (Figure 5-19). Thoroughly clean and dry these parts in the same way as was done for valve body parts. Test each valve in its bore in the governor housing. They should move freely in their bores without sticking or binding.

Also check the valves for any signs of burning or scoring and replace them if necessary. Inspect the springs for a loss of tension and burning marks and replace if necessary.

Reassembly

To reassemble the governor, place the spring around the secondary valve and insert them into the secondary valve bore. Then insert the retaining pin into the governor housing pin holes. Now install the primary valve into the governor housing.

▲ **WARNING:** Never interchange components of the primary and secondary governors. Also note that the flat facets of the primary valve must face outward when it is installed.

If the governor assembly was removed from the governor support and parking gear, be sure to tighten the bolts to specifications with a torque wrench. After assembly, install the governor and torque the bolts to specifications. Some transmissions use a drive ball on the output shaft which locks the governor to the shaft (Figure 5-20). Make sure it is in place when installing the governor.

✓ **SERVICE TIP:** On Honda AS, AK, CA, and F4 model transmissions that are governor controlled, a complaint of no reverse and/or wrong gear starts can be caused by either stuck valves in the valve body or high governor pressure at a stop. To identify the problem, measure governor pressure. Locate the governor tap for the model you are working on and install a pressure gauge into the tap. Check the governor pressure at a stop and

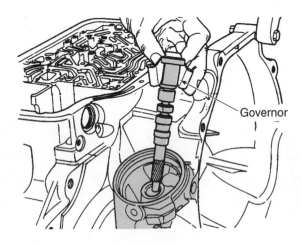

Figure 5-19 After the retaining cover has been removed, the governor assembly can be pulled out of its bore. (Reprinted with the permission of Ford Motor Co.)

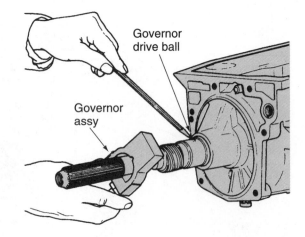

Figure 5-20 Governor drive ball in output shaft (Reprinted with the permission of Ford Motor Co.)

on the road when the problem occurs. If the gauge reads less than 2 psi, the problem is sticky valves in the valve body. If the gauge reads 2 psi or more, then high governor pressure is the problem. Either the governor valve is sticking or something else is allowing a higher pressure to leak into the governor circuit. Whether the governor valve appears to be sticky or not, use 600 grit or finer sandpaper to polish the valve. Be careful not to round the corners of the valve. Make sure the governor is correctly assembled. There is a small tube that must be installed into the governor shaft. This tube separates line pressure from governor pressure. If it's left out, line pressure will flow directly into the governor circuit, causing high governor pressure.

Oil Pump Service

The oil pump (Figure 5-21) of an automatic transmission should be carefully inspected during any transmission overhaul and especially when low line pressure was measured during a pressure test.

Classroom Manual
Chapter 5, page 101

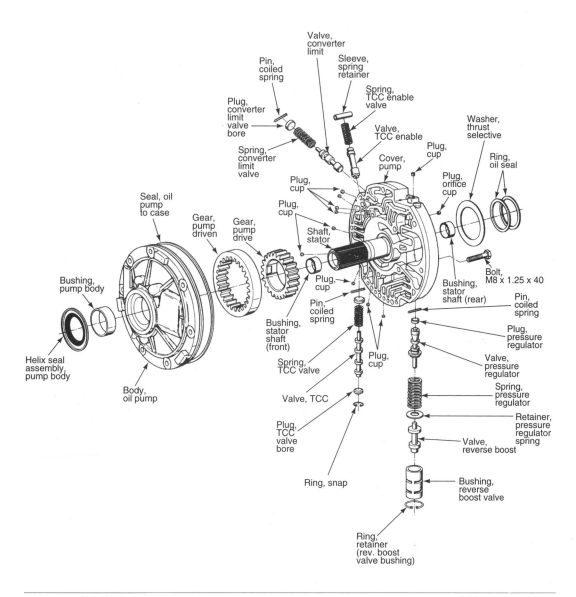

Figure 5-21 A typical gear-type oil pump (Courtesy of the Hydra-Matic Division of General Motors Corp.)

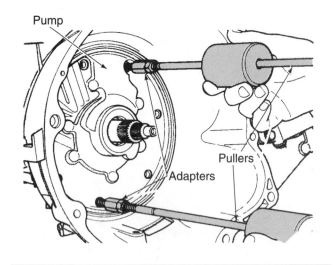

Figure 5-22 Using a puller to remove an oil pump from a transmission case (Courtesy of the Chrysler Corp.)

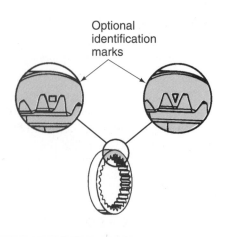

Figure 5-23 Identification marks on oil pump gear sets (Courtesy of the Oldsmobile Motor Division of General Motors Corp.)

Carefully remove the oil pump assembly. Some transmissions require the use of a special puller to remove the oil pump from the transmission case (Figure 5-22).

Disassembly

Before disassembling the pump, mark the alignment of the gears. If acceptably worn gears are reinstalled in a position other than their wear pattern, excessive noise will result. When the pump halves have been separated, look at the relationship between the inner and outer gears. Most gears will have a mark on them indicating the top side of the gear (Figure 5-23).

If no mark is present, you should use a nondestructive type of marker to be sure you install the gears in the same position they were originally in (Figure 5-24). This ensures that the converter drive hub will mate correctly with the inner gear.

Another way of assuring proper position of the gears is to observe the existing wear pattern on the gears as they are being removed and make a note of this for use during reassembly. If the inner gear is installed upside down, there will not be enough free movement, which could cause one or more of the following problems: a broken inner gear, broken flexplate, pump cover scoring, or broken transmission case ears. The proper procedure for disassembling an oil pump is shown in Photo Sequence 5. This procedure is given as an example. Always refer to the procedures for the specific transmission you are working on.

Figure 5-24 Marking the location of the outer and inner gears so that they can be properly meshed during reassembly (Courtesy of the Hydra-Matic Division of General Motors Corporation)

Photo Sequence 5
Proper Procedure for Disassembling an Oil Pump

P5-1 Remove the front pump bearing race, front clutch thrust washer, gasket, and O-ring. Inspect the pump bodies, pump shaft, and ring groove areas.

P5-2 Unbolt and separate the pump bodies.

P5-3 Mark the gears with machinist bluing ink or paint before removing them so that the gears will remain in the same relationship during reassembly.

P5-4 Inspect the gears and all internal surfaces for defects and visible wear.

P5-5 Measure between the outer gear and the pump housing crest 0.003 to 0.006 in. or 0.08 to 0.15 mm. Replace the pump if the measurements exceed specifications.

P5-6 Measure between the outer gear teeth and the crescent. Clearance should be 0.006 to 0.007 in. or 0.15 to 0.18 mm. Replace the pump if clearance exceeds 0.008 in.

P5-7 Place the pump flat on the bench and with a feeler gauge and straightedge, measure between the gears and the pump cover. The clearance should be no more than 0.002 in. (0.05 mm). Replace the pump if the measurement exceeds specifications.

P5-8 Put the halves back together.

P5-9 Torque the securing bolts to specifications. Replace all O-rings and gaskets.

Inspection

Inspect the pump bore for scoring on both its bottom and sides. A converter that had a tight fit at the pilot hub could hold the converter drive hub and inner gear too far into the pump, causing cover scoring. A front pump bushing that has too much clearance may allow the gears to run off center, causing them to wear into the crescent and/or the sides of the pump body.

 SERVICE TIP: Some service manuals will call for a maximum front bushing wear of 0.004 in.; however, a dimension of 0.0005 to 0.0015 in. will avoid wear, as well as keep the pressure losses at the bushing to a minimum.

The stator shaft should be inspected for looseness in the pump cover (Figure 5-25). This can be done while you check for interference inside a torque converter. The shaft's splines and bushing should also be carefully looked at. If the splines are distorted, the shaft and the pump cover should be replaced. Because the bushings control oil flow through the converter and cooler, their fit must be checked. Bushings must be tight inside the shaft and provide the input shaft with a 0.0005 to 0.003 in. clearance.

Inspect the gears and pump parts for deep nicks, burrs, or scratches. Examine the pump housing for abnormal wear patterns. The fit of each gear into the pump body, as well as the centering effect of the front bushing, controls oil pressure loss from the high-pressure side of the pump to the low-pressure input side. Scoring or body wear will greatly reduce this sealing capability.

SERVICE TIP: All pumps, valve bodies, and cases should be checked for warpage and should be flat filed to take off any high spots and burrs prior to reassembly.

On positive displacement pumps, use a feeler gauge to measure the clearance between the outer gear and the pump pocket in the pump housing (Figure 5-26). Also, check the clearance between the outer pump gear teeth and the crescent (Figure 5-27), and between the inner gear teeth and the crescent. Compare these measurements to the specifications. Use a straightedge and feeler gauge to check gear side clearance (Figure 5-28) and compare the clearance to the specifications. If the clearance is excessive, replace the pump.

Variable displacement vane-type pumps require different measuring procedures. However, the inner pump rotor-to-converter drive hub fit is checked in the same way as described for the

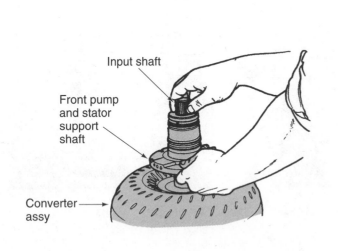

Figure 5-25 Checking the play between the splines of the oil pump and the input shaft (Reprinted with the permission of Ford Motor Co.)

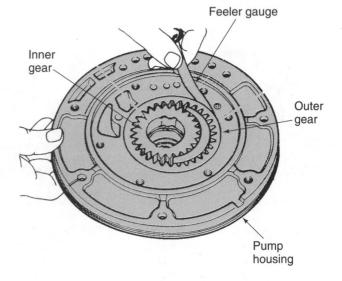

Figure 5-26 Measuring oil pump gear-to-pocket clearance (Courtesy of Chrysler Corp.)

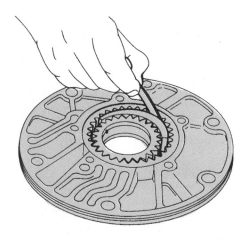

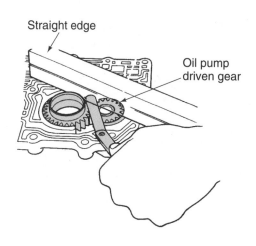

Figure 5-27 Measuring the clearance between the outer gear and the crescent (Courtesy of Nissan Motors Co., Ltd.)

Figure 5-28 Measuring the side clearance of the pump's gears (Courtesy of Honda Motor Co.)

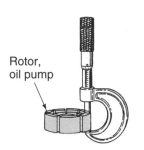 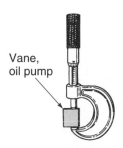 

ROTOR SELECTION		VANE SELECTION		SLIDE SELECTION	
THICKNESS (mm)	THICKNESS (in.)	THICKNESS (mm)	THICKNESS (in.)	THICKNESS (mm)	THICKNESS (in.)
17.593 - 17.963	.7068 - .7072	17.943 - 17.961	.7064 - .7071	17.983 - 17.993	.7080 - .7084
17.963 - 17.973	.7072 - .7076	17.961 - 17.979	.7071 - .7078	17.993 - 18.003	.7084 - .7088
17.973 - 17.983	.7076 - .7080	17.979 - 17.997	.7078 - .7085	18.003 - 18.013	.7088 - .7092
17.983 - 17.993	.7080 - .7084			18.013 - 18.023	.7092 - .7096
17.993 - 18.003	.7084 - .7088				

Figure 5-29 Vane-type oil pump measurements and selection chart (Courtesy of the Buick Motor Division of General Motors)

other pumps. The pump rotor, vanes, and slide are originally selected for size during assembly at the factory (Figure 5-29). Changing any of these parts during overhaul can destroy this sizing and possibly the body of the pump. You must maintain the original sizing if any parts are found to need replacement. These parts are available in select sizes for just this reason.

 SERVICE TIP: Some early pump vanes cracked as a result of excessive load. These pumps did not have enough vanes to handle the hydraulic loads encountered during operation. Updated pumps with more vanes cured the problem.

The vanes are subject to edge wear, as well as cracking and subsequent breakage. The outer edge of the vanes should be rounded, with no flattening (Figure 5-30). These pumps have an aluminum body and cover halves; therefore, any scoring indicates that they should be replaced.

Inspect the reaction shaft's seal rings. If the rings are made of cast iron, check them for nicks, burrs, scuffing, or uneven wear patterns, and replace them if they are damaged (Figure 5-31).

Gear-type and rotor-type pumps are positive displacement pumps.

Classroom Manual
Chapter 5, page 104

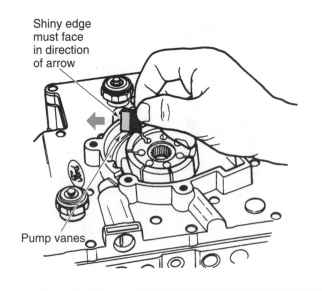

Shiny edge must face in direction of arrow

Pump vanes

Figure 5-30 Placement of vanes in the pump's rotor (Reprinted with the permission of Ford Motor Co.)

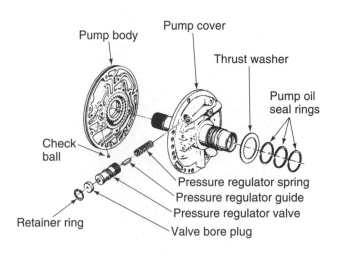

Pump body

Pump cover

Thrust washer

Pump oil seal rings

Check ball

Pressure regulator spring

Pressure regulator guide

Pressure regulator valve

Retainer ring

Valve bore plug

Figure 5-31 Location of oil seal rings on a typical oil pump assembly (Courtesy of the Oldsmobile Division of General Motors Corp.)

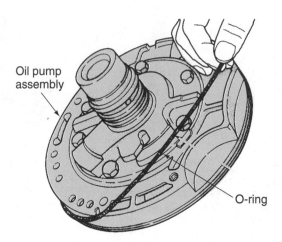

Oil pump assembly

O-ring

Figure 5-32 Outer oil pump seal (Courtesy of Chrysler Corp.)

Make sure the rings are able to rotate in their grooves. Check the clearance between the reaction shaft support ring groove and the seal ring. If the seal rings are the Teflon full-circle type, cut them out and use the required special tool to replace them.

The outer area of most pumps utilizes a rubber seal (Figure 5-32). Check the fit of the new seal by making sure the seal sticks out a bit from the groove in the pump. If it does not, it will leak. The seal at the front of the pump is always replaced during overhaul. Most of these seals are the metal-clad lip seal type. Care must be taken to avoid damage to the seating area when removing the old seal.

Check the area behind the seal to be sure the drainback hole is open to the sump. If this hole is clogged, the new seal will possibly blow out. The drainback hole relieves pressure behind the seal. A loose fitting converter drive hub bushing can also cause the front pump seal to blow out.

Reassembly

The pump seal can be installed with a hammer and seal driver. Apply some RTV sealant around the outside surface of the seal case when installing the seal. Place some transmission fluid in the

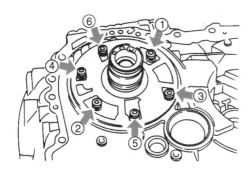

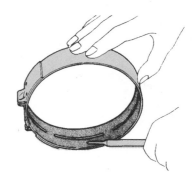

Figure 5-33 Oil pump bolts must be torqued in the sequence recommended by the manufacturer. (Courtesy of Nissan Motors Co., Ltd.)

Figure 5-34 The friction material of the bands must be carefully inspected. (Courtesy of Nissan Motors Co., Ltd.)

pocket of the pump housing and install the gears into the housing according to their alignment marks. Align and install the reaction shaft support and tighten the bolts to the specified torque.

Place some petroleum jelly in two or three spots around the oil pump gasket and position it onto the transmission case. Tighten the pump attaching bolts to specifications in the specified order (Figure 5-33). After the bolts are tight, check the rotation of the input shaft. If the shaft does not rotate, disassemble the transmission to locate the misplaced thrust washer.

Band Service

Servicing of bands and their components includes inspection of the bands, as well as the drums that the bands are wrapped around. Before the introduction of overdrive automatic transmissions, most bands operated in a free condition during most driving conditions. This means the band was not applied in the cruising gear range. However, many overdrive automatic transmissions use a band in the overdrive cruise range, which puts an additional load on the band and subsequently causes additional wear on the band. For this reason, a thorough inspection of the bands is very important (Figure 5-34).

Classroom Manual
Chapter 5, page 146

The bands in a transmission will be either single or double wrap, depending on the application. Both types can be the heavy-duty cast iron type or the normal strap type. The friction material used on clutches and bands is quite absorbent. This characteristic can be used to tell if there is much life left in the lining. Simply squeeze the lining with your fingers to see if any fluid appears. If fluid appears, this tells you the lining can still hold fluid and has some life left in it. It is hard to tell exactly how long the band will last, but at least you have an indication that it is still useable. Strap or flex-type bands should never be twisted or flattened out. This may crack the lining and lead to flaking of the lining.

Inspection

Band failures found during overhaul are easy to spot. Look for chipping, cracks, burn marks, glazing, and nonuniform wear patterns and flaking. If any of these defects are apparent, the band should be replaced.

Also inspect brake band friction material for wear. If the linings show wear, carefully check the band struts (Figure 5-35), levers, and anchors for wear. Replace any worn or damaged parts. Look at the linings of heavy-duty bands to see if the lining is worn evenly. A twisted band will show tapered wear on the lining. If the friction material is blackened, this is caused by an excessive buildup of heat. High heat may weaken the bonding of the lining and allow the lining to come loose from the metal portion of the band.

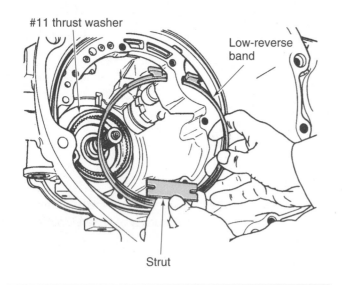

#11 thrust washer

Low-reverse band

Strut

Figure 5-35 While removing the band, carefully inspect the band strut. (Courtesy of Chrysler Corp.)

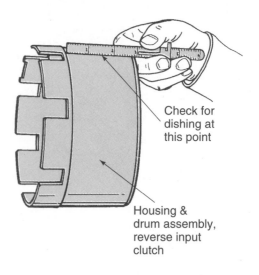

Check for dishing at this point

Housing & drum assembly, reverse input clutch

Figure 5-36 Check the flatness of the drum with a straightedge. (Courtesy of the Hydra-Matic Division of General Motors Corp.)

The drum surface should be checked for discoloration, scoring, glazing, and distortion. The drums will be either iron castings or steel stampings. Cast iron drums that are not scored can generally be restored to service by sanding the running surface with 180-grit emery paper in the drum's normal direction of rotation. A polished surface is not desirable on cast iron drums.

The surface of the drum must also be flat (Figure 5-36). This is not usually a problem with a cast iron drum, but it can affect the stamped steel-type drum. It is possible for the outer surface of the drum to dish outward during its normal service life. This is a common problem on the GM 4L60 and should be inspected on any transmission that has a stamped steel band surface. Check the drum for flatness across the outer surface where the band runs. Any dishing here will cause the band to distort as it attempts to get a full grip on the drum. Distortion of the band weakens the bond of the friction material to the band and will cause early failure due to flaking off of the friction lining. A dished stamped steel drum should be replaced. Check the service manual for maximum allowable tolerances.

Band Adjustments

After the band assembly has been installed in the transmission housing and around its drum, the band needs to be adjusted. Band adjustment is also part of a "transmission tune-up" on some models. Many transmissions have provisions for externally adjusting the band running clearance (Figure 5-37). Some transmissions have no provisions for band adjustment other than selectively sized servo apply pins and struts.

To set the running clearance on transmissions that have an adjustment screw, loosen the locknut on the adjuster screw and back it off about five turns. Backing off the locknut allows for tightening the screw to a specified torque, which simulates a fully applied band (Figure 5-38). The amount of required torque varies with different manufacturers, but it gives the same result.

After torquing, the adjuster screw is backed off a number of turns as specified by the manufacturer. To hold the adjustment, the locknut is generally tightened to 30–35 ft.lb., while the adjuster screw is held stationary. The timing of band application has a lot to do with how the shift feels to the driver. This is one of the reasons there are so many different tightening torques on the various transmissions. The torque setting and number of turns the adjuster is backed off provides the proper clearance and grip for the many different types of bands used. Additionally, the pitch of the threads on the adjuster screws varies even on the same type of transmission. This is why you cannot use a single adjustment sequence for all transmissions.

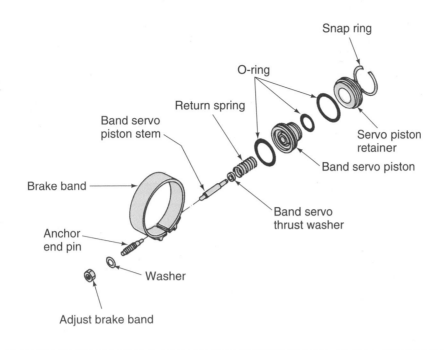

Figure 5-37 A complete band/servo assembly with an adjusting screw (Courtesy of Nissan Motors Co., Ltd.)

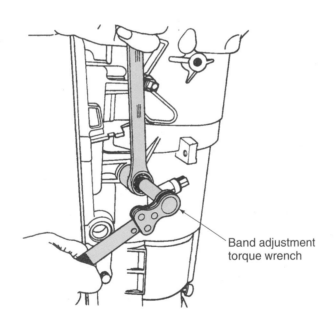

Figure 5-38 Adjusting a band with a torque wrench and a box-end wrench (Reprinted with the permission of Ford Motor Co.)

Not all screw-type band adjusters are on the exterior of the transmission case. Torqueflite transmissions have the low-reverse band adjustment inside the oil pan. This requires the removal of the pan to make band adjustments. This may seem inconvenient, but since the low-reverse band is normally applied at idle conditions, not much lining wear occurs. Therefore, the low-reverse band does not require adjustment as often as the intermediate band, and it can be done during a fluid and filter change when the pan is removed.

Adjusting the bands
may or may not be
part of a scheduled
maintenance
program. Some
manufacturers
recommend periodic
adjustments; others
recommend
adjustment only
after the
transmission has
been overhauled.
Some transmissions
do not have a band
adjustment.

As you can see, band adjustment is not a difficult task. However, gaining access to the adjuster screw may require the removal of some other linkages, wiring, or even cooler lines that could be in your way. Be sure any components you moved out of the way are restored to their original position after the band adjustment is finished.

CAUTION: It is important that band adjustment procedures be followed exactly as outlined by the manufacturer. Serious damage to the transmission can result if the procedure is not followed.

Servo and Accumulator Service

On some transmissions, the servo and accumulator assemblies are serviceable with the transmission in the vehicle (Figure 5-39). Others require the complete disassembly of the transmission. Internal leaks at the servo or clutch seal will cause excessive pressure drops during gear changes.

Before disassembling a servo or any other component, carefully inspect the area to determine the exact cause of the leakage. Do this before cleaning the area around the seal. Look at the path of the fluid leakage and identify other possible sources. These sources could be worn gaskets, loose bolts, cracked housings, or loose line connections.

Inspect the outside area of the seal. If it is wet, determine if the oil is leaking out or if it is merely a lubricating film of oil. When removing the servo, continue to look for the causes of the leak. Check both the inner and outer parts of the seal for wet oil, which means leakage. When removing the seal, inspect the sealing surface, or lips, before washing. Look for unusual wear, warping, cuts and gouges, or particles embedded in the seal.

Band servos and accumulators are basically pistons with seals in a bore held in position by springs and retaining snap-rings (Figure 5-40). Remove the retaining rings and pull the assembly from the bore for cleaning. Check the condition of the piston and springs (Figure 5-41). Cast iron seal rings may not need replacement, but rubber and elastomer seals should always be replaced.

Accumulators

Begin the disassembling of an accumulator by removing the accumulator plate snap-ring. After removing the accumulator plate, remove the spring and accumulator pistons. If rubber seal rings are installed on the piston, replace them whenever you are servicing the accumulator. Lubricate

Classroom Manual
Chapter 5, page 147

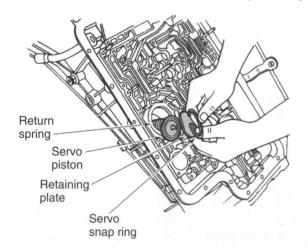

Return
spring

Servo
piston

Retaining
plate

Servo
snap ring

Figure 5-39 The servos in some transmissions are serviceable while the transmission is in the vehicle. The servos are contained in their own bores. (Reprinted with the permission of Ford Motor Co.)

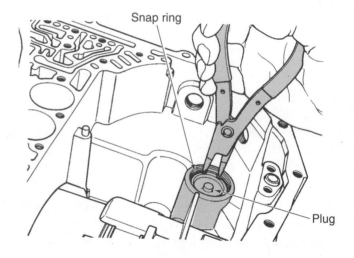

Snap ring

Plug

Figure 5-40 Servos and accumulators are normally retained by a snap ring. (Courtesy of Chrysler Corp.)

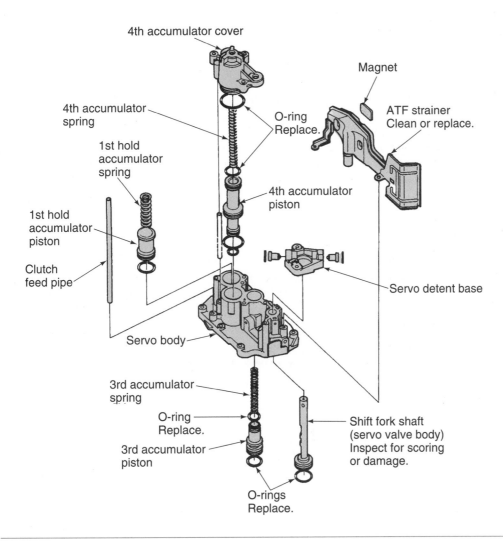

4th accumulator cover

Magnet

4th accumulator
spring

O-ring
Replace.

ATF strainer
Clean or replace.

1st hold
accumulator
spring

4th accumulator
piston

1st hold
accumulator
piston

Clutch
feed pipe

Servo detent base

Servo body

3rd accumulator
spring

O-ring
Replace.

Shift fork shaft
(servo valve body)
Inspect for scoring
or damage.

3rd accumulator
piston

O-rings
Replace.

Figure 5-41 A servo/accumulator assembly (Courtesy of Honda Motor Co.)

the new accumulator piston ring and carefully install it on the piston. Lubricate the accumulator cylinder walls and install the accumulator piston and spring. Then reinstall the accumulator plate and retaining snap-ring.

Many accumulator pistons can be installed upside down. This results in free travel of the piston or too much compression of the accumulator spring. Note the direction of installation of the piston during the teardown process, as you will not always find a good picture when you need to reinstall the accumulator. Because the movement of the accumulator has an effect on shift feel, correct installation is critical. It is quite common for manufacturers to mate servo piston assemblies with accumulators. This takes up less space in the transmission case and, because they have the same basic shape, can reduce some of the machining during manufacture.

Servos

A servo is disassembled in a similar fashion. The servo's piston, spring, piston rod, and guide should be cleaned and dried. Check the servo piston for cracks, burrs, scores, and wear. Servo pistons may be made of either aluminum or steel. Aluminum pistons should be carefully checked for cracks and their fit on the guide pins. Cracked pistons will allow for a pressure loss and being loose on the guide pin may allow the piston to bind in its bore. Whether it is a steel or aluminum piston, the seal groove should be free of nicks or any imperfection that might pinch or bind the seal. Clean up any problems with a small file or scraper. A side clearance of 0.003 to 0.005 in. is required.

Classroom Manual
Chapter 5, page 147

Seals

Most original equipment servo seals are of the Teflon type. These seals will exhibit no feeling of drag in the bore due to the slipperiness of the Teflon. A majority of replacement transmission gasket sets will supply cast iron hook-end seal rings, in place of the Teflon seals. When they are installed, the cast iron seals will have a noticeable drag as the servo piston is moved through the bore. This is not a problem, and in some cases may even improve operation.

Check the cast iron seal rings to make sure they are able to turn freely in the piston groove. These seal rings are not typically replaced unless they are damaged, so carefully inspect them.

Some servo pistons have a molded-on rubber seal. This type is usually replaced during overhaul since the rubber is subject to the same deterioration as other seals of similar construction.

Inspection

Inspect the servo or accumulator spring for possible cracks. Also check where the spring rests against the case or piston. The spring may wear a groove into the aluminum, so be sure the piston or case material has not worn too thin. Some servos utilize a steel chafing plate in order to eliminate this problem.

Inspect the servo cylinder for scores or other damage. Move the piston rod through the piston rod guide and check for freedom of movement. Check the band servo components for wear and scoring. Replace all other components as necessary, then reassemble the servo assembly.

Assembly

Multipart servo/accumulators must be assembled in the correct sequence. Incorrect assembly can result in dragging bands and harsh shifting. If the servo pin has Teflon seals, these will need to be cut off with a knife and a new Teflon seal installed with the correct sleeve and sizing tools.

When reassembling the servo, lubricate the seal ring with ATF and carefully install it on the piston rod. Lubricate and install the piston rod guide with its snap ring into the servo piston. Then install the servo piston assembly, return spring, and piston guide into the servo cylinder. Some servos are fitted rubber lip seals which should be replaced (Figure 5-42). Lubricate and install the new lip seal. On spring-loaded lip seals, make sure that the spring is seated around the lip and that the lip is not damaged during installation.

Selective Servo Apply Pins

Transmissions without a band adjusting screw use selective servo apply pins that maintain the correct clearance between the band and the drum. This type of servo apply pin must be checked for

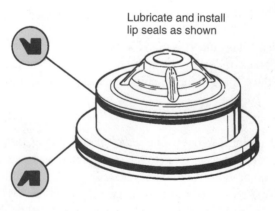

Lubricate and install lip seals as shown

Figure 5-42 Proper installation of servo piston seals (Courtesy of the Hydra-Matic Division of General Motors Corp.)

correct length during overhaul to ensure proper stroking action of the servo piston, as well as shift timing. The travel needed to apply a band relates to the timing of band application. Although the valving and orificing determines true shift timing, the adjustment of the servo pin completes the job by providing the correct clearance of the band to the drum. To select the correct apply pin, most transmissions require the use of special tools (Figure 5-43). The tools are used to simulate a fully applied band and allow the technician to determine the correct length of pin (Figures 5-44 and 5-45).

If a replacement band, drum, or even case has been used, an apply pin check must be made. This is the only way to be certain you have the correct length pin. Since these pins are selective, you will have to start by checking the length of the pin already in the transmission. If it is too long or too short, a new length pin will be necessary. These pins are available at the Parts Departments of most dealerships and suppliers for automatic transmission rebuilders.

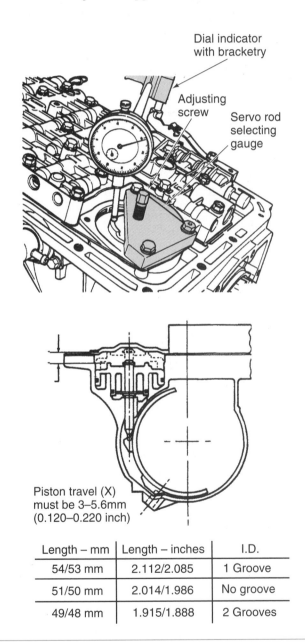

Piston travel (X) must be 3–5.6mm (0.120–0.220 inch)

Length – mm	Length – inches	I.D.
54/53 mm	2.112/2.085	1 Groove
51/50 mm	2.014/1.986	No groove
49/48 mm	1.915/1.888	2 Grooves

Figure 5-43 Using special tools to determine correct apply pin length (Reprinted with the permission of Ford Motor Co.)

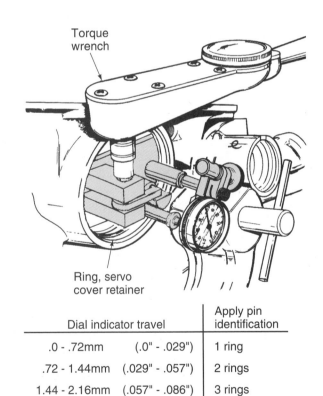

Dial indicator travel		Apply pin identification
.0 - .72mm	(.0" - .029")	1 ring
.72 - 1.44mm	(.029" - .057")	2 rings
1.44 - 2.16mm	(.057" - .086")	3 rings
2.16 - 2.88mm	(.086" - .114")	Wide band

Figure 5-44 Procedure for checking servo pin length on most GM transmissions (Courtesy of the Hydra-Matic Division of General Motors Corp.)

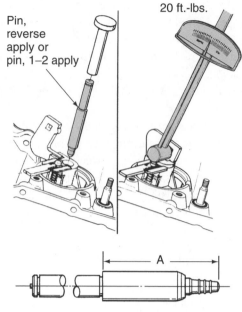

Pin, reverse apply or pin, 1–2 apply

20 ft.-lbs.

Reverse band apply pin	
Identification	Dimension A
2 wide bands	70.86-71.01mm (2.789"-2.795")
3 grooves & wide band	71.91-72.06mm (2.831"-2.837")
2 grooves & wide band	72.96-73.11mm (2.872"-2.878")
1 groove & wide band	74.01-74.16mm (2.913"-2.919")
No groove	75.03-75.18mm (2.953"-2.959")
1 groove	76.08-76.23mm (2.995"-3.001")
2 groove	77.13-77.28mm (3.036"-3.042")
3 groove	78.18-73.33mm (3.077"-3.083")
4 groove	79.20-79.35mm (3.118"-3.124")

1–2 band apply pin	
Identification	Dimension A
1 ring & wide band	56.24-56.39mm (2.214"-2.220")
1 ring	57.23-57.38mm (2.253"-2.259")
2 rings	58.27-58.42mm (2.294"-2.299")
3 rings	59.31-59.46mm (2.335"-2.340")
Wide band	60.34-60.49mm (2.375"-2.381")
2 rings & wide band	61.34-61.49mm (2.414"-2.420")

Classroom Manual
Chapter 5, page 150

Many transmission rebuilders replace all friction discs and steel plates as a part of the overhaul process. This can actually save time because it eliminates the inspection time and there is no doubt of whether any discs or plates were of questionable condition.

Figure 5-45 Procedure for checking servo pin length on other GM transmissions (Courtesy of the Buick Motor Division of General Motors Corp.)

Clutch Pack Assemblies

Two types of multiple-disc clutches will be found in the transmission: rotating drum (Figure 5-46) and case grounded (Figure 5-47). Both are serviced in the same way but require slightly different inspection procedures.

✓ **SERVICE TIP:** New friction discs should not be used with used steel plates unless the steel plates are deglazed.

All friction disc and steel plate packs are held in place by snap rings. These snap rings may be selective in thickness and must be kept with the clutch pack during disassembly. It is common to

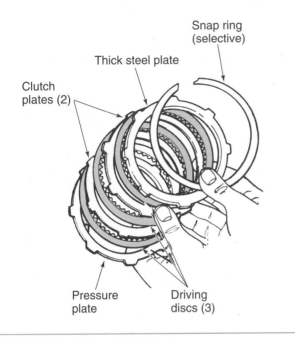

Figure 5-46 Clutch assembly from a rotating drum-type clutch (Courtesy of Chrysler Corp.)

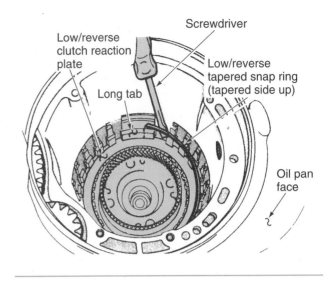

Figure 5-47 A case ground clutch assembly (Courtesy of Chrysler Corp.)

use the same diameter snap ring on more than one clutch in a single transmission. However, the rings can be of differing thickness. This thickness variation can be used to set the clearance in the clutch pack. The snap rings may also have a distinct shape which will only be effective when used in the correct groove.

An example of the complete disassembly, inspection, assembly, and clearance checks is shown in Photo Sequence 6. Always refer to the manufacturer's recommendations for doing the same thing on a particular transmission.

Disassembly

Using a screwdriver, remove the large clutch retaining plate snap ring and remove the thick steel clutch pressure plate (Figure 5-48). Now remove the clutch pack. Some clutch packs have a wave snap ring, which must be looped through the clutch pack until the steel plates can be removed.

Special Tools

Snap-ring pliers

Feeler gauge

OSHA-approved air nozzle

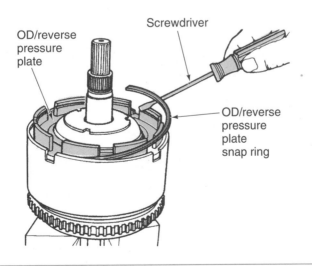

Figure 5-48 Removing large snap ring to disassemble the clutch pack (Courtesy of Chrysler Corp.)

Photo Sequence 6
Proper Procedure for Disassembling, Inspecting, Assembling, and Clearance Checking Direct Clutch

P6-1 Set the direct clutch on the bench.

P6-2 Pry out the snap ring.

P6-3 Remove the pressure plate and the clutch pack assembly.

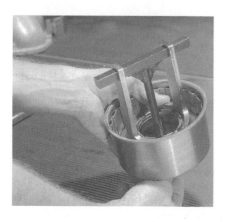

P6-4 Install the piston compressor and compress the piston.

P6-5 Remove the snap ring.

P6-6 Remove the piston compressor, the spring retainer assembly, and the piston.

P6-7 Install new seals on the piston.

P6-8 Check the movement of the check ball.

P6-9 Lubricate the outer piston seal with petroleum jelly.

Proper Procedure for Disassembling, Inspecting, Assembling, and Clearance Checking Direct Clutch (continued)

P6-10 Reinstall the piston.

P6-11 Place the piston spring retainer into the drum.

P6-12 Install the compressor and snap ring.

P6-13 Alternately install friction and steel discs.

P6-14 Install the pressure plate and snap ring.

P6-15 Check the clearance with a feeler gauge.

P6-16 Air test the assembly.

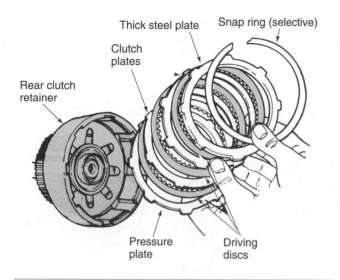

Figure 5-49 When removing the clutch discs from the drum, keep them in the order that they were in. (Courtesy of Chrysler Corp.)

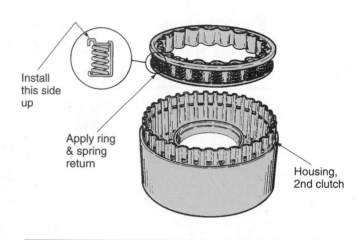

Figure 5-50 Retainer ring for captive return springs (Courtesy of the Buick Motor Division of General Motors Corp.)

Once the snap ring is removed, the backing plate will be the first steel plate you will take out. This plate, like the snap ring, may be a selective part. Since the backing plate is thicker than the other steel plates in the clutch pack, it is easy to spot. Now remove the remainder of the friction discs and steel plates, keeping them in the order they were in the drum (Figure 5-49).

Using a clutch spring compressor tool, compress the clutch return springs. Then using snapring pliers, remove the clutch hub retainer snap ring, retaining plate, and springs. Most retainers have small tabs that prevent you from removing the snap ring without using a compressor (Figure 5-50).

There are many types of clutch spring compressor tools available. The many different locations and depths the clutch springs may have in their drums dictate having more than one simple compressor (Figures 5-51 and 5-52).

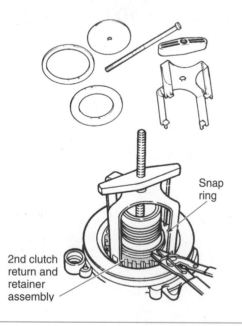

Figure 5-51 A clutch compressor tool for a GM 4T80E transaxle (Courtesy of Kent-Moore Tools)

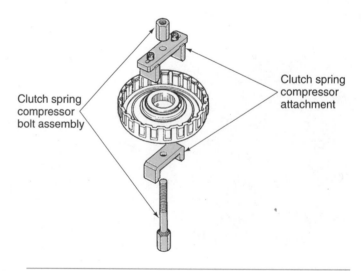

Figure 5-52 A clutch compressor tool for Honda transaxles (Courtesy of Honda Motor Co.)

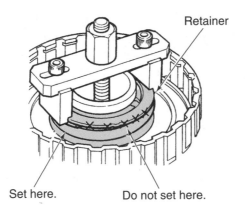

Retainer

Set here.

Do not set here.

Figure 5-53 Proper position of the spring compressor will prevent damage to the piston and seals. (Courtesy of Honda Motor Co.)

With the compressor installed on the spring retainer (Figure 5-53), compress the spring and retainer just enough to allow the snap ring to be removed. Pushing the retainer down too much may bend or distort it.

CAUTION: Be careful when compressing the springs. Careless procedures can allow the springs and retainer to fly into your face.

With the retainer removed, a single, large coil spring or multiple small coil springs will be exposed. Note the number and placement of the multiple springs for use during assembly.

If no coil springs were encountered when the clutch pack was removed, then a Belleville or disc-type return spring is used (Figure 5-54). This type of clutch will have a heavy steel plate with one rounded side found in the bottom of the clutch pack. This heavy plate is a pressure plate. The released position of these springs has very little or no outward force on the retaining snap ring; therefore, the snap ring can be removed without the use of a compressor. The snap ring for this application could be selective, so remember which snap ring goes where.

Some clutches can have two or more snap rings, so pay attention to their placement in the clutch. Additionally, a wavy snap ring is sometimes used to retain the Belleville spring. These are used to give some cushion to the application of the clutch.

Classroom Manual
Chapter 5, page 153

Belleville-type return springs are usually found in the forward clutch.

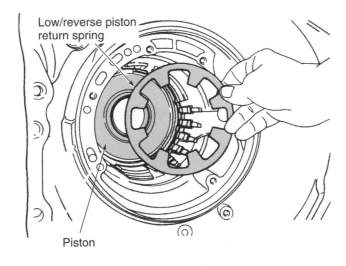

Low/reverse piston return spring

Piston

Figure 5-54 A clutch assembly fitted with a Belleville spring (Courtesy of Chrysler Corp.)

For easy removal of the piston from its drum, mount the clutch on the oil pump. Then lift the piston out of the bore. If it will not come out, the drum can be inverted and slammed squarely against a hardwood bench top. This will dislodge a tight piston, but it will not damage the drum.

Another way to remove the piston is to charge the apply circuit of the clutch with compressed air. Use an air nozzle with a rubber tip and apply air pressure to pop out the piston from the drum. Air is blown in either at the feed hole in the drum or by placing the drum on the clutch support and using its normal feed passage (Figure 5-55).

> ■ **CAUTION:** Air pressure should be reduced to 25–30 lb. to avoid expelling the piston at a high rate of speed, which could cause injury.

With the piston removed, take note of the types of seals used and their position. This is important with lip seals, as the lip will face toward the direction the fluid comes from. Lip seals installed in the wrong direction will not hold pressure. Seals are generally replaced during overhaul. The reuse of old seals is not recommended.

Inspection and Cleaning

Once a clutch assembly has been taken apart, you may wish to inspect the clutch components or continue to disassemble the remainder of the clutch units in the transmission. If you choose the latter, make sure you keep the parts of each clutch separate from the others.

Clean the components of the clutch assembly. Make sure all clutch parts are free of any residue of varnish, burned disc facing material, or steel filings. Take special care to wash out any foreign material from the inside of drums and the hub disc splines. If left in, the material can be washed out by the fresh transmission fluid and sent through the transmission. This can ruin the rebuild.

The clutch splines must be in good shape with no excessively rounded corners or shifted splines. Test their fit by trial fitting three new clutch discs on the splines. Move the discs up and down the splines to check for binding. If they bind, this can cause dragging of the discs during a time when they should be free-floating. Replace the hubs if the discs drag during this check. Check the spring retainer. It should be flat and not distorted at its inner circumference. Check all springs for height, cracks, and straightness (Figure 5-56). Any springs that are not the correct height or that are distorted, should be replaced. Many retainers have the springs attached to them by crimping. This speeds up production at the assembly line. Turning this type of retainer upside down is a quick

---

Pistons located in the transmission case can be removed by blowing compressed air into the apply passage in the case.

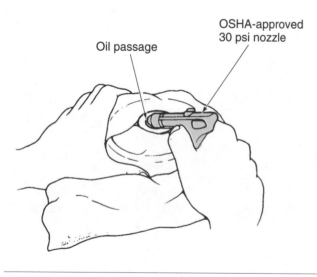

Oil passage
OSHA-approved
30 psi nozzle

Figure 5-55 Wrap a rag around the clutch drum when using air to pop the piston out of the drum. (Courtesy of Honda Motor Co.)

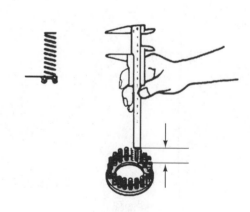

Figure 5-56 Check the height of each one of the return springs. (Courtesy of Mazda Motors)

check of spring length. Closely examine the Belleville spring for signs of overheating or cracking, and replace it if it is damaged.

The steel plates should be checked to be sure they are flat and not worn too thin. Check all steels against the thickest one in the pack or a new one. Most steels will have an identifying notch or mark on the outer tabs. If the plates pass inspection, remove the polished surface finish and the steels are ready for reuse. The steel plates should also be checked for flatness by placing one plate on top of the other and checking the space on the inside and outside diameters. Clutch plates must not be warped or cone-shaped. Also, check the steel plates for burning and scoring, and for damaged driving lugs. Check the grooves inside the clutch drum and check the fit of the steel plates, which should travel freely in the grooves.

Close inspection of the friction discs is simple. The discs will show the same types of wear as bands will. Disc facings should be free of chunking, flaking, and burnt or blackened surfaces. Discs that are stripped of their facing have been overheated and subject to abuse. In some cases, the friction discs and steel plates can be welded together. This occurs when the facing comes off the disc due to a loosening of the facing's bonding because of extreme heat. As the facing comes off, metal-to-metal contact is made between the discs and steel plates. The friction involved as the clutch tries to engage causes them to fuse together. This may lock the clutch in an engaged condition. Depending on which clutch is affected, driveability problems can include: drives in neutral, binds up in reverse (forward clutch seized), starts in direct drive, binds up in second (high-reverse clutch seized), and other problems that are not that common.

If the discs do not show any signs of deterioration, squeeze each disc to see if fluid is still trapped in the facing material. If fluid comes to the surface, the disc is not glazed. Glazing seals off the surface of the disc and prevents it from holding fluid. Holding fluid is basic to proper disc operation. It allows the disc to survive engagement heat, which would burn the facing and cause glazing. Fluid stored in the friction material cools and lubricates the facings as it transfers heat to the steel plates and also carries heat away as some oil is spun out of the clutch pack by centrifugal force. This helps avoid the scorching and burning of the disc.

Clutch discs must not be charred, glazed, or heavily pitted. If a disc shows signs of flaking friction material or if the friction material can be scraped off easily, replace the disc.

A black line around the center of the friction surface also indicates that the disc should be replaced. Examine the teeth on the inside diameter of each friction disc for wear and other damage (Figure 5-57).

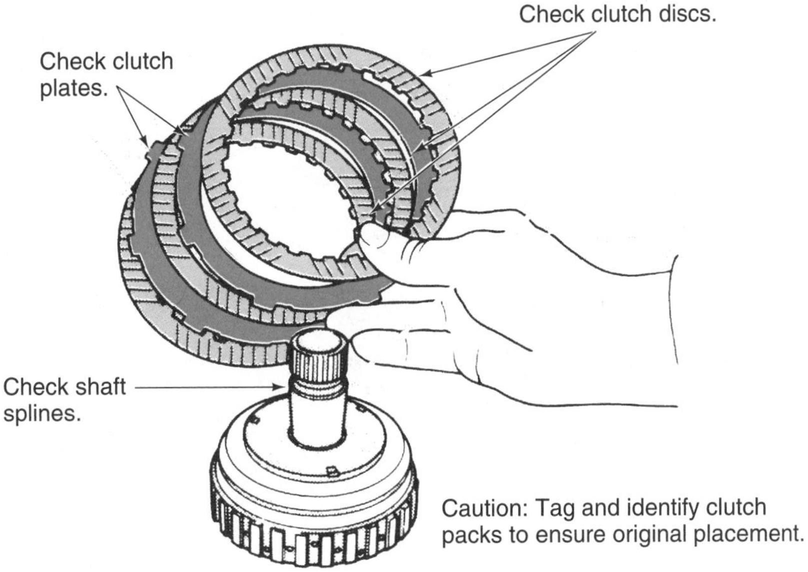

The steel plates are commonly called "steels" by the trade.

Classroom Manual
Chapter 5, page 150

Check clutch discs.

Check clutch plates.

Check shaft splines.

Caution: Tag and identify clutch packs to ensure original placement.

Figure 5-57 Carefully examine the clutch discs. (Courtesy of Chrysler Corp.)

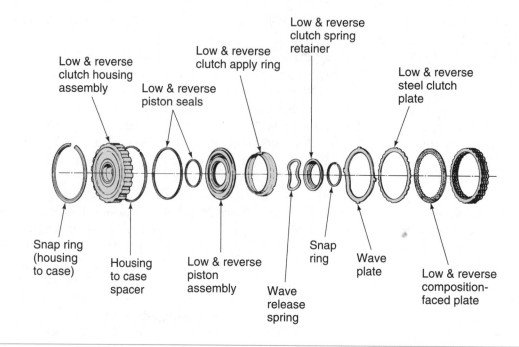

Figure 5-58 A clutch assembly with a waved plate (Courtesy of the Oldsmobile Division of General Motors Corp.)

Wave plates are used in some clutch assemblies to cushion the application of the clutch (Figure 5-58). These should be inspected for cracks and other damage. Never mix wave plates from one clutch assembly to another. As an aid in assembly, most wave plates will have different identifying marks.

The clutch pistons (Figure 5-59) are checked for cracks, warpage, and fit in their bore. Carefully examine the seal ring grooves and inside diameter of the piston for cracks, nicks, and burrs. Groove wear can be accelerated by excessive pump pressures. The excess pressure forces the seal rings against the sides of the grooves so hard that they cannot turn. If the bores of the clutch are severely grooved, a stuck pressure regulator valve could be the problem.

The reverse side of the pump cover is the clutch support that incorporates seal rings for the fluid circuits leading to the clutch drums. The seal rings fit loosely into the grooves in the clutch

Classroom Manual
Chapter 5, page 151

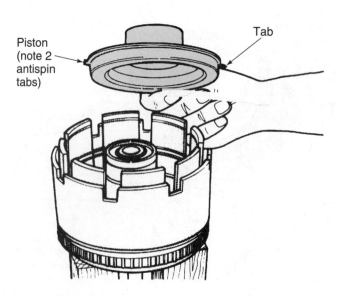

Figure 5-59 A clutch piston with antispin tabs (Courtesy of Chrysler Corp.)

support and rely on pump oil pressure to push them against the side of the groove to make the seal. The seal rings should be checked for side play in the grooves and for proper fit into the drum. Check the grooves for burrs, step wear, or pinched groove conditions. It should be noted that these seals rotate with the drum. Any condition that hinders rotation will cause the ring seals to bind, resulting in drum wear. This can destroy the drum, if it is not corrected.

Carefully inspect aluminum pistons, which may have hairline cracks that will cause pressure leakage during use. This could cause clutch slipping as a result of lost hydraulic force pushing the piston against the discs. If burned discs were found in the clutch pack, be sure to check for cracks in the piston.

Stamped steel pistons have replaced most aluminum pistons because they are cheaper to produce. Aluminum pistons require a casting process followed by machining, whereas stamped pistons can be made by the thousands and may only require a simple spot weld operation to prepare them for use. Stamped pistons show cracking more easily than aluminum. Also, look for any separation where spot welding is used on these pistons.

It is common to find annular check balls in clutch pistons (Figure 5-60). These balls allow for an air release from the bore while it is being filled with ATF and for a quick release of pressure when the clutch is released. Inspect each check ball to be sure it is free in its bore. Even after a thorough cleaning in solvent, the check balls may not be free. A fine wire can be pushed into the bore, followed by a spray of carburetor cleaner, to remove any stubborn deposits. The bore should then be blown out with compressed air.

A check ball may also be located in the drum and should be checked. However, it is often very difficult to use the same cleaning techniques as used on pistons. A quick way to determine if the check ball is free is to shake the clutch drum to hear the relief check ball rattle. If the check ball does not rattle, replace the drum.

The ability of the check ball to seal is also important. To check how well the ball seats in its bore, pour ATF into the bore. Observe the other end of the bore. If fluid leaks out, the ball is not seating and the piston or drum should be replaced.

Examine the outside surface of the drum for glazing. Glazing can be removed with emery cloth. Also check the drum's cylinder walls for deep scratches and nicks.

Inspect the front clutch bushing for wear and scores. If the bushing is worn, replace it. Also inspect any bushings found in the clutch drums for excessive wear, scoring, or looseness in the drum bore. Replace as needed.

Clutch Pack Reassembly

Begin assembly of a clutch unit by gathering the new seals and other new parts that may be necessary. Prior to installation, all clutch discs and bands are to be soaked in the type of transmission

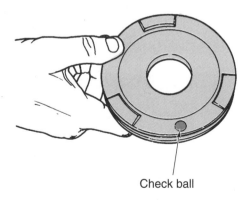

Check ball

Figure 5-60 A clutch piston with an annular check ball (Reprinted with the permission of Ford Motor Co.)

fluid that will be used in the transmission. The minimum soak time is one-half hour. Be sure that all discs are submerged in the fluid and that both sides are coated.

⚠ **WARNING:** Use caution when installing the piston to prevent damage to the seals. Be careful not to stretch the seals during installation.

Before fitting the new rubber seals to the piston, check them against the old seals. This will ensure correct sizing and shape of the new seal. Most overhaul kits include more seals than are required to complete the job. This is because changes in transmission design may dictate the use of a seal or gasket of different design or size. Therefore, both the old design seal and the new design seal are included in the kit. The rebuilder must check to be sure the correct seal is being used. This also holds true for cast iron and Teflon ring seals.

✓ **SERVICE TIP:** Using the hook-end, cast iron, high-reverse, stator support seals from a C-4 on a C-5 will result in no reverse and no high gear because the inner bore of the C-5 clutch drum is larger than that of the C-4. Doing this would be building in a failure that, if not detected during an air check, would require the removal of the transmission to correct.

Once the correct seals are chosen from the kit, they can be installed on the piston (Figure 5-61). Remember to position lip seals so they face the direction that the fluid pressure comes into the drum.

Seals should be lubricated with automatic transmission fluid, trans-gel, or petroleum jelly. Never use chassis lube, "white lube," or motor oil. These will not melt into the transmission fluid as it heats up, but will clog filters and orifices or cause valves and check balls to stick.

Manufacturers switched to Teflon seals for these positions because they helped reduce the wear at the bore in the drum. During an overhaul, they may be replaced with hook-end-type steel rings.

Care must be taken to be sure the ends of scarf-cut Teflon seals are installed correctly. These rings must also be checked for fit into the drum. They should have a snug, but not too tight, fit into the bore.

Some manufacturers use an endless type of Teflon seal. These seals must be installed with special sleeves and pushing tools to avoid overstretching the seal. The seals are first pushed over the installing sleeve to their location in the groove, then a sizing tool is slipped over the sleeve and seal ring to fit the seal to the groove (Figure 5-62).

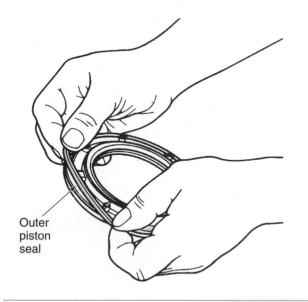

Figure 5-61 Install new seals around the outside of the piston before reassembling the clutch. (Reprinted with the permission of Ford Motor Co.)

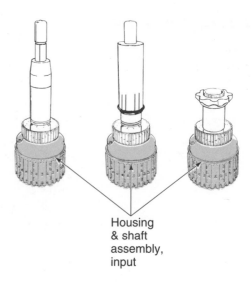

Figure 5-62 Tools and procedure for installing solid seals onto an input shaft (Courtesy of the Hydra-Matic Division of General Motors Corp.)

⚠️ **WARNING:** Never assemble a clutch assembly when it is dry. Always lubricate its components thoroughly with clean ATF.

Assemble the piston, being careful not to allow the seal to kink or become damaged during installation. The piston can now be installed. Several methods may be used to aid piston installation. Lathe-cut seals can be helped into their bores by using a thin feeler gage mounted on a handle. These are available from most automotive tool suppliers. Pistons with lathe-cut seals are installed by first positioning the piston in the bore of the clutch drum. Then slowly work the piston down in the bore until resistance is felt. Using the feeler-type seal installer, work your way around the outer circumference of the seal, using a downward action followed by a clockwise pulling motion as you push the seal back into the groove in the piston.

Occasionally, a piston will not allow access to the outer seal area. A large chamfered edge is at the top of the seal bore to allow the seal to be worked into the bore without the help of any special tools. The piston can be installed by rotating the piston as you push down. Use even pressure to avoid binding the piston or cocking the piston in its bore. Uneven pressure can also cause the ring seal to be pushed out and/or to tear.

☑️ **SERVICE TIP:** Some rebuilders use a wax stick to coat lathe-cut seals for installation. This is available under the trade name "Door Ease." Its original use was to stop squeaks on rubber door bumpers and latches. If Door Ease is used, do not coat the drum bore with ATF. The Door Ease works fine by itself.

Pistons with lip seals require a more delicate installation. The lips can be bent back or torn unless proper caution is taken during installation. The basic shape of a lip seal makes it necessary to use an installation tool (Figure 5-63). The lip must be pushed back towards the piston body in order to allow the seal to enter the bore. Lip seals will often stick in snap ring grooves as you try to slip the piston and seals into the drum. Piano wire installers can be used to roll lip seals back away from the snap ring grooves or the bore of the drum (Figure 5-64). The round cross section of the wire prevents cutting or tearing of the seal lip during installation.

Some rebuilders use an old credit card in place of the feeler gauge. Its wide surface and plastic edge will not cut into the rubber seals.

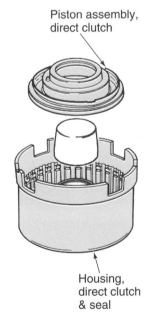

Piston assembly, direct clutch

Housing, direct clutch & seal

Figure 5-63 Using a seal protector to install a piston into a drum without damaging the seals (Courtesy of the Hydra-Matic Division of General Motors Corp.)

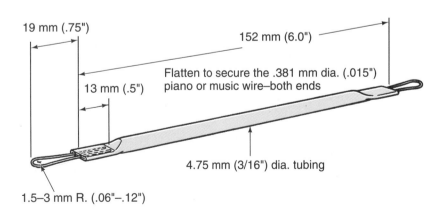

19 mm (.75")

152 mm (6.0")

13 mm (.5")

Flatten to secure the .381 mm dia. (.015") piano or music wire–both ends

4.75 mm (3/16") dia. tubing

1.5–3 mm R. (.06"–.12")

Figure 5-64 Piano wire-type piston installation tool (Courtesy of the Oldsmobile Division of General Motors Corp.)

While holding the piston as squarely in the drum as you can, work the tool around the lip to allow the seal to enter into the bore or around the center of the drum. Do not apply too much downward force as the seal lip is worked into the bore. The piston will fall into place after the lip is fully inserted.

Pistons with multiple seals require special care to avoid damaging the other seals as you work on one seal. Multiple-seal piston installation is made simple by using plastic ring seal installers. These rings compress the seals back into the piston grooves so they will not hang up. Two installers are often used at the same time.

Regardless of the type of tool used to install the piston seals, always take your time to avoid tearing or rolling the new seals during installation.

When working with stamped steel pistons, install the ring spacer on the top side of the piston, making sure it has the correct thickness for the application.

Once the piston seals enter the bore, push the piston all the way down until it stops. Then lift the piston up slightly. Push it back down to the bottom of the bore to be sure it is all the way down. After the piston is installed, rotate the piston by hand to ensure that there is no binding.

Reassemble the springs and retainer after the piston has been installed. Place the single- or multiple-spring set onto the piston using the pockets provided to locate them. Be sure the loose spring sets are spaced as they were during teardown. Set the retainer plate over the springs. Captive spring retainers are simply set on the piston and require no other setup.

Position the spring compressor on the retainer plate and compress the springs. Be careful not to allow the retainer to catch in the snap ring groove while compressing the spring. This will bend or distort the plate, making it unsuitable for use. Remember to compress the springs only enough to get the snap ring into its groove. Use snap ring pliers to expand or contract the snap ring. Once the snap ring is installed and fully seated, release the compressor.

Clutches with a Belleville return spring will not require a compressor for assembly. The Belleville spring is merely laid on the piston and centered in the bore. A wire ring is sometimes inserted on the piston where the Belleville spring touches the piston. This is used to prevent the steel spring from chafing the aluminum piston. If left out, there will be too much endplay and piston damage will occur.

The large snap ring that retains the Belleville spring is now inserted in its groove. This snap ring may be either flat or of wavy construction. There may also be a plastic spacer ring used under the snap ring to center the spring. Be certain the snap ring seats firmly against the drum. Then install the pressure plate on top of the Belleville spring.

Continue clutch assembly by stacking the clutch pack. Begin by installing the dish plate, with the dish facing outward. If a wavy plate or cushion spring is used, it will normally be installed next to the piston. Alternately stack the steels and friction discs until the correct number of plates has been installed. Place the backing plate (the thickest steel plate) on top and install the retainer ring in its groove. Assemble the remaining clutch packs.

SERVICE TIP: Used steels can be deglazed by sanding their faces with 180-grit emery paper, which can produce a dull surface on the plates. Glass bead blasting and abrasive shot tumbling are used by automatic transmission rebuilding shops to refinish polished but still usable steel plates.

As mentioned before, all models of a transmission do not always have the same number of discs and steels in their clutches. Most overhaul kits supply enough discs and steels to rebuild all models. They are likely to have more discs and steels than are required. This is why a technician should always note how many discs and steels are in each pack while disassembling the transmission.

Trying to fit all of the supplied discs and steels may result in clutches with no freeplay or no room for the snap ring. If these problems come up while stacking the pack, refer to a service manual. It may list the number of discs and steels for the model you are building. After the plates are installed,

Loose springs are used mostly with aluminum pistons.

position the retainer plate and install the retaining snap ring. Proceed by measuring the clearance between the plate and snap ring.

⚠️ **WARNING:** Steel plates and friction discs from the same transmission may look the same, but may have different thicknesses. Compare the new plates and discs to the ones removed from each clutch during teardown. Incorrect disc and steel thicknesses can cause buildup headaches, if you are not aware of this fact. The new disc facings should match the type that was removed in both thickness and grooving, if correct clutch engagement is to be maintained.

✔️ **SERVICE TIP:** There are three different direct clutch counts for the A500 and A518. The only reliable way to determine the correct number of clutches for the transmission you are working on is to measure the snap ring groove height in the drum. If the distance from the groove to the top of the drum is 0.485 inch, there should be 5 clutches and 4 steels. If the measurement is 0.350 inch, there should be 6 clutches and 5 steel. If the measurement is 0.100 inch, 8 clutches and 7 steels should be installed.

Clearance Checks

The clearance check of a clutch pack is critical for correct transmission operation. Excessive clearance causes delayed gear engagements, while too little clearance causes the clutch to drag. Adjusting the clearance of multiple-disc clutches can be done with the large outer snap ring in place.

With the clutch pack and pressure plate installed, use a feeler gauge to check the distance between the pressure plate and the outer snap-ring (Figure 5-65). Clearance can also be measured between the backing plate and the uppermost friction disc. If the clutch pack has a waved snap ring, place the feeler gauge between the flat pressure plate and the wave of the snap ring farthest away from the pressure plate. Compare the distance to specifications. Attempt to set pack clearance to the smallest dimension shown in the chart.

Clearances can also be checked with a dial indicator and hook tool (Figure 5-66). The hook tool is used to raise one disc from its downward position and the amount that it is able to move is recorded on the dial indicator. This represents the clearance.

Another way to measure clearance is to mount the clutch drum on the clutch support and use 25–35 psi of compressed air through the oil pump body channels to charge the clutch. Clearance

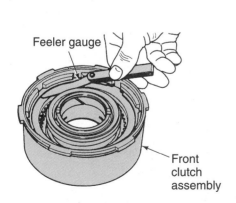

Figure 5-65 Measuring clutch clearance with a feeler gauge (Courtesy of Chrysler Corp.)

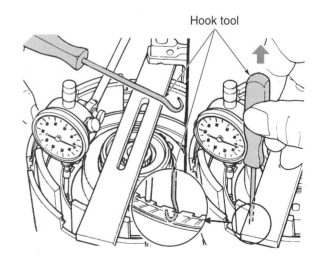

Figure 5-66 Using a hook tool to measure clutch clearance (Courtesy of Chrysler Corp.)

	Service limit	
1st	0.65–0.85 mm	(0.026–0.033 in)
2nd	0.65–0.85 mm	(0.026–0.033 in)
3rd	0.40–0.60 mm	(0.016–0.024 in)
4th	0.40–0.60 mm	(0.016–0.024 in)
Low-hold	0.80–1.00 mm	(0.031–0.039 in)

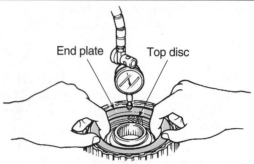

Figure 5-67 Dial indicator setup for measuring clutch clearances (Courtesy of Honda Motor Co.)

can be measured by mounting a dial indicator so that it reads backing plate movement as the air forces the piston to apply the clutch (Figure 5-67).

If the clearance is greater than specified, install a thicker snap ring to take up the clearance. If the clutch clearance is insufficient, install a thinner snap ring.

Another way to adjust clutch clearance is to vary the thickness of the clutch pressure plate. By using a pressure of the desired thickness, you can obtain adequate clutch clearance (Figure 5-68).

Air Testing

After the clearance of the clutch pack is set, perform an air test on each clutch. This test will verify that all of the seals and check balls in the hydraulic component are able to hold and release pressure.

Air checks can also be made with the transmissions assembled. This is the absolute best way to check the condition of the circuit, because there are very few components missing from the circuit. The manufacturers of different transmissions have designed test plates that are available to test different hydraulic circuits (Figure 5-69). Testing with the transmission assembled also allows for testing of the servos.

P/N	Plate no.	Thickness mm (in)
22551–PX4–003	1	2.1 (0.082)
22552–PX4–003	2	2.2 (0.086)
22553–PX4–003	3	2.3 (0.090)
22554–PX4–003	4	2.4 (0.094)
22555–PX4–003	5	2.5 (0.098)
22556–PX4–003	6	2.6 (0.102)
22557–PX4–003	7	2.7 (0.106)
22558–PX4–003	8	2.8 (0.110)
22559–PX4–003	9	2.9 (0.114)

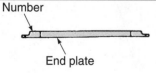

Figure 5-68 A typical chart of the various pressure plate thicknesses available to correct clutch clearances (Courtesy of Honda Motor Co.)

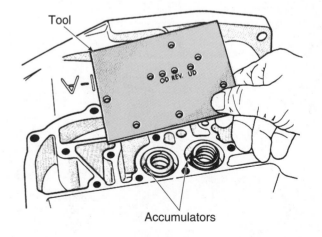

Figure 5-69 A manufacturer's air pressure test plate (Courtesy of Chrysler Corp.)

To test a clutch assembly, install the oil pump assembly with its reaction shaft support over the input shaft and slide it into place on the front clutch drum. When the clutch drums are mounted on the oil pump, all components in the circuit can be checked. If the clutch cannot be checked in this manner, blocking off apply ports with your finger and applying air pressure through the other clutch apply port will work.

> ✓ **SERVICE TIP:** FWD transaxles that utilize a chain to connect the torque converter output to the transaxle input shaft do not have a common shaft for the clutches and the oil pump. Therefore, it is impossible to test the drums from these transaxles while they are mounted on the oil pump. Instead, the drums should be tested while they are mounted on the front sprocket support. These are tested in the same way as any other drum.

The 4T60 and AXOD transmissions are common transaxles that have a chain link and an off-center input shaft.

Invert the entire assembly and place it in an open vise or transmission support tool. Then air test the circuit using the test hole designated for that clutch (Figure 5-70). Be sure to use low-pressure compressed air (25–35 psi) to avoid damage to the seals. Higher pressures may blow the rubber seals out of the bore or roll them on the piston.

While applying air pressure, you may notice some escaping air at the metal or Teflon seal areas. This is normal, as these seals have a controlled amount of leakage designed into them. There should be no air escaping from the piston seals. The clutch should apply with a dull but positive thud. It should also release quickly without any delay or binding. Examine the check ball seat for evidence of air leakage.

One-way Clutch Assemblies

Because they are purely mechanical in nature, one-way clutches are relatively simple to inspect and test. The durability of these clutches relies on constant fluid flow during operation. If a one-way clutch has failed, a thorough inspection of the hydraulic fluid feed circuit to the clutches must be made to determine if the failure was due to fluid starvation. The rollers and sprags ride on a wave of fluid when they overrun. Since most of these clutches spend most of their time in the overrunning state, any loss of fluid can cause rapid failure of the components.

Sprags, by design, produce the fluid wave effect as they slide across the inner and outer races, making them somewhat less prone to damage. Rollers, due to their spinning action, tend to throw off fluid, which allows more chance for damage during fluid starvation. During the check of the hydraulic circuit, take a look at the feed holes in the races of the clutch. Use a small-diameter wire and spray carburetor cleaner or brake cleaner to be certain the feed holes are clear. Push the

Shop Manual
Chapter 5, page 154

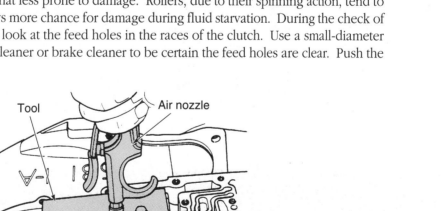

Tool   Air nozzle

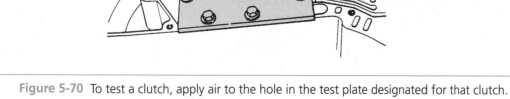

Figure 5-70 To test a clutch, apply air to the hole in the test plate designated for that clutch. (Courtesy of Chrysler Corp.)

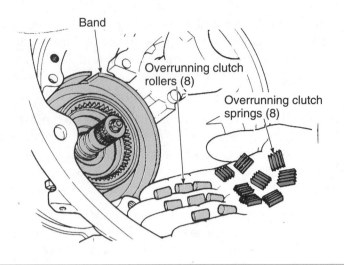

Figure 5-71 Inspect all parts of a roller-type one-way clutch. (Courtesy of Chrysler Corp.)

wire through the feed holes and spray the cleaner into them. Blowing through them with compressed air after cleaning is recommended.

Roller Clutches

Classroom Manual
Chapter 5, page 154

Roller clutches should be disassembled to inspect the individual pieces (Figure 5-71). The surface of the rollers should have a smooth finish with no evidence of any flatness. Likewise, the race should be smooth and show no sign of brinnelling, as this indicates severe impact loading. This condition may also cause the roller clutch to "buzz" as it overruns.

All rollers and races that show any type of damage or surface irregularities should be replaced. Check the folded springs for cracks, broken ends, or flattening out. All of the springs from a clutch assembly should have approximately the same shape. Replace all distorted or otherwise damaged springs. The cam race of a roller clutch may show the same brinnelled wear due to impact overloading. The cam surface, like the smooth race, must be free of all irregularities.

Sprag Clutches

Sprag clutches (Figure 5-72) cannot easily be disassembled; therefore, a complete and thorough inspection of the assembly is necessary. Pay particular attention to the faces of the sprags. If the faces are damaged, the clutch unit should be replaced. Sprags and races with scored or torn faces are an indication of dry running and require the replacement of the complete unit.

Once the one-way clutches are ready for installation, verify that they overrun in the proper direction (Figure 5-73). In some cases, it is possible to install these clutches backwards, which would cause them to overrun and lock in the wrong direction. This would result in some definite driveability problems. To make sure you have installed the clutch in the correct direction, determine the direction of lockup before installing the clutch.

Transmission Cooler Service

Classroom Manual
Chapter 5, page 100

Vehicles equipped with an automatic transmission can have an internal or external transmission fluid cooler, or both (Figure 5-74). The basic operation of either type of cooler is that of a heat exchanger. Heat from the fluid is transferred to something else, such as a liquid or air. Hot ATF is sent from the transmission to the cooler, where it has some of its heat removed, then the cooled ATF returns to the transmission.

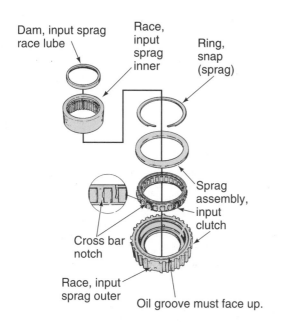

Dam, input sprag race lube

Race, input sprag inner

Ring, snap (sprag)

Sprag assembly, input clutch

Cross bar notch

Race, input sprag outer

Oil groove must face up.

Figure 5-72 A typical sprag one-way clutch assembly (Courtesy of the Hydra-Matic Division of General Motors Corp.)

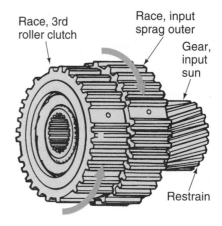

Must freewheel in direction of arrows and hold in opposite direction.

Race, 3rd roller clutch

Race, input sprag outer

Gear, input sun

Restrain

Figure 5-73 A sprag and/or roller clutch functional check (Courtesy of the Hydra-Matic Division of General Motors Corp.)

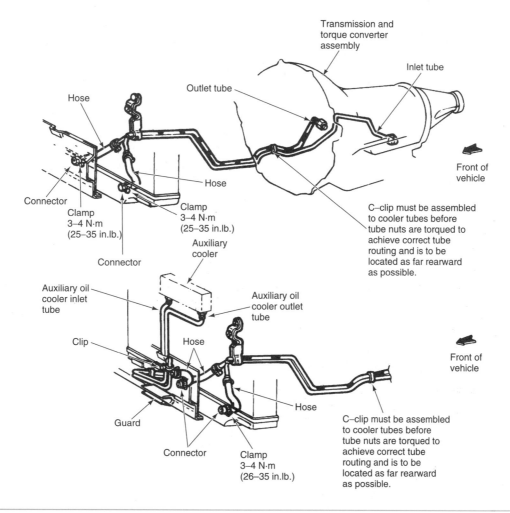

Transmission and torque converter assembly

Inlet tube

Outlet tube

Hose

Hose

Connector

Clamp 3–4 N·m (25–35 in.lb.)

Clamp 3–4 N·m (25–35 in.lb.)

Connector

Auxiliary cooler

Front of vehicle

C–clip must be assembled to cooler tubes before tube nuts are torqued to achieve correct tube routing and is to be located as far rearward as possible.

Auxiliary oil cooler inlet tube

Auxiliary oil cooler outlet tube

Clip

Hose

Front of vehicle

Guard

Connector

Clamp 3–4 N·m (26–35 in.lb.)

Hose

C–clip must be assembled to cooler tubes before tube nuts are torqued to achieve correct tube routing and is to be located as far rearward as possible.

Figure 5-74 Typical transmission coolers (Reprinted with the permission of Ford Motor Co.)

Internal coolers are located inside the engine's radiator. Heated ATF travels from the torque converter to a connection at the radiator. Inside the radiator is a small internal cooler, which is sealed from the liquid in the radiator. ATF flows through this cooler and its heat is transferred to the liquid in the radiator. The ATF then flows out of a radiator connection, back to the transmission.

● **CUSTOMER CARE:** This type of ATF cooler is less efficient as engine temperatures increase; therefore, it is recommended that an external cooler be installed in addition to the internal one, when the vehicle is used during conditions where the engine will be operating at high temperatures, to provide extra cooling for the transmission.

External coolers are mounted outside the engine's radiator, normally just in front of it. Air flowing through the cooler removes heat from the fluid before it is returned to the transmission.

The engine's cooling system is the key to efficient transmission fluid cooling. If anything affects engine cooling, it will also affect ATF cooling. The engine's cooling system should be carefully inspected whenever there is evidence of ATF overheating or a transmission cooling problem.

If the problem is the transmission cooler, examine it for signs of leakage. Check the dipstick for evidence of coolant mixing with the ATF. Milky fluid indicates that engine coolant is leaking into and mixing with the ATF because of a leak in the cooler. At times, the presence of ATF in the radiator will be noticeable when the radiator cap is removed, as ATF will tend to float to the top of the coolant. A leaking transmission cooler core can be verified with a leak test.

External cooler leaks are evident by the traces or film buildup of ATF around the source of the leak. It usually takes little time to determine the source of the leakage. However, internal coolers present a little bit more difficulty and a leak test should be conducted. To do this, place a catch pan under the fittings that connect the cooler lines to the radiator. Then, disconnect and plug both cooler lines from the transmission (Figure 5-75). Tightly plug one end of the cooler and apply compressed air (50–75 psi) into the open end of the cooler.

▪ **CAUTION:** Be sure the radiator is cool to the touch. Hot coolant may spray on you if the system is hot. The hot coolant can cause serious injuries. Wearing safety goggles is a good precaution when working around hot liquids.

▪ **CAUTION:** Wear safety goggles when working with compressed air.

▪ **CAUTION:** Do not apply more than 75 psi to the cooler. Extreme pressures can cause the cooler to explode.

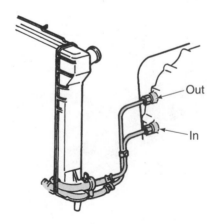

Figure 5-75 Before testing the cooling system, identify which line is the in line and which is the out line. (Reprinted with the permission of Ford Motor Co.)

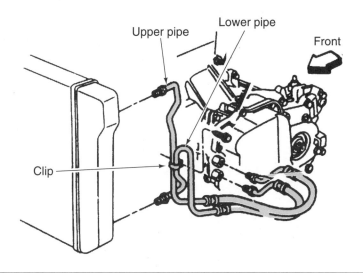

Figure 5-76 Typical routing of transmission oil cooler lines for a transaxle (Courtesy of the Chevrolet Motor Divivision of General Motors Corp.)

Carefully check the coolant through the top of the radiator. If there is a leak in the transmission cooler, bubbles will be apparent in the coolant. If it leaks, the cooler must be replaced.

The inability of the transmission fluid to cool can also be caused by a plugged or restricted fluid cooling system. If the fluid cannot circulate, it cannot cool. To verify that the transmission cooler is plugged, disconnect the cooler return line from the radiator or cooler (Figure 5-76). Connect a short piece of hose to the outlet of the cooler and allow the other end of the hose to rest inside an empty container. Start the engine and measure the amount of fluid that flows into the container after 20 seconds. Normally, one quart of fluid should flow into the container. If less than that fills the container, a plugged cooler is indicated.

 WARNING: A clogged cooler will result in the subsequent failure of a new or rebuilt transmission.

A tube-type transmission cooler can be cleaned by using cleaning solvent or mineral spirits and compressed air. A fin-type cooler, however, cannot be cleaned in this same way. Therefore, normal procedure includes replacing the radiator (which includes the cooler). In both cases, the cooler lines should be flushed to remove any debris.

If the cooler is plugged, disconnect the cooler lines at the transmission. Apply compressed air through one port of the cooler, then to the other port. The air should blow any debris out of the cooler. Always use low air pressure, no more than 30 psi. Higher pressures may damage the cooler. If little air passes through the cooler, the cooler is severely plugged and may need to be replaced.

If the vehicle has two coolers, pull the inlet and outlet lines from the side of the transmission and, using a hand pump, pump mineral spirits into the inlet hose until clear fluid pours out of the outlet hose. To remove the mineral spirits from the cooler, release some compressed air into the inlet port, then pump one quart of ATF into the cooler.

SERVICE TIP: A transmission cooler is a good hiding place for debris when there has been a major transmission failure. Blowing out the cooler and lines will not always remove all of the dirt. Install a temporary auxiliary oil filter in the cooler lines to catch any leftover debris immediately after rebuilding a transmission.

Check the condition of the cooler lines from their beginning to their ends (Figure 5-77). A line that has been accidently damaged while the transmission has been serviced will reduce oil flow through the cooler and shorten the life of the transmission. If the steel cooler lines need to be replaced, use only double-wrapped and brazed steel tubing. Never use copper or aluminum tubing to replace steel tubing. The steel tubing should be double-flared and installed with the correct fittings.

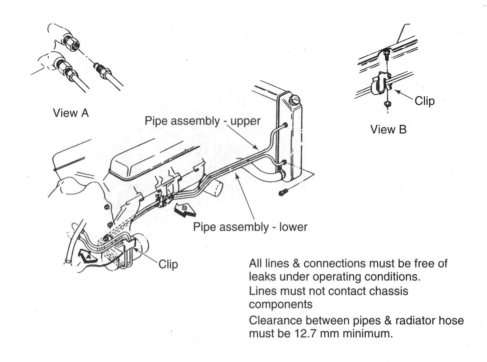

View A

Pipe assembly - upper

Clip

View B

Pipe assembly - lower

Clip

All lines & connections must be free of leaks under operating conditions.
Lines must not contact chassis components
Clearance between pipes & radiator hose must be 12.7 mm minimum.

Figure 5-77 Typical routing of oil cooler lines for a transmission (Courtesy of the Oldsmobile Division of General Motors Corp.)

CASE STUDY

A car dealer called to see if the Automotive Department at the college could help with a diagnostic problem. His customer had a new Ford pickup with a C-5 transmission. The owner complained of a buzzing noise while driving. The technician at the dealership disassembled the transmission and installed an overhaul kit. The next day, the customer picked up his truck. He came back the next day, slightly annoyed, with the same complaint of buzzing while driving. The same technician once again disassembled the transmission and installed another overhaul kit. This time the technician road tested the truck after installing the kit. The noise was still there. By this time, both the owner of the truck and the technician were not very happy. After two "rebuilds," the transmission still had the same noise.

The transmission was brought to the college and was checked out on the transmission "dyno." In the forward ranges, the buzzing was only apparent in intermediate and high gear ranges. There was no buzzing in low or reverse gear ranges. The students, using a power flow chart, determined that the only gear train member not in motion in low and reverse was the rear carrier. Because the rear carrier is held by the one-way clutch, we decided to disassemble the transmission to check the clutch. The inner race of the roller clutch had a flat spot from an apparent machining mishap during manufacture. Every time a roller went past the flat spot, it clicked into the flat, producing the buzz at road speeds. We installed a new roller inner race, which was supplied by the dealer, and, like magic, the buzz was gone.

Why were students able to determine the problem while an experienced technician could not find it? Very simply, the technician did not understand the basic theories of transmission operation. The road test very clearly showed that the buzzing only occurred in second and third gears. A basic knowledge of the Simpson gear train and/or knowing how to read and use a flow chart would have helped the technician diagnose the problem. Apparently this technician, unfortunately like many others, only knew how to "put a kit in it" and hope that the noise would go away. Just as a point of interest, according to the service manager, the technician loaded up his tools and quit when he learned that students, armed with a power flow chart and a shop manual had diagnosed and repaired the problem that had him stumped.

Terms to Know

Cap	Intercooler	Steels
Fin-type cooler	Mineral spirits	Tube-type cooler
Flow chart	Plug	VFM
Heat exchanger	Reverse stall test	

ASE Style Review Questions

1. *Technician A* says all parts of the valve body should be soaked in mineral spirits before reassembling.
   Technician B says a lintfree rag is a must when wiping down valves.
   Who is correct?
   A. A only **C.** Both A and B
   B. B only **D.** Neither A nor B

2. While removing scratches in a valve:
   Technician A uses fine emery cloth to remove the scratch.
   Technician B uses a sandblaster or glass bead machine to polish the surface of the valve.
   Who is correct?
   A. A only **C.** Both A and B
   B. B only **D.** Neither A nor B

3. *Technician A* says overtorquing the hold-down bolts of the valve body can cause the valves to stick in their bore.
   Technician B says flat filing the surface of the valve body will allow the valve body to seal properly and will therefore allow the valves to move freer in their bores.
   Who is correct?
   A. A only **C.** Both A and B
   B. B only **D.** Neither A nor B

4. *Technician A* says the vanes in a variable-displacement oil pump should be replaced if the outer edges are worn flat.
   Technician B says all parts of vane-type pumps are selectively sized.
   Who is correct?
   A. A only **C.** Both A and B
   B. B only **D.** Neither A nor B

5. While discussing a pressure test:
   Technician A says this test is the most valuable diagnostic check for slippage in one gear.
   Technician B says the test can identify the cause of late or harsh shifting.
   Who is correct?
   A. A only **C.** Both A and B
   B. B only **D.** Neither A nor B

6. *Technician A* says all adjustable bands have their locknut and adjusting screw on the outside of the transmission case.
   Technician B says some bands do not have adjusting screws.
   Who is correct?
   A. A only **C.** Both A and B
   B. B only **D.** Neither A nor B

7. *Technician A* says transmissions originally equipped with Teflon seals must be refitted with Teflon seals during an overhaul.
Technician B says a press is needed for the installation of Teflon seals.
Who is correct?
   A. A only
   B. B only
   C. Both A and B
   D. Neither A nor B

8. While discussing the results of an oil pressure test:
Technician A says when the fluid pressures are high, internal leaks, a clogged filter, low oil pump output, or a faulty pressure regulator valve are indicated.
Technician B says if the fluid pressure increased at the wrong time, an internal leak at the servo or clutch seal is indicated.
Who is correct?
   A. A only
   B. B only
   C. Both A and B
   D. Neither A nor B

9. *Technician A* says an air test can be used to check servo action.
Technician B says an air test can be used to check for internal fluid leaks.
Who is correct?
   A. A only
   B. B only
   C. Both A and B
   D. Neither A nor B

10. While checking clutch discs:
Technician A says the steel plates should be replaced if they are worn flat.
Technician B says the friction discs should be squeezed to see if they can hold fluid. If they hold fluid and look okay, they are serviceable.
Who is correct?
   A. A only
   B. B only
   C. Both A and B
   D. Neither A nor B

Table 5-1 ASE TASK

Perform pressure tests and determine needed repairs.

Problem Area	Symptoms	Possible Causes	Classroom Manual	Shop Manual
LOW PRESSURE	Pressure low at idle in all ranges	1. Defective EGR	38	33
		2. Low fluid level	98	33
		3. Clogged filter	99	34
		4. Loose oil tubes, valve body, or regulator	106	105
		5. Internal fluid leaks	106	99
		6. Stuck pressure regulator valve	106	105
		7. Defective oil pump	101	113
		8. Defective modulator	117	46
		9. Missing check ball	106	106
	Pressure low in reverse	Applied clutches or servos leak	147	122
	Pressure low in all forward ranges, but normal in reverse	Applied clutches or servos leak	147	122
	Pressure good at idle in all ranges, but low at low vacuum	1. Defective vacuum modulator	117	46
		2. Defective throttle linkage assembly or valve	118	49
		3. Defective oil pump	101	113
		4. Clogged filter	99	34
		5. Internal leaks	106	99
	Pressure low in a particular range	Leakage in the applied clutches or servos	147	122
	Pressure low in neutral and park	Valve body leaks	106	105
	Incorrect governor pressure	1. Stuck governor weights	115	111
		2. Stuck governor valve	115	111
	Pressure low in all throttle positions	1. Clogged oil cooler or lines	100	142
		2. Internal leaks	106	99
		3. Worn oil pump	101	113
HIGH PRESSURE	Pressure high at idle in all ranges	1. Defective EGR	38	33
		2. Defective vacuum modulator	117	46
		3. Leaking vacuum lines	117	46
		4. Defective throttle valve and/or linkage	118	49
		5. Defective pressure regulator or boost valve	106	105

Disassembly and Assembly of Common Transmissions

Upon completion and review of this chapter, you should be able to:

❏ Disassemble, service, and reassemble a Chrysler Torqueflite 36RH transmission.

❏ Disassemble, service, and reassemble a Chrysler 41TE transaxle.

❏ Disassemble, service, and reassemble a Chrysler 42LE transaxle.

❏ Disassemble, service, and reassemble a Ford A4LD transmission.

❏ Disassemble, service, and reassemble a Ford ATX transaxle.

❏ Disassemble, service, and reassemble a Ford AXOD transaxle.

❏ Disassemble, service, and reassemble a GM 4L60 transmission.

❏ Disassemble, service, and reassemble a GM 3T40 transaxle.

❏ Disassemble, service, and reassemble a GM 4T60 transaxle.

❏ Disassemble, service, and reassemble Honda CA, F4, and G4 transaxles.

❏ Disassemble, service, and reassemble a Mazda GF4A-EL transaxle.

❏ Disassemble, service, and reassemble a Nissan L4N71B/E4N71B transmission.

Basic Tools

Basic mechanic's tool set
Appropriate service manual
Transmission holding fixture
Torque wrench
Rubber-tipped air nozzle
Pan of clean ATF
Feeler gauge

This chapter contains many photo sequences that show the step-by-step procedures for overhauling common transmissions and transaxles. Although a particular model of transmission was chosen for the photographs and other variations of the same transmission are also used, the sequences given in this chapter are for educational purposes and do not include the many procedural changes dictated by the different models.

These sequences also do not include detailed procedures for rebuilding subassemblies, such as the valve body and clutch packs. Detailed information on these, as well as other subassemblies, is given in other chapters of this manual.

Chrysler Torqueflite Transmissions

☑ **SERVICE TIP:** Since 1992, Chrysler has been implementing a new system of identification for its transmissions. This system is a four-character identification system in which each character represents a feature or characteristic of the transmission. The first character relates to the number of forward speeds in the transmission. The second character indicates relative input torque capacity: 0 to 2 (light to heavy duty) for passenger cars and 0 to 7 for trucks. The third character indicates whether the vehicle is FWD or RWD and relates to the position of the engine in the drive line: R indicates rear wheel drive; T indicates transverse engine mounting in a FWD vehicle; L indicates longitudinal engine mounting in a FWD vehicle; and A indicates all wheel drive. The fourth character describes the type of control the transmission relies on: E indicates electronic controls and H indicates hydraulic controls. For the next few years, Chrysler's transmissions will be referred to by their old or new number. Following is a list of the new names and old names for Chrysler transmissions:

New Name	Old Name
30RH	A-904
30TH	A-404

31RH	A-998
31TH	A-413/670
32RH	A-999
36RH	A-727
37RH	A-727 (diesel)
40RH	A-500
41AE	A-604 (AWD)
41TE	A-604
42LE	none
42RE	none
42RH	A-500
46RH	A-518

Chrysler Corporation introduced the Torqueflite transmission in 1956. This transmission was the first modern three-speed automatic transmission with a torque converter and the first to use the Simpson two-planetary compound gear train. Nearly all Torqueflite-based transmissions and transaxles use a rotor-type oil pump and all use a Simpson gear train.

The original Torqueflite was called the A-466 and it had a cast iron case with separate aluminum castings for the bellhousing and the extension housing. Late-model Torqueflites have a one-piece aluminum housing that incorporates the bellhousing and a bolt-on extension.

There are two basic versions of the three-speed Torqueflite transmission: the A-904 and the A-727. The 904 is the light-duty version, while the 727 is the heavy-duty version. The A-727 has a larger case than the A-904 and was originally designed to be used behind large displacement engines.

The A-904 Torqueflite is typically used behind small displacement engines on RWD vehicles. Many vehicles equipped with this transmission are also fitted with a lockup torque converter. The A-998 and A-999 transmissions are newer versions of the A-904. The A-998 was designed specifically for the 318 cid (5.2-L) V-8, while the A-999 was used behind the 360 cid (5.8-L) V-8. All three designs look very similar and share the same basic components.

All the Torqueflite-based transmissions and transaxles used today have two multiple-disc clutches, an overrunning clutch, two servos and bands, and two planetary gear sets to provide three forward gear ratios and a reverse ratio. The two multiple-disc clutches are called the front and rear clutch packs. The servos and bands are also referred to by their location, front and rear, or by their function, kickdown and low/reverse.

In 1989, Chrysler introduced the A-500 and A-518, which are four-speed automatic transmissions used exclusively in the mid-size Dodge Dakota and full-size Dodge RAM pickups and vans. The first three forward gear ratios are the same as the A-904 and A-727 Loadflite units. Fourth gear is provided by a separate planetary gear set and controlled by an overdrive clutch, direct clutch, and overrunning clutch in an overdrive assembly attached to the rear of the transmission. The A-500 is actually a modified A-999 three-speed transmission. The A-518 is a modified A-727 and is designed for heavy-duty use.

These transmissions have extra long extension housings, which hold the additional planetary gear set and an additional shaft. This shaft serves as the output shaft and the three-speed output shaft became an intermediate shaft linking the output from the Simpson gear train to the overdrive assembly.

To control the operation of the overdrive planetary gear set, two multiple-disc clutches (direct and overdrive clutches) and a one-way overrunning clutch are used. The intermediate shaft is locked to the output shaft whenever the one-way clutch is locked, providing for direct drive.

Photo Sequence 7 goes through the procedure for overhauling a 36RH transmission. The overhaul procedures for other Torqueflite transmissions are similar.

Classroom Manual
Chapter 6, page 165

Special Tools
Dial indicator and holding fixture
Clutch compressor tool
Oil pump puller
Seal remover/installer

Photo Sequence 7
Typical Procedure for Overhauling a Torqueflite 36RH (727) Transmission

P7-1 Check and record the endplay of the input shaft. Then unbolt and remove the oil pan and gasket.

P7-2 Unbolt and remove the oil filter. Then unbolt and remove the valve body by first disconnecting the parking lock rod from the manual lever and unbolting the TCC solenoid.

P7-3 Remove the accumulator piston spring and lift the piston from the case.

P7-4 Remove the parking lock rod. Then unbolt and remove the extension housing. There is a cover plate in the bottom of the extension housing that, once removed, will allow access to the output shaft's snap ring. This ring must be expanded while pulling the extension housing off the case.

P7-5 Tighten the front band adjustment screw to prevent damage to the discs during disassembly.

P7-6 Remove the oil pump bolts and pull the oil pump out of the case with special pullers.

P7-7 Loosen the front band adjustment screw and remove the band strut and anchor.

P7-8 Remove the front clutch assembly, with the input shaft, from the case.

P7-9 Remove the front planetary gear set.

Typical Procedure for Overhauling a Torqueflite 36RH (727) Transmission (continued)

P7-10 Loosen the rear band adjustment screw and remove the band strut and band. Then remove the output shaft with the governor attached to it.

P7-11 Compress the return spring for the front servo and remove the retaining snap ring.

P7-12 Remove the servo assembly from the case.

P7-13 Compress the return spring for the rear servo and remove the retaining snap ring.

P7-14 Remove the servo assembly from the case.

P7-15 Replace all servo piston seals.

P7-16 Reinstall the servo assemblies after careful inspection.

P7-17 Inspect the planetary gear set and drive shells. Replace components as needed.

P7-18 Inspect the bands and their struts and anchors. Replace as needed.

P7-19 Disassemble and inspect the clutch units. Replace parts as needed. Remember to allow the discs to soak in ATF before assembling the clutch pack.

P7-20 Replace the pistons' oil seal rings in the clutch assemblies.

P7-21 Reassemble the clutch packs. Then check the clearances.

P7-22 Unbolt the stator support from the oil pump housing. Separate the support from the oil pump.

P7-23 Check the wear of the gears and the pump itself. Replace parts as needed.

P7-24 Replace the pump seals and reassemble the pump. Torque the bolts to specifications.

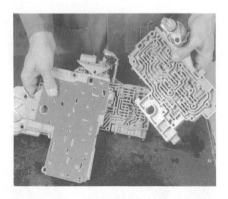

P7-25 Disassemble, clean, and inspect the valve body.

P7-26 Bolt the valve body together and tighten the bolts to specifications.

P7-27 Install the output shaft with the governor into the case. Then install the rear band and anchor assembly and tighten the adjustment screw.

P7-28 Install the rear planetary gear set and drum. Make sure all thrust washers are being installed in their correct locations.

P7-29 Install the front clutch assembly and the front band assembly. Then tighten the adjustment screw.

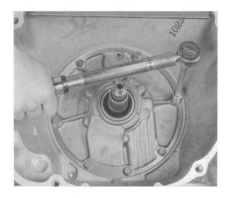

P7-30 Assemble the servos with new seals. Then install the oil pump and torque bolts in the proper sequence and to the proper torque.

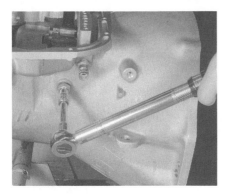

P7-31 Adjust both bands according to specifications.

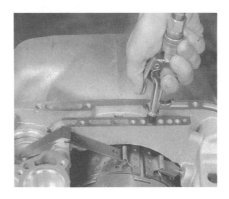

P7-32 Air test the entire transmission before installing the oil filter and oil pan.

Chrysler Torqueflite Transaxles

In 1978, the basic Torqueflite transmission was modified for use as an FWD transaxle. Torqueflite transaxles contain the same basic parts as the A-904 transmission, with the addition of a transfer shaft, final drive gears, and differential unit.

Many different transaxle models (A-404, 413, 415, 470, 604, 670, and 42LE) have been used since then, most designed for use with a particular engine. All three-speed Torqueflite-based transaxles have two multiple-disc clutches, an overrunning clutch, two servos and bands, and two planetary gear sets. The four-speed models use two planetary gear sets and five multiple-disc clutches. Current models of these transaxles have electronically controlled shifting and lockup torque converters. The 41TE (A604) is commonly used. The overhaul procedure for this transaxle is shown in Photo Sequence 8.

Classroom Manual
Chapter 6, page 165

Special Tools
Dial indicator and holding fixture
Clutch compressor tool
Oil pump puller
Seal remover/installer
Output shaft holding tool

Photo Sequence 8
Typical Procedure for Overhauling a 41TE (A604) Transaxle

P8-1 Remove all electrical switches from the outside of the transaxle case.

P8-2 Remove the solenoid assembly and gasket.

P8-3 Remove the torque converter.

P8-4 Loosen and remove the oil pan's bolts. Then remove the oil pan.

P8-5 Remove the oil filter.

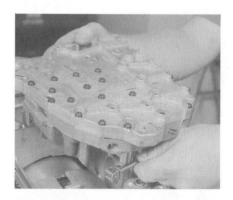

P8-6 Loosen and remove the valve body attaching bolts. Then push the park rod rollers away from the guide bracket and remove the valve body assembly. Pull straight up on the valve body, as the manual shaft is attached to it.

P8-7 Remove the retaining snap ring for the accumulator. Then remove the accumulator assembly.

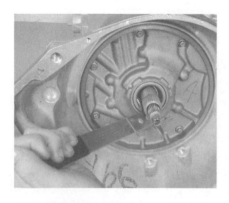

P8-8 Remove the oil pump seal, using the correct seal puller.

P8-9 Loosen and remove the oil pump's attaching bolts. Then, using the correct puller, remove the oil pump. While pulling the pump, push in on the input shaft.

P8-10 Pull out the input shaft and clutch assembly.

P8-11 Remove and discard the oil pump gasket. Then remove the caged needle bearing assembly from the input shaft.

P8-12 Remove the front sun gear assembly.

P8-13 Twist and pull on the front carrier and rear ring gear assembly to remove it.

P8-14 Remove the rear sun gear with its thrust bearing.

P8-15 Using the proper clutch compressor, remove the 2–4 clutch snap ring and clutch retainer. Mark the alignment of the retainer with the return spring located below it. Then remove the return spring.

P8-16 Remove the clutch pack and tag it as the 2–4 pack.

P8-17 Remove the low/reverse tapered snap ring from the case. Follow the recommended sequence while prying the snap ring out of the case.

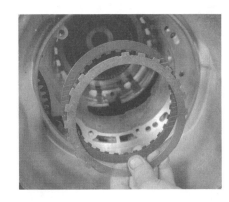

P8-18 Remove the low/reverse reaction plate and one friction disc.

P8-19 Remove the low/reverse reaction plate flat snap ring. Then remove the clutch pack. Tag the pack as the low/reverse clutch.

P8-20 Loosen and remove the rear cover bolts. Then remove the rear cover.

P8-21 Remove the transfer shaft gear nut and washer. Then, using a gear puller, remove the transfer gear.

P8-22 Remove the transfer gear selective shim and the bearing retainer. Then remove the transfer shaft bearing snap ring.

P8-23 Remove the transfer shaft and the output shaft gear bolt and washer. Then, using a gear puller, remove the output gear and selective shim.

P8-24 Remove the rear carrier assembly.

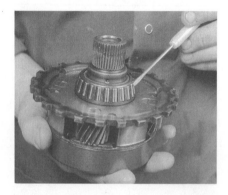

P8-25 After careful inspection of all parts, begin reassembly by installing the rear carrier assembly.

P8-26 Install the output gear, selective shim, washer, and bolt. While holding the shaft with a special holding tool, tighten the bolt to specifications.

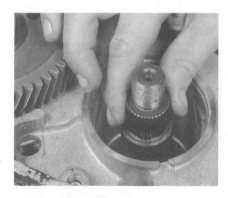

P8-27 Install the transfer shaft.

P8-28 Install the transfer shaft bearing snap ring, selective shim, and bearing retainer. Then install the transfer gear, washer, and nut. While holding the shaft with a special tool, tighten the nut to specifications.

P8-29 Apply a 1/8-inch bead of RTV on the rear cover. Then install the rear cover and tighten the rear cover bolts to specifications.

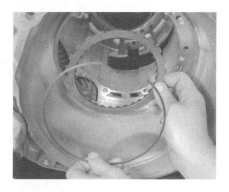

P8-30 Install the low/reverse clutch pack, the reaction plate flat snap ring, the friction disc, and, low/reverse reaction plate. Then install the low/reverse tapered snap ring into the case. Follow the recommended sequence while installing the snap ring into the case.

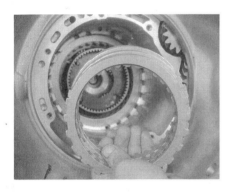

P8-31 Install the 2–4 clutch pack and return spring.

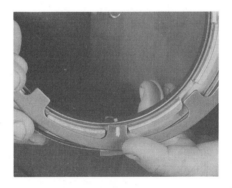

P8-32 Using the proper clutch compressor, install the 2–4 clutch snap ring and clutch retainer. Make sure the alignment marks on the spring and retainer are matched.

P8-33 Install the rear sun gear with its thrust bearing. Then install the front carrier and rear ring gear assembly by twisting and pushing on it.

P8-34 Install the front sun gear assembly.

P8-35 Install the new oil pump gasket. Then install the oil pump. Tighten the oil pump's attaching bolts to specifications.

P8-36 Install the oil pump seal using the correct seal installing tool.

Photo Sequence 8
Typical Procedure for Overhauling a 41TE (A604) Transaxle (continued)

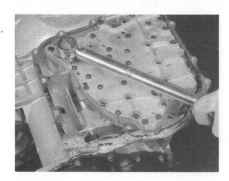

P8-37 Install the accumulator assembly and retaining snap ring. Then install the valve body assembly. Push the park rod rollers away from the guide bracket while positioning the valve body. Install and tighten the valve body retaining bolts to specifications.

P8-38 Install a new oil filter. Then install the oil pan with a new gasket. Tighten the oil pan bolts to specifications.

P8-39 Install the torque converter, the solenoid assembly with a new gasket, and all electrical switches on the outside of the transaxle case.

Classroom Manual
Chapter 6, page 165

Special Tools
Dial indicator and holding fixture
Clutch compressor tool
Oil pump puller
Seal remover/installer
Output shaft holding tool

With the introduction of new mid-size FWD cars in 1993, Chrysler introduced the 42LE, which is a longitudinally mounted transaxle. Although this transaxle is similar to the 41TE, its location in the vehicle requires different overhaul procedures. Photo Sequence 9 shows the step-by-step procedures for overhauling this new design of transaxle.

Photo Sequence 9
Typical Procedure for Overhauling a 42LE Transaxle

P9-1 Remove all of the electrical switches from the outside of the transaxle case.

P9-2 With the transaxle standing up on end and the torque converter removed, measure the endplay of the transaxle.

P9-3 Remove the electrical wiring harness.

P9-4 Remove the oil pan retaining bolts, oil pan, and oil pan gasket.

P9-5 Remove the oil filter.

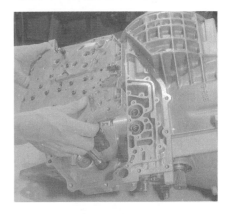

P9-6 Remove the valve body.

P9-7 Remove the retaining snap ring for the accumulator cover, and remove the accumulator and spring.

P9-8 Remove the solenoid assembly from the valve body.

P9-9 Remove the snap ring and long stub shaft from the transaxle.

P9-10 Index the inner and outer differential adjusters at the case.

P9-11 Remove the outer lock bracket and back out the adjuster one complete revolution.

P9-12 Place the seal protector on the shaft and carefully remove the differential cover.

Photo Sequence 9
Typical Procedure for Overhauling a 42LE Transaxle (continued)

P9-13 Remove the differential carrier and ring gear assembly.

P9-14 Remove the inner adjuster lock bracket and inner adjuster.

P9-15 Remove the rear chain cover.

P9-16 Remove the chain snubber guide.

P9-17 Remove the snap rings and wave washers from the output and transfer shafts.

P9-18 Install the special chain spreader tool between the chain sprockets and remove the chain and sprockets as a unit.

P9-19 Remove the transfer shaft nut, rear cone, rear cup, oil baffle, rear shim, transfer shaft, and transfer shaft seals. Follow the procedures in the service manual for removing the staked sections of the nut flange. Use the special tool to hold the nut while you turn the transfer shaft to loosen the nut.

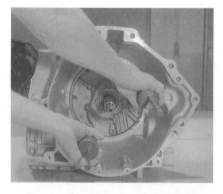

P9-20 Remove the oil pump attaching bolts, then using the proper puller, remove the oil pump.

P9-21 Remove and discard the oil pump gasket. Then remove the bypass valve from the case.

P9-22 Remove the input shaft clutch assembly.

P9-23 Remove the caged needle bearing from the input shaft. Note its orientation as it comes off so that it can be properly installed during reassembly.

P9-24 Remove the thrust washer from the front sun gear.

P9-25 Twist and pull on the front carrier and rear annulus assembly to remove it. Then remove the rear sun gear with its thrust washer.

P9-26 Using the proper clutch compressor, remove the 2–4 clutch snap ring and clutch retainer.

P9-27 Remove the 2–4 clutch retainer and return spring.

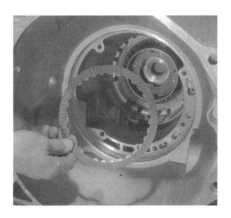

P9-28 Remove the lower reaction plate.

P9-29 Remove the 2–4 clutch pack.

P9-30 Remove the low/reverse tapered snap ring following the sequence in the service manual to pry it out of the case.

Photo Sequence 9
Typical Procedure for Overhauling a 42LE Transaxle (continued)

P9-31 Remove the low/reverse reaction plate.

P9-32 Remove the low/reverse reaction plate flat snap ring.

P9-33 Use an arbor press to press the output shaft from the case.

P9-34 Using an arbor press, press the output shaft into the case.

P9-35 Install the low/reverse reaction plate flat snap ring.

P9-36 Install the low/reverse reaction plate.

P9-37 Follow the recommended procedure for installing the low/reverse tapered snap ring.

P9-38 Install the 2–4 clutch pack.

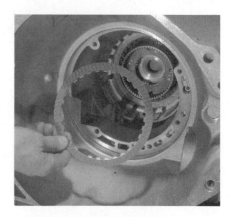

P9-39 Install the lower reaction plate.

P9-40 Install the return spring and 2–4 clutch retainer.

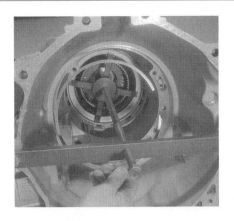

P9-41 Using the proper clutch compressor, install the 2–4 clutch retainer and snap ring.

P9-42 Install the rear rear sun gear with its thrust washer and then seat the rear annulus and front carrier in proper position.

P9-43 Install the front sun gear.

P9-44 Install the front sun gear thrust washer on the front sun gear.

P9-45 Install the caged needle bearing on the input shaft so that the inside diameter elbow flange is seated against the rear sun gear. Use some petroleum jelly to hold it in place.

P9-46 Install the input shaft clutch assembly.

P9-47 Install a new oil pump gasket on the oil pump.

P9-48 Install a new oil pump gasket. Then, install the bypass valve and oil pump. Tighten the oil pump's attaching bolts to specification.

P9-49 Install the transfer shaft nut, rear cone, rear cup, oil baffle, rear shim, transfer shaft, and transfer shaft seals. Use the special tool to hold the nut while you turn the transfer shaft to tighten the nut. Follow the procedures in the service manual for staking the nut.

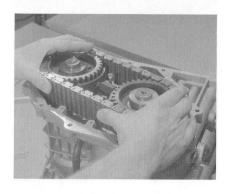

P9-50 Install the chain and sprockets as a unit using the special chain spreader tool.

P9-51 Install the wave washers and snap rings on the output and transfer shafts.

P9-52 Install the chain snubber guide.

P9-53 Install the rear chain cover and torque the mount bolts to specification.

P9-54 Install the inner adjuster and lock bracket. Torque the lock bracket mount bolt to specification.

P9-55 Install the differential carrier and ring gear assembly.

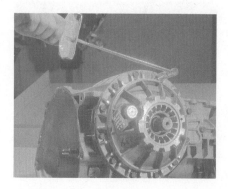

P9-56 Install the differential cover being careful to protect the seals using the special tool. Torque the mount bolts to the specified value.

P9-57 Install the outer adjuster lock and torque its bolt to specification.

P9-58 Install the long stub shaft and snap ring.

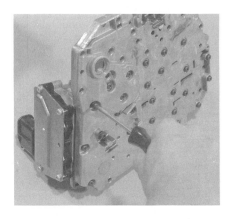

P9-59 Install the solenoid assembly on the valve body.

P9-60 Install the accumulator spring, accumulator, O-ring, and snap ring.

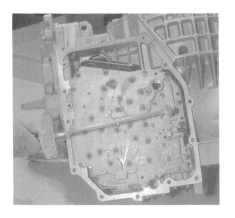

P9-61 Install the valve body.

P9-62 Install the oil filter.

P9-63 Install a new oil pan gasket and then install the oil pan.

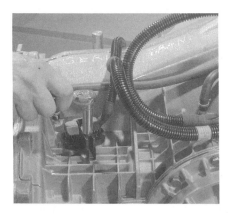

P9-64 Install the wiring harness.

P9-65 Install the electrical switches on the outside of the transaxle case. Tighten them to specified values.

Ford Motor Company Transmissions

Ford Motor Company began to use the Simpson gear train with the introduction of the C-4 transmission in 1964. Previous transmissions were based on the Ravigneaux design, as are some of their current models. Ford Motor Company has five basic transmissions that use the Simpson gear train and are quite similar to each other: the C-3, C-4, C-5, C-6, and A4LD. The A4LD is a four-speed transmission, whereas the others are three-speed transmission. All of these units use a gear and crescent-type oil pump as the source of all hydraulic pressures.

In appearance, these transmissions are similar and all have a separate and removable extension housing. The major external difference between the models is the size and the reinforcements of the case. The bellhousings of the C-3, C-4, and C-5 are removable from the transmission case, whereas the bellhousing of the C-6 is part of the case's casting.

To control planetary gear action, these transmissions use two multiple-disc clutches, two bands, and one overrunning clutch. The two clutches are referred to as the high, front, or reverse/high and the rear or forward clutches. The two bands are the intermediate, front, or kickdown and low/reverse or rear units. The exception to this is the C-6, which uses a clutch in place of a band for low and reverse.

The A4LD is a modified version of the C-3. Its housing was enlarged and an additional planetary gear set was added in front of the Simpson gear set to provide for an overdrive fourth gear. The transmission consists of a torque converter, planetary gear train, three multiple-disc clutch packs, three bands, a one-way clutch, and a hydraulic control system. The overhaul procedures for this transmission are shown in Photo Sequence 10.

The AOD is a four-speed transmission with an overdrive and it uses the Ravigneaux gear train. Another feature of the AOD is the split-torque-type torque converter. This arrangement consists of two input shafts driven by the converter. One of these shafts relays torque into the transmission through the normal hydraulic action of a converter and the other shaft provides a mechanical link from the engine to the gear train when the transmission is operating in third and fourth gears. The AOD uses four multiple-disc clutches, two one-way clutches, and two bands to obtain the various gear ranges.

Classroom Manual
Chapter 6, page 171

Special Tools

Dial indicator and holding fixture
Clutch compressor tool
Oil pump puller
Oil pump alignment tool
Seal remover/installer
Tapered punch

Photo Sequence 10
Typical Procedure for Overhauling an A4LD Transmission

P10-1 Remove the torque converter.

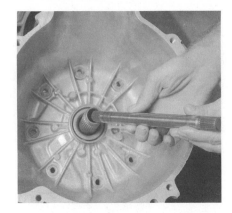

P10-2 Remove the input shaft. Note that the two splined ends of the shaft are different.

P10-3 Remove the bellhousing retaining bolts, then the bellhousing and oil pump as an assembly. Also remove the thrust washer and gasket.

P10-4 Separate the oil pump from the bellhousing.

P10-5 Remove the steel plate from the housing.

P10-6 Remove the oil pan bolts, then the oil pan.

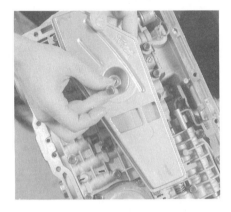

P10-7 Remove the retaining bolt for the filter screen. Remove the filter screen. Then remove the detent spring under the screen.

P10-8 Disconnect the two wires at the TCC solenoid.

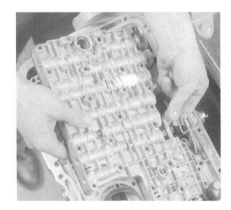

P10-9 Remove the valve body retaining bolts. While moving the valve body away, unlock and remove the selector lever connecting link. Then remove the valve body and gasket. Note and mark the valve body bolts. They are different lengths.

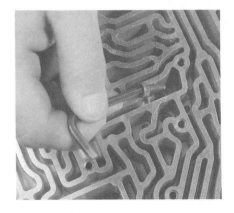

P10-10 Remove the Allen head retaining bolt holding the center support from within the case's worm tracks.

P10-11 Remove the bolts retaining the extension housing to the case. Then remove the extension housing, parking pawl, and pawl spring.

P10-12 Remove the governor retaining bolts, then the governor.

P10-13 Loosen the overdrive band locknut and back off the adjustment screw. Discard the locknut.

P10-14 Remove the anchor and apply struts.

P10-15 Lift out the clutch assembly and band. Mark the band as "overdrive" and label the anchor end of the band.

P10-16 Lift out the overdrive one-way clutch and planetary assembly.

P10-17 Remove the center support retaining snap ring.

P10-18 Remove the overdrive apply lever and shaft. Then remove the overdrive control bracket from the valve body side of the case. The overdrive apply lever shaft is longer than the intermediate apply lever shaft.

P10-19 Remove and mark the thrust washer from the top of the center support.

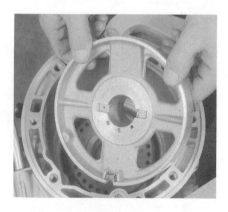

P10-20 Carefully remove the center support bearing by prying on it evenly and upwardly.

P10-21 Remove and mark the thrust washer from below the center support.

P10-22 Loosen the intermediate band locknut and back off the adjusting screw. Discard the locknut. Remove the anchor and apply struts. Then remove the reverse/high and forward clutch assembly.

P10-23 Remove the intermediate band. Mark the band as "intermediate" and label the anchor end of the band.

P10-24 Remove the forward planetary gear assembly with its thrust washer. Mark the thrust washer.

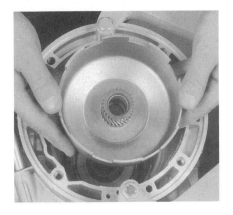

P10-25 Remove the sun gear shell.

P10-26 Remove the planet and ring gear assembly.

P10-27 Remove the large snap ring from the carrier of the reverse planetary gear assembly.

P10-28 Remove the reverse planetary gear assembly with its thrust washer. Mark the thrust washer.

P10-29 Remove the small snap ring on the output shaft and remove the output shaft ring gear.

P10-30 Remove the low/reverse drum and one-way clutch assembly.

P10-31 Remove the low/reverse servo from the valve body side of the case.

P10-32 Remove the low/reverse band. Then remove and mark the thrust washer.

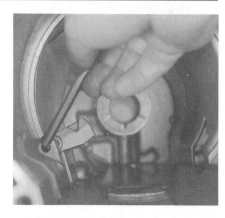

P10-33 Remove the intermediate band apply lever and shaft.

P10-34 Turn the transmission so that the output shaft points upward. Then lift out the output shaft.

P10-35 Remove the park gear/collector body assembly from the rear of the case. Then remove and mark the thrust washer.

P10-36 Remove the vacuum modulator and throttle valve retaining bolt. Then remove the modulator, throttle valve actuator rod, and throttle valve.

P10-37 Remove the intermediate and overdrive servos' cover snap rings, covers, pistons, and springs. Mark the covers as to which is the overdrive servo cover.

P10-38 Remove the neutral/safety switch and kickdown lever nut, lever, and O-ring.

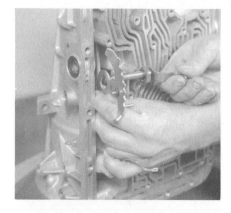

P10-39 Remove the linkage centering pin, manual lever nut, manual lever, internal kickdown lever and park pawl rod, and detent plate assembly. Then remove the lever shaft oil seal.

Typical Procedure for Overhauling an A4LD Transmission (continued)

P10-40 Remove the TCC and 3–4 shift solenoid connector. To remove the connector, a tab on the outside of the case must be depressed while the connector is pulled.

P10-41 Before installing the center support into the case, install new high clutch seals onto the support hub. You must size these seals. Failure to do so will cause the seals to be cut or rolled over during installation. Use the overdrive brake drum for sizing. Carefully rotate the center support while inserting it into the drum housing. Apply a liberal amount of petroleum jelly to the center support hub and seals.

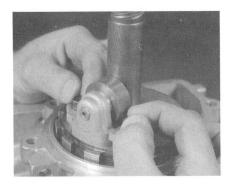

P10-42 Make sure the center support is seated fully into the overdrive drum. Set the assembly aside to allow the assembly to stand for several minutes. This will allow the seals to seat in their grooves.

P10-43 Position the thrust washer in the rear of the case. Install the collector body and output shaft. Then carefully install the governor onto the collector body. Tighten the bolts to specifications.

P10-44 Position the thrust washer into the case from the front. Then install the low-reverse brake drum, output shaft ring gear and snap ring. Position the reverse planet assembly and thrust washers.

P10-45 Install the snap ring into the drum to hold the planet assembly. Then install the low-reverse band as marked at disassembly. Replace the piston's O-ring and install the low-reverse servo piston assembly to hold the band in position.

P10-46 Replace the piston's O-ring and install the intermediate servo piston assembly, piston cover, and snap ring. Replace the piston's O-ring and install the overdrive servo piston assembly, piston cover, and snap ring.

P10-47 Install the intermediate servo apply lever and shaft into the case. Then install the complete forward clutch and reverse-high clutch assemblies. Rotate the transmission so that the output shaft points downward. Install the intermediate band, apply strut, and anchor strut. Temporarily install the input shaft.

P10-48 Using a depth gauge, check the transmission's rear endplay to determine the amount of space between the thrust washer surface of the overdrive center support and the intermediate brake drum. Select the proper sized thrust washer required to obtain endplay specifications. Install the thrust washer and retain it with petroleum jelly. *Note*: If the average reading during this check is outside of specifications, this indicates improper reassembly, missing parts, or some of the parts out of specifications. Correct the problem before continuing reassembly. If within specifications, remove the depth gauge and input shaft.

P10-49 Insert the input shaft with its short splines facing down, through the center support and into the splines of the forward clutch cylinder. Carefully place the center support into the case. Do not seat it into the intermediate brake drum, but make sure it is in line with the retaining bolt hole in the worm tracks.

P10-50 Without applying force on the center support, gently wiggle the input shaft, allowing the center support to slide into the intermediate brake drum with its own weight. When it is fully seated, position the thrust washer on top of the center support.

P10-51 Install the snap ring to retain the center support. The ends of the snap ring should be positioned in the wide shallow cavity located in the five o'clock position. Then install the Allen head bolt that retains the center support to the case. *Note*: Two types of snap rings are used. One has no notches and the other has a notch on its outer and inner diameters. The outer-diameter notch should be positioned on the left.

P10-52 Install the sun gear and support into the overdrive planetary assembly and one-way clutch. Make sure the needle bearing race is centered inside the planetary assembly.

P10-53 Install the overdrive planetary assembly and one-way clutch into the case. Install the overdrive drum assembly, overdrive bracket, apply lever, and shaft. Then install the overdrive band, apply strut, and anchor strut.

Typical Procedure for Overhauling an A4LD Transmission (continued)

P10-54 Make sure the needle bearing race in the overdrive planetary assembly is centered and the overdrive clutch is fully seated. Then place the selective washer on top of the overdrive clutch drum and temporarily install the pump assembly (without its gasket) into the case.

P10-55 Using a dial indicator, check the endplay. If the endplay exceeds limits, replace the selective washer with one that will allow for proper endplay. When the endplay is correct, remove the oil pump and the selective thrust washer.

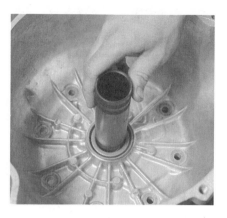

P10-56 Install a new oil pump seal and position the separator plate onto the converter housing. Position the pump assembly onto the separator plate and converter housing. Install the retaining bolts fingertight.

P10-57 With the recommended tool, align the pump in the converter housing to prevent seal leakage, pump gear breakage, or bushing failure.

P10-58 Install and tighten the retaining bolts. Then remove the alignment tool. Install the input shaft into the pump and install the converter into the pump gears. Rotate the converter to check for free movement. Then remove the converter and input shaft.

P10-59 Coat the converter housing gasket with petroleum jelly and position it onto the transmission case. Install the seal ring onto the converter housing and position the selected thrust washer onto the rear of the pump. With the converter housing and pump aligned to the transmission case, install the retaining bolts with new washers and tighten them to specifications.

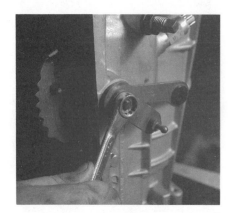

P10-60 Install a new locknut on the overdrive band adjusting screw. Tighten the adjusting screw to 10 ft. lb. and back it off exactly two turns. Then tighten the locknut to 35–45 ft. lb. Repeat this procedure for the intermediate band. Perform air pressure tests to ensure proper transmission operation.

P10-61 Install the shift lever oil seal. Then install the internal shift linkage, external manual control lever, and centering pin. Then tighten the nut to specifications. Install the O-ring, kickdown lever, and retaining nut. Install the neutral/safety switch. Install the converter clutch solenoid, throttle valve, rod, vacuum diaphragm, retaining clamp, and bolt.

P10-62 Using tapered punches, align the valve body to the separator plate and gasket. Use petroleum jelly to hold the gasket in place, then tighten the retaining bolts to specifications. Attach and lock the link to the manual valve. Carefully ease the valve body into the case and install and tighten the retaining bolts. Make sure the correct length bolts are installed in their proper location.

P10-63 Connect the converter clutch solenoid wires and 3–4 shift solenoid wires. Then install the reverse servo piston assembly and spring. Make sure the piston rod is correctly seated into the reverse band apply end. Then install the correct servo rod and cover, using a new servo cover gasket. Then tighten the attaching bolts. Install new O-rings on the filter screen. Then install the filter screen and the oil pan with a new gasket. Tighten the pan bolts to specifications in two steps.

P10-64 Install the parking pawl and return spring into the extension housing. Then install the extension housing with a new gasket. Make sure the parking pawl rod is fully seated. Then install and tighten the extension housing bolts. Finish assembly by installing the input shaft and torque converter. Make sure the torque converter is properly seated.

Ford Transaxles

The Ford ATX transaxle uses a Ravigneaux planetary gear train. It also uses a gear-type oil pump that is driven indirectly by the torque converter. Because the oil pump is located at the opposite end of the transaxle, as is the torque converter, a drive shaft is used to rotate the oil pump.

The transaxle has been continuously updated and modified to fit different applications. The most obvious change made to the transaxle is the torque converter. The ATX has been available with three different designs of torque converters: the conventional three-element converter, a converter with centrifugal lockup, and a split torque converter. Photo Sequence 11 covers the overhaul procedure for an ATX transaxle.

Classroom Manual
Chapter 6, page 193

Special Tools
Dial indicator and holding fixture
Clutch compressor tool
Oil pump puller
Seal remover/installer
Servo cover compressor tool
Depth micrometer and fixture

Photo Sequence 11
Typical Procedure for Overhauling an ATX Transaxle

P11-1 Unbolt and remove the oil pan and gasket. Remove the filter, filter gasket, fill tube, and oil pump drive shaft. Discard all gaskets.

P11-2 Remove the throttle lever return spring. Then remove the control and main oil pressure regulator baffle plates. Remove the valve body bolts. Mark their location and length for reassembly. Then remove the valve body and governor filter.

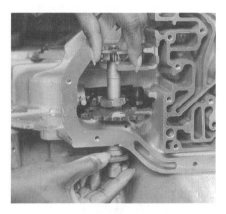

P11-3 Remove the manual valve retaining pin from the case. Then remove the manual valve linkage.

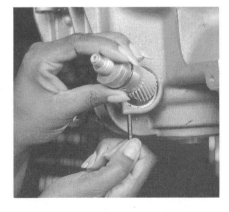

P11-4 Remove the speedometer driven gear retaining pin from the case. Then remove the speedometer gear.

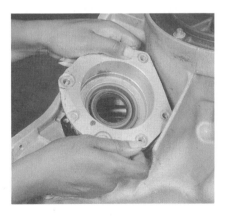

P11-5 Unbolt and remove the differential retainer.

P11-6 Lift the differential upward and out of the case.

Photo Sequence 11
Typical Procedure for Overhauling an ATX Transaxle (continued)

P11-7 Remove and discard the oil pump retaining bolts and washers. Using the correct puller, remove the oil pump assembly from the case. Make sure the selective thrust washer under the oil pump is also removed. Remove and discard the pump gasket.

P11-8 Remove the thrust bearing from the intermediate clutch. Then remove the clutch assembly.

P11-9 Remove the thrust bearing from the direct clutch and remove the direct clutch assembly.

P11-10 Remove the intermediate clutch hub and ring gear assembly. Then remove the thrust washer from the planetary gear assembly.

P11-11 Remove the reverse clutch snap ring. Then remove the reverse clutch pack.

P11-12 Remove the reverse clutch piston.

P11-13 Pry the reverse clutch cylinder up and remove it from the case.

P11-14 Using the proper tool, remove the servo retaining ring. Then remove the tool and the servo.

P11-15 Remove the low/intermediate band.

Typical Procedure for Overhauling an ATX Transaxle (continued)

P11-16 Remove the sun gear and drum assembly.

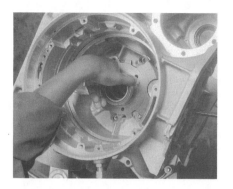

P11-17 Using a screwdriver, pry the transfer housing from the idler gear shaft. Then remove the transfer housing.

P11-18 Remove the input gear's needle bearing assembly and the thrust bearing.

P11-19 Use an Allen wrench to hold the idler gear shaft and remove the nut from the shaft.

P11-20 Gently tap the shaft and remove the idler gear and shaft from the case.

P11-21 Clean the threads of the idler gear shaft and install a new O-ring. Then position the idler gear and shaft into the case. Install the input gear's thrust bearing and caged needle bearing. Then install the input gear.

P11-22 Install the thrust bearing onto the input gear. Then position the transfer housing in the case. Make sure the housing is fully seated on its alignment dowels. Then install new transfer housing bolts and tighten them to specifications.

P11-23 Install the thrust washer onto the transfer housing. Then install the sun gear and drum assembly.

P11-24 Install the intermediate band. Make sure the band lug engages with the strut.

Photo Sequence 11
Typical Procedure for Overhauling an ATX Transaxle (continued)

P11-25 Using the proper tool, install the servo assembly. Then install the servo retaining ring.

P11-26 Position the reverse clutch cylinder in the case and install the reverse clutch piston into the cylinder using a seal protector.

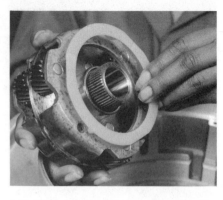

P11-27 Install the thrust washer onto the planetary gear assembly. Use petroleum jelly to hold it in place. Install the planetary assembly onto the sun gear.

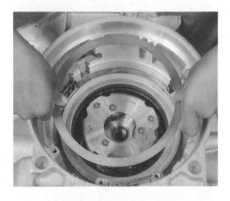

P11-28 Install the reverse clutch return spring and holder assembly. Then install the clutch pack wave spring.

P11-29 Install the thrust bearing and align the oil pump gasket and oil pump.

P11-30 Using the correct tool fixture and depth micrometer, check the endplay. Select the correct-sized thrust washer to ensure proper endplay.

P11-31 Install the thrust washer onto the pump. Use petroleum jelly to hold it in place. Then install the oil pump into the case.

P11-32 Install new pump retaining bolts and washers. Tighten the bolts to specifications.

P11-33 Install the differential unit with differential bearing spacer shim.

Typical Procedure for Overhauling an ATX Transaxle (continued)

P11-34 Position the differential retainer with a new gasket into the case. Then tighten the retaining bolts to specifications.

P11-35 Install a new seal on the speedometer gear retainer and install the speedometer driven gear and retaining pin in the case.

P11-36 Connect the manual valve linkage and install the manual valve retaining pin into the case.

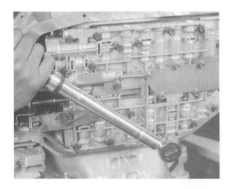

P11-37 Install the governor filter and valve body and tighten the bolts to specifications. Make sure the different length bolts are in their correct location. Install the control and main oil pressure regulator baffle plates. Then install the throttle lever return spring.

P11-38 Install a new filter and filter seal. Then install the fill tube and pump drive shaft.

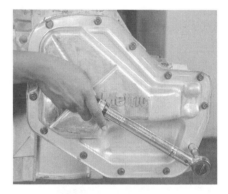

P11-39 Install the oil pan with a new gasket. Tighten the bolts to specifications.

P11-40 Assembly is complete with the installation of the torque converter. Make sure the converter is fully engaged with the oil pump drive shaft.

Special Tools

Dial indicator and hold-
ing fixture
Clutch compressor tool
Oil pump puller
Seal remover/installer
Servo cover compressor
tool
Depth micrometer and
fixture
Seal protector

.The AXOD four-speed transaxle doesn't use a Ravigneaux or a Simpson gear set. It relies on two simple planetary units that operate in tandem. The planetary carriers of each planetary unit are locked to the other planetary unit's ring gear. Each planetary set has its own sun gear and set of planetary pinion gears. The AXOD uses four multiple-disc clutches, two band assemblies, and two one-way clutches to control the operation of the planetary gear set. All of the multiple-disc clutches are applied hydraulically and released by several small coil springs when hydraulic pres-sure is diverted from the clutch's piston. The overhaul procedure for this transaxle is shown in Photo Sequence 12.

Photo Sequence 12
Typical Procedure for Overhauling an AXOD Transaxle

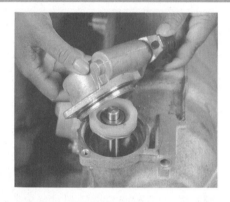

P12-1 Remove the torque converter and mount the transaxle in a vertical position on a holding fixture. Then drain the fluid.

P12-2 Move the transaxle to a horizontal position. Remove the two governor cover bolts. Then remove the cover and seal. Discard the seal. Lift the governor, the speedometer drive gear assembly, and the bearing from the case.

P12-3 Remove the bolts from the over-drive servo cover. Then mark the align-ment of the cover and remove it, the piston assembly, and spring. Discard the O-ring from the cover. Then remove the low-intermediate servo cover bolts. Mark the alignment of the cover and remove it, the piston assembly, and spring. Remove and discard the gasket.

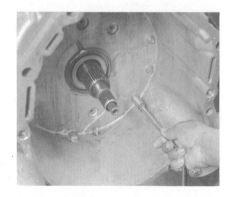

P12-4 Remove the neutral safety switch bolts and remove the switch.

P12-5 Remove the dipstick tube attach-ing bolt and pull the tube from the case. Then, remove the chain cover bolts from inside the torque converter housing.

P12-6 Remove the valve body cover bolts, valve body cover, and gasket. Discard the gasket.

P12-7 Unplug the electrical connectors from the pressure switches and solenoid. Use both hands to do this and do not pull on the wires; pull on the connector. Compress the tabs on both sides of the five-pin bulkhead connector from inside the chain cover and remove the connector and wiring.

P12 8 Using a 9-mm wrench on the flats of the manual shaft, rotate the shaft clockwise so that the manual valve is all the way in. Note the location, according to length, of the oil pump and valve body bolts. Then remove the bolts. Do not remove the two bolts that hold the oil pump and valve body together. The oil pump cover bolts should also not be removed at this time.

P12-9 Push in the T.V. plunger. Pull the pump and valve body assembly outward. Rotate the valve body clockwise and remove the manual valve link from the manual valve and disconnect it from the detent lever. Now remove the oil pump and valve body assembly.

P12-10 Remove the throttle valve bracket bolts and remove the bracket. Pull the oil pump drive shaft out of the case and remove and discard the Teflon seals from the shaft.

P12-11 Place the transaxle in a vertical position. Then remove and discard the circlip for the left-hand output shaft. Remove the shaft's seal with a seal remover.

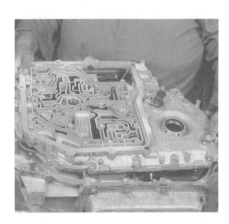

P12-12 Mark the location and length of the chain cover bolts. Then remove them and the chain cover. Discard the gasket.

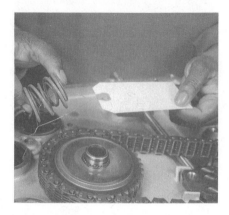

P12-13 Mark and remove the metal thrust washers from the chain cover.

P12-14 Simultaneously lift out both sprockets with the chain assembly. Mark and remove the thrust washers from the drive and driven sprocket supports. Inspect the drive sprocket support bearing to determine if it needs replacing.

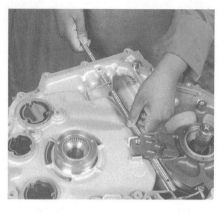

P12-15 Remove the lockpin and roll pins from the manual shaft. Be careful not to damage the machined surfaces. Slide the shaft out of the case. Then pry the seal from the case.

P12-16 Use a straightedge and note whether the machined bolt hole surface of the driven sprocket support is above or below the case's machined surface. Remember this for reassembly. Then remove the driven sprocket support assembly. It may be necessary to back out the reverse clutch anchor bolt.

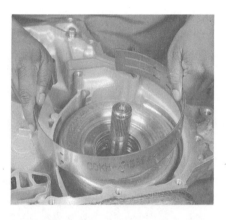

P12-17 Remove the Teflon seals from the support assembly. Mark and remove the thrust washers and needle bearing. Then remove the plastic overdrive band retainer and the overdrive band.

P12-18 Lift the front sun gear and shell assembly out of the case.

P12-19 Remove the oil pan cover bolts and the cover. Discard the gasket. Remove the reverse apply tube/oil filter bolt, bracket, and oil filter screen.

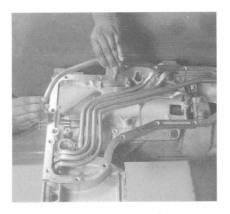

P12-20 Remove the tube retaining bracket bolts and brackets. *Note*: For complete transaxle disassembly, the reverse apply tube must be removed prior to removing the reverse clutch. The rear lube tube *must* also be removed and the seal replaced whenever the differential is removed.

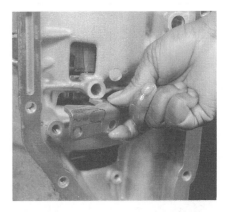

P12-21 Remove the park rod abutment bolts. Remove the roll pin for the park pawl shaft. Then, using a magnet, remove the park pawl shaft, park pawl, and return spring.

P12-22 Place the transaxle in a horizontal position. Grasp the outer diameter of the reverse clutch cylinder with your fingertips and slide the assembly out of the case.

P12-23 Place the transaxle in a vertical position. Grasp the front planetary shaft and lift out both the front and rear planetary assemblies. Lift out the low-intermediate drum and sun gear assembly. Then remove the low-intermediate band.

P12-24 Using a screwdriver inserted through the side of the case, remove the snap ring for the final drive assembly from the case. Lift out the final drive assembly using the output shaft.

P12-25 Remove the final drive ring gear, thrust washer, and needle bearing. Remove and discard the rear lube tube seal by tapping it toward the inside of the case. Then using a seal installer, install the converter oil seal and right-hand output shaft seal.

P12-26 Place the transaxle in a vertical position. Install the needle bearing over the case boss with its flat side facing up and outer lip facing down.

P12-27 Install the final drive ring gear with its external splines up. Lightly tap the ring gear to fully seat it in the case.

P12-28 Reassemble the governor drive gear, differential assembly, final drive sun gear, parking gear, needle bearings, rear planetary support, and thrust washer.

P12-29 Lower the final drive assembly into the case. Install the snap ring and align it with the low-intermediate band anchor pin.

P12-30 Mount a dial indicator with its plunger on the end of the output shaft. Check end clearance. If the clearance is not within specifications, replace the thrust washer with one that will bring the clearance to specifications. Available thicknesses are: 0.045–0.049" (1.15–1.25 mm), Orange; 0.055–0.059" (1.40–1.50 mm) Purple; and 0.064–0.069" (1.65–1.75 mm) Yellow.

Typical Procedure for Overhauling an AXOD Transaxle (continued)

P12-31 Install the park pawl, return spring, park pawl shaft, and locator pin. Make sure the park pawl engages the park gear and returns freely. Install the park rod actuating lever and park rod into the case. Install the park rod abutment and start the abutment bolts. Push in the park pawl and locate the rod between the pawl and the abutment.

P12-32 Using a 3/8" drift, gently install the lube tube seal flush against the rear case support. Install the low-intermediate band and align the anchor pin pocket with the anchor pin. Install the low-intermediate drum and sun gear assembly.

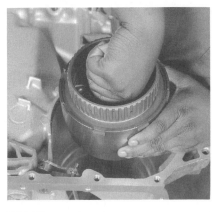

P12-33 Reassemble the ring gear and shell assembly, rear planetary, needle bearing, front planetary, and snap ring. Carefully slide the planetary assembly over the output shaft.

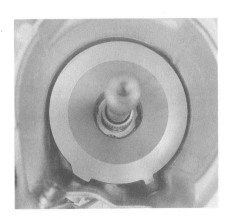

P12-34 Lower the reverse clutch into the case and start clutch plate engagement. Align the clutch cylinder anchor pin pocket with the anchor pin case hole. Use the intermediate clutch hub to complete clutch plate engagement and fully seat the reverse clutch. Rotate the planet with the hub to engage splines.

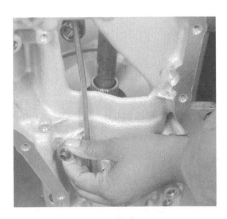

P12-35 Start the reverse anchor pin bolt, but do not tighten it. Reassemble the forward, direct, and intermediate clutch assembly. Lower the assembly into the case. Align the shell and sun gear splines with those in the forward planetary.

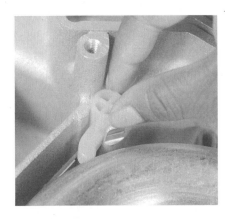

P12-36 Install the overdrive band into the case. Install the plastic retainer with the crosshairs facing up.

P12-37 Check the drive sprocket end clearance to determine required thrust washer thicknesses. To do this, you must first determine if the machined bolt hole surfaces on the driven sprocket support are above or below the case machined surface. If they are above the surface, place a depth micrometer on the machined bolt hole surface and measure the distance to the case's machined surface. If they are below the case's surface, place the depth micrometer on the case's surface and measure the distance to the machined bolt hole surface. Install the correct thrust washer. Repeat this procedure for the drive sprocket.

P12-38 Tap the seal for the manual shaft into the case. Start the manual shaft through the seal and slide the manual detent lever onto the shaft. Then slide the shaft through the park rod actuating lever and tap it into the bore in the case. Install a new lock pin through the case hole. Make sure it is aligned with the groove in the shaft. Install new roll pins.

P12-39 Align the tabs of the thrust washers and install them onto the drive and driven sprocket supports. Lubricate and install the cast iron sealing ring onto the input shaft. Install the chain over the sprockets.

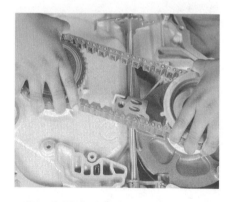

P12-40 Lower the chain assembly onto the sprocket supports. Rotate the supports while lowering the chain assembly to ensure that the sprockets are fully seated on the supports.

P12-41 Install and align the thrust washers into the chain cover. Install a new chain cover gasket onto the cover. Then install the correct accumulator springs in their proper location. Carefully apply downward pressure to overcome accumulator spring pressure, and start two of the chain cover bolts. Position these bolts 180 degrees apart.

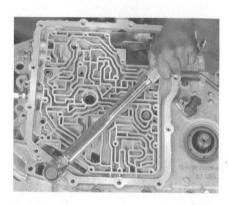

P12-42 Start the remaining chain cover bolts and tighten them in sequence and to specifications. The input shaft should now have little endplay and should rotate freely. If it does not rotate freely, remove the chain cover and check for a damaged cast iron seal.

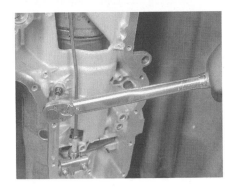

P12-43 Tighten the park rod abutment bolts, reverse anchor pin bolt, and locknut to specifications. Lightly tap the lube tubes until they are fully seated in their bores. Install the tube retaining brackets.

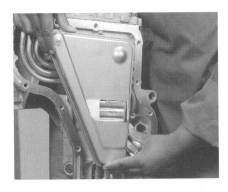

P12-44 Install the O-rings onto the oil filter and press the filter into the case. Install the reverse apply tube/oil filter bracket and bolt. Install the oil pan with a new gasket. Tighten the bolts to specification.

P12-45 Install new Teflon seals onto the pump drive shaft and install the shaft.

P12-46 Install the T.V. bracket and tighten the bracket bolts to specification. Connect the manual valve link to the detent lever.

P12-47 Start the oil pump and valve body assembly over the pump shaft and connect the manual valve link to the manual valve.

P12-48 Install the oil pump and valve body assembly.

P12-49 Install the bulkhead connector and other electrical connectors. Install the neutral start switch. With the manual shaft in neutral, align the switch using a 0.089" drill bit. Then tighten the switch to specification.

P12-50 Install the valve body cover with a new gasket. Then tighten the bolts to specification. Install the dipstick tube grommet and dipstick tube in the case.

General Motors Transmissions

Most General Motors transmissions and transaxles are based on the Simpson gear train. The THM 375, 400, and 475 transmissions are three-speed, heavy-duty transmissions generally used in full-sized RWD cars and trucks. The 425 is based on the same design as the others, but it was modified for use in FWD vehicles with a longitudinally placed engine. The 425 was replaced by the 325. The 325 and 425 transmissions are not classified as transaxles because the differential and final drive gears are not built into the transmission case; a separate final drive unit is bolted to the transmission to drive the front wheels.

The THM 250 and 250C are light-duty units used primarily in compact and intermediate models. The 350 and 350C are medium-duty units that were used in intermediate and full-sized cars and light trucks. The 375B is a heavy-duty version of the 350. It has an additional direct clutch for increased torque capacity. The primary difference between a 250 and 350 transmission is that a 250 uses a band as the second gear holding device, while a 350 uses a band plus a multiple-disc clutch and a one-way roller clutch as holding devices during second gear operation.

The 200C, 200-4R, 325, and 325-4L are light- to medium-duty transmissions. The 200, 200C, and 325 are three-speed units and the 200-4R and 325-4L have four forward speeds.

Most General Motors transmissions are fitted with gear-type oil pumps. Exceptions to this are late-model 200-4R and 700-R4 transmissions, which use a variable displacement vane-type oil pump. Although there are differences between the various models of GM transmissions with a Simpson gear train, most use three multiple-disc clutches, one band, and a single one-way roller clutch to provide the various gear ratios. Each clutch is applied hydraulically and released by several small coil springs. Two exceptions to this should be noted. The 350/350C has four multiple-disc clutches instead of three. And the 200/200C uses a waved spring for clutch release in place of small coil springs.

The 200-4R and 325-4L use the same multiple-disc clutches as the 200 with additional multiple-disc and one-way clutches. The forward, direct, and fourth clutches are released by several small coil return springs and the drive and low/reverse clutches are released by wave plate springs.

The 700-R4 (4L60) is used in most full-size GM cars and trucks and in some heavy-duty and high-performance vehicles. It is a four-speed unit consisting of two planetary gear sets, five multiple-disc clutches, one sprag clutch, one roller clutch, and a band. The five multiple-disc clutches are released by several small coil springs. The 700-R4 has a one-piece case casting that incorporates the bellhousing. The extension housing is a separate casting bolted to the rear of the case. The overhaul procedures for this transmission are outlined in Photo Sequence 13.

Classroom Manual
Chapter 6, page 186

Special Tools
Dial indicator and holding fixture
Clutch compressor tool
Oil pump puller
Seal remover/installer
Servo cover compressor tool
Output shaft support tool

Photo Sequence 13
Typical Procedure for Overhauling a 4L60 Transmission

P13-1 Remove the torque converter.

P13-2 Mount the transmission in a holding fixture and position it so the oil pan is facing up. Position a servo cover compressor on two oil pan bolts. Then compress the servo cover and remove the retaining ring, servo cover, and O-ring.

P13-3 Remove the 2–4 servo assembly. Servo pin length should be checked prior to disassembly. This helps to identify any wear on the 2–4 band and/or reverse input drum.

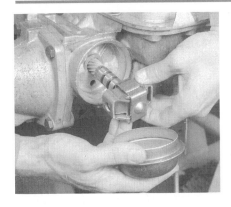

P13-4 Remove the governor cover and O-ring. Then remove the governor assembly.

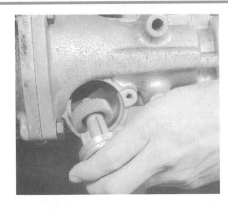

P13-5 Remove the speed sensor assembly or the speedometer driven gear. This type of transmission will be equipped with one or the other.

P13-6 Remove the extension housing bolts. Then pull the housing away from the case.

P13-7 Remove the oil pan bolts, pan, and gasket.

P13-8 Remove the oil filter.

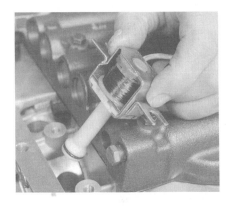

P13-9 Remove the solenoid retaining bolts. Then remove the solenoid and O-ring.

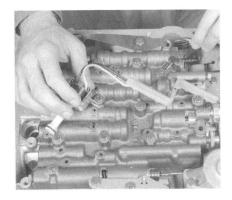

P13-10 Remove the wiring harness. Note the location and position for reference during reassembly.

P13-11 Loosen and remove the valve body and auxiliary valve body attaching bolts. Mark the location of each bolt as they are different lengths. Also, mark the location of any check balls you encounter.

P13-12 Carefully remove the 1–2 accumulator cover retaining bolts, cover and pin assembly, piston, seal, and spring.

P13-13 Remove the spacer plate, and note the location of the check balls and filter.

P13-14 Remove the 3–4 accumulator spring, piston, and pin. Note the location of the spacer plate, gasket, check balls, and filters.

P13-15 Remove the oil pump retaining bolts. Then, using the correct puller, remove the oil pump.

P13-16 Remove the oil pump seal and gasket. Then remove the reverse input clutch-to-pump thrust washer from the pump.

P13-17 Remove the reverse and input clutch assembly by lifting it out with the input shaft.

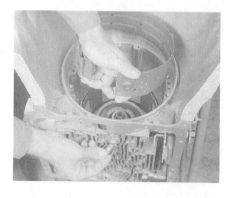

P13-18 Remove the 2–4 band anchor pin. Then remove the band from the case.

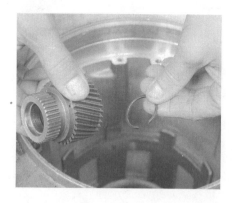

P13-19 Remove the input sun gear. Install an output shaft support tool onto the output shaft, then remove the input carrier-to-output shaft retaining ring.

P13-20 Remove the input carrier and output shaft. Then remove the input carrier thrust washer from the reaction carrier shaft.

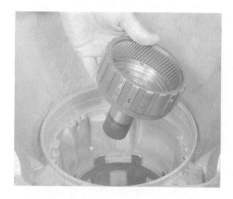

P13-21 Remove the input ring gear and the reaction carrier shaft.

P13-22 Remove the reaction sun shell and thrust washer. Then remove the sun shell-to-clutch race thrust washer and support-to-case retaining ring.

P13-23 Remove the spring retainer from the low/reverse support. Then remove the reaction sun gear, low/reverse clutch race, clutch rollers, support assembly, and reaction carrier assembly.

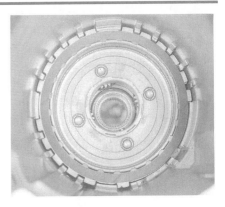

P13-24 Remove the low/reverse clutch assembly.

P13-25 Remove the reaction ring gear and bearing assembly. Then remove the reaction ring gear support-to-case bearing assembly.

P13-26 Remove the parking lock bracket retaining bolts and the lock bracket.

P13-27 Remove the parking pawl shaft, parking pawl, and return spring by using a screw extractor to remove the shaft plug from the case.

P13-28 Using the correct spring compressor, compress the low/reverse clutch spring retainer and remove the retaining ring and spring assembly.

P13-29 Remove the low/reverse piston assembly by applying air to the apply passage.

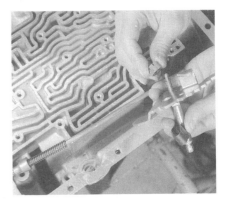

P13-30 Remove the manual shaft nut, shaft, and retainer. Then remove the parking lock actuator assembly and inner detent lever.

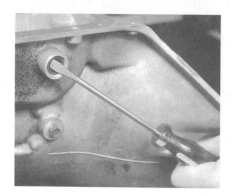

P13-31 Pry the shaft seal from the case.

P13-32 Install a new shaft seal into the case. Then install the parking lock actuator assembly and detent lever. Install the manual shaft, nut, and retainer.

P13-33 Install new seals onto the low/reverse piston and coat them with petroleum jelly.

P13-34 With the transmission in a vertical position, install the piston into the transmission case. Make sure the piston is properly aligned and fully seated.

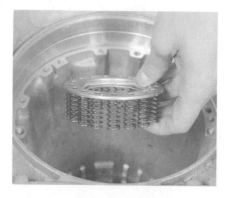

P13-35 Install the clutch springs. Using the spring compressor, compress the springs past the ring groove in the case. Then install the retaining ring.

P13-36 Coat the bearing assembly with petroleum jelly and install it into the case. Then install the reaction ring gear and support.

P13-37 Install the bearing onto the support. Then install the oil deflector and reaction carrier assembly into the case.

P13-38 Install the clutch pack with the correct number of plates.

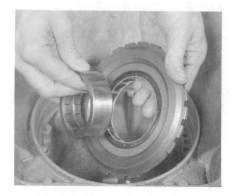

P13-39 Install the low/reverse support into the case. Then install the inner race by pushing down while rotating it until it is fully engaged.

Typical Procedure for Overhauling a 4L60 Transmission (continued)

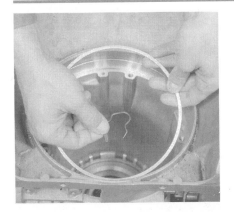

P13-40 Install the spring retainer. Then install the low/reverse retainer ring.

P13-41 Install the snap ring onto the reaction sun gear. Then install the sun gear into the reaction carrier.

P13-42 Install the thrust washer onto the low/reverse clutch race and install the reaction sun gear shell onto the sun gear.

P13-43 Install the thrust washer onto the reaction sun gear shell, making sure the thrust washer tangs are positioned in the slots in the shell.

P13-44 Install the input ring gear and reaction carrier shaft into the sun gear shell. The carrier shaft splines must engage with the reaction carrier.

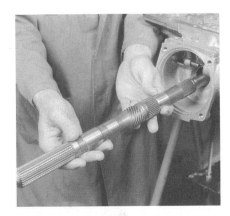

P13-45 Install the thrust washer onto the reaction carrier shaft. Then install the output shaft into the transmission case.

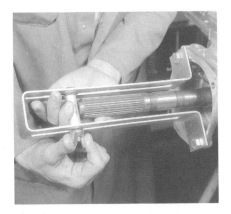

P13-46 Install the output shaft support tool so that the shaft is positioned as upward as possible. Then install the input carrier assembly onto the output shaft.

P13-47 Install a new retaining ring onto the output shaft. Then remove the output shaft support tool.

P13-48 Install the input sun gear.

P13-49 Install the selective thrust washer and bearing assembly onto the input housing.

P13-50 Install new input shaft seals, then size them according to the procedures given in the service manual.

P13-51 Position the reverse input assembly onto the input clutch assembly. Then install the reverse and input clutch assemblies into the case as a single unit. Align the 3–4 clutch plates of the input assembly with the input ring gear.

P13-52 Install the 2–4 band into the case. Align the anchor pin end with the case pin hole and install the anchor pin into the case. Make sure the anchor pin lines up with the end of the band. Then install the 2–4 servo assembly into the case and index the apply pin onto the end of the band. Install the servo cover with a new O-ring.

P13-53 Check endplay. Install the correct thickness of thrust washer at the rear of the oil pump. Then install the oil pump. Make sure all holes, especially the filter and pressure regulator holes, are lined up.

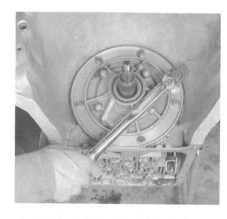

P13-54 Install the retaining bolts and torque them to specifications.

Typical Procedure for Overhauling a 4L60 Transmission (continued)

P13-55 Install the 3–4 accumulator piston pin into the case. Then install the piston seal and piston onto the pin. Install the accumulator spring, check balls, filters, retainer, and the spacer plate and gasket.

P13-56 Install the 1–2 accumulator spring, oil seal ring, piston, cover, and bolts. Tighten the bolts to specifications.

P13-57 Install the valve body and auxiliary valve body. Then install the retaining bolts and tighten them to specifications.

P13-58 Install the speedometer gear or speed sensor rotor onto the output shaft. Then install a new O-ring into the output shaft sleeve. Install the extension housing and torque the bolts to specifications. Using a seal installer, install a new oil seal in the extension housing.

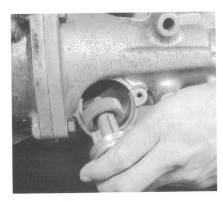

P13-59 Install the speedometer driven gear or the speed sensor into the extension housing. Then install all external electrical connectors. Install the manual shift lever and the torque converter.

General Motors Transaxles

The THM 125 and 125C are three-speed automatic transaxles designed for light-duty use. The torque converter does not directly drive the transaxle; rather, the turbine shaft drives two sprockets and a chain, which transfers engine torque to the gear train. The 125/125C (3T40) models also use a vane-type oil pump, which is mounted to the valve body and is driven indirectly, through a drive shaft, by the torque converter. They also use three multiple-disc clutches, one band, and a single one-way roller clutch to provide the various gear ratios. Each clutch is applied hydraulically and released by several small coil springs. The recommended overhaul procedure for this transaxle is shown in Photo Sequence 14.

Classroom Manual
Chapter 6, page 180

Special Tools

Dial indicator and holding fixture
Clutch compressor tool
Oil pump puller
Seal remover/installer
Servo cover compressor tool
Output shaft support tool

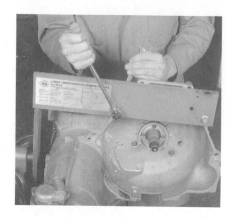

P14-1 Remove the torque converter. Place the transaxle on a holding fixture and position the unit so that the right axle end is down to drain the fluid.

P14-2 Remove the speed sensor housing bolts, housing, and oil seal. Then remove the speed sensor, speed sensor rotor, and governor assembly. Or remove the speedometer driven gear mount retainer bolt and retainer, then pull the driven gear from the governor housing. Remove the governor cover bolts, cover, and O-ring.

P14-3 Remove the oil pan bolts. Then remove the oil pan and oil strainer.

P14-4 Remove the reverse oil pipe retaining bracket-to-servo cover bolt. Remove the retaining servo cover bolts and lift off the intermediate servo cover and gasket. Then remove the intermediate servo assembly.

P14-5 Remove the snap ring and remove the intermediate band apply pin from the intermediate servo piston. Then remove the third accumulator exhaust valve and spring. *Note*: The apply pin is selective size.

P14-6 Disconnect the low/reverse oil pipe, oil pipe seal backup washer, and O-ring. Then remove the low/reverse cup plug.

P14-7 Remove the dipstick stop and parking lock bracket from above the parking pawl and parking pawl actuator rod.

P14-8 Rotate the final drive unit until both ends of the output shaft retaining ring are showing.

P14-9 With a C-ring removal tool, loosen the ring from the output shaft. Rotate the shaft 180 degrees and remove the ring from the shaft. Then remove the output shaft.

P14-10 Remove the valve body mounting bolts. Then remove the throttle lever and bracket assembly retaining bolts from the valve body. Lift off the throttle lever and bracket assembly with the throttle valve cable link.

P14-11 Remove all auxiliary valve body bolts except the lower left bolt. This bolt is only removed when it is necessary to separate the auxiliary valve body from the main valve body. Remove the remaining control valve bolts and lift the control valve and oil pump assembly away from the valve body.

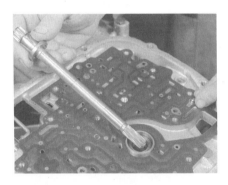

P14-12 Remove the check ball from the direct clutch passage on the spacer plate. Then lift out the oil pump drive shaft.

P14-13 Remove the spacer plate and gaskets. Then remove the other check balls from the case cover. Mark the location of all check balls.

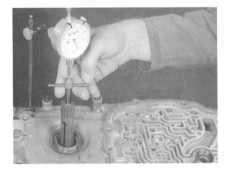

P14-14 Using the correct tools and setup, measure the input shaft-to-case cover endplay. The endplay is adjusted by a selective snap ring on the input shell.

P14-15 Disconnect the manual valve rod from the manual valve.

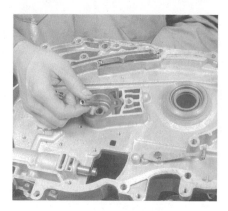

P14-16 Remove the transmission case cover bolts and separate the case cover from the case. Once separated, lay the case cover with the 1–2 accumulator facing up. This prevents the loss of the 1–2 accumulator pin.

P14-17 Remove the 1–2 accumulator spring, piston, and gasket. Then remove the case cover-to-drive sprocket thrust washer and driven sprocket thrust bearing assembly.

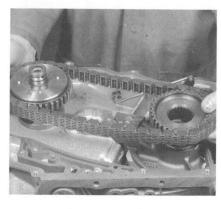

P14-18 Remove and discard the O-ring from the input shaft. Then remove the drive sprocket, driven sprocket, and chain as an assembly with the selective fit thrust washers.

P14-19 Remove the detent lever-to-manual shaft pin and the manual shaft-to-case pin. Then pull the manual lever from the case. Remove the manual valve rod, detent lever assembly, and park lock actuator rod.

P14-20 Remove the driven sprocket support and thrust washer from the direct clutch assembly.

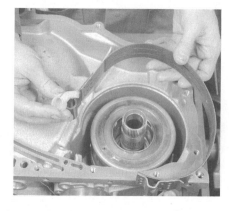

P14-21 Remove the intermediate band anchor hole plug. Then remove the intermediate band assembly.

Typical Procedure for Overhauling a 3T40 Transaxle (continued)

P14-22 Remove the input shaft with the direct and forward clutch assembly. As the shaft is being pulled out, separate the clutch assemblies. Then remove the thrust washers and the input ring gear.

P14-23 Remove the input carrier assembly, the thrust washers for the input ring and sun gears, the sun gear, and the input drum.

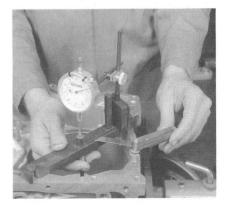

P14-24 Measure the sun gear-to-input drum endplay to check the fit of the selective snap ring. The snap ring is positioned on the reaction sun gear. If the endplay exceeds specifications, replace the snap ring with one of the correct thickness during reassembly.

P14-25 Measure the low/reverse clutch housing-to-low roller clutch race thrust washer endplay. This washer is located between the clutch housing and the one-way clutch assembly. If the endplay exceeds specifications, replace it with one of the correct thickness during reassembly.

P14-26 Remove the reaction sun gear. Then remove the low/reverse clutch housing-to-case snap ring. Now with the correct tool, remove the low/reverse housing.

P14-27 Remove the low/reverse clutch housing-to-case spacer ring from its groove in the case. Then lift the final drive sun gear shaft and the reaction gear set assembly out of the case.

P14-28 Remove the roller clutch and reaction carrier assembly from the final drive sun gear shaft.

P14-29 Measure the final drive-to-case endplay to select the correct thickness of thrust washer for reassembly. The thrust washer is located between the differential carrier and the carrier case.

P14-30 Remove the final drive ring gear spacer-to-case snap ring. Then, using the appropriate tool, pull the final drive unit from the case.

P14-31 Remove the final drive differential-to-case thrust washer and the differential carrier-to-case thrust roller bearing from the case.

P14-32 Prior to assembly, lubricate all bearings, bushings, and seals with clean ATF. Install the correct-sized differential and carrier thrust washers. Use petroleum jelly to keep them in place.

P14-33 Install the final drive ring gear spacer. Make sure the opening in the spacer aligns with the parking pawl opening in the case.

P14-34 Install the final drive spacer-to-case snap ring. Then install the reaction sun gear into the case.

P14-35 Install the low/reverse backing plate and clutch plates into the case. Then install the low/reverse wave plate and housing-to-case spacer ring.

P14-36 Install the low/reverse clutch housing into the case. Make sure the oil feed bores in the housing line up with those in the case.

Typical Procedure for Overhauling a 3T40 Transaxle (continued)

P14-37 Install the selective snap ring onto the reaction sun gear. Then install the sun gear onto the final drive sun gear shaft. Rotate the reaction sun gear while pushing down on the low/reverse clutch housing until the housing drops below the snap ring groove in the case.

P14-38 Install the thick low/reverse clutch housing-to-case snap ring. Then install the input drum onto the reaction sun gear. Install the tanged thrust washers for the input sun and ring gears. The thrust washer for the ring gear is larger than the one for the sun gear.

P14-39 Install the input pinion carrier onto the input sun gear. Then put the ring gear over the carrier.

P14-40 Install the input shaft-to-input ring gear thrust washer onto the forward and direct clutch assembly. Then install the clutch assembled into the case, making sure they are fully seated.

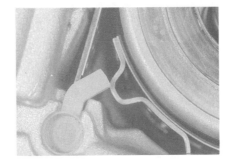

P14-41 Install the intermediate band. Install the anchor hole plug.

P14-42 Install the driven sprocket support-to-direct clutch housing thrust washer. Then install the support.

P14-43 Install the manual shaft and parking lock actuator rod into the case. Then install the detent lever on the manual shaft and push the manual shaft into place. Install the manual shaft-to-detent lever and the manual shaft-to-case retaining pins.

P14-44 Assemble the drive and driven sprockets with their chain link. Then install the assembly with the appropriate thrust washers.

P14-45 Install the case cover-to-driven sprocket roller bearing thrust washer.

P14-46 Install the 1–2 accumulator piston and spring into the case. Then install the inner and outer case-to-cover gaskets and case cover. Install and tighten the case cover bolts according to specifications.

P14-47 Connect the manual valve rod to the manual valve. Install the check balls into the case cover.

P14-48 Install the check ball into the direct clutch passage on the spacer plate. Then install the oil pump shaft into the bore in the case cover.

P14-49 Install the control valve body and tighten the bolts to specification. Then install the valve body wiring harness. Connect the lever link for the T.V. bracket and install the bracket onto the valve body.

P14-50 Install the valve body cover with a new cover. Tighten bolts to specifications.

P14-51 Rotate the transaxle so that the oil pan is facing up. Then install the output shaft into the transaxle. Rotate the final drive to allow the retaining ring to be installed in its groove in the output shaft.

P14-52 Install the parking lock bracket and dipstick stop. Then install a new low/reverse oil pipe seal assembly, O-ring backup washer, and the O-ring for the end of the oil pipe. Install the oil pipe and retainer bracket.

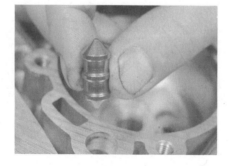

P14-53 Install the intermediate servo piston assembly with new seals. Install the third accumulator exhaust valve and spring into the check valve bore.

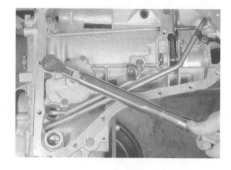

P14-54 Install the reverse oil pipe bracket-to-oil pipe and servo cover, the servo cover, and the servo cover bolts. Install a new oil strainer and O-ring.

Typical Procedure for Overhauling a 3T40 Transaxle (continued)

P14-55 Install the oil pan with a new oil pan gasket and tighten the bolts to specifications.

P14-56 Install the governor assembly with a new O-ring on its cover. Install the speedometer gear or speed sensor rotor into the governor cover. Then tighten the governor cover bolts to specifications. Install the torque converter.

Classroom Manual
Chapter 6, page 203

Special Tools

Dial indicator and holding fixture
Clutch compressor tool
Oil pump puller
Seal remover/installer
Servo cover compressor tool
Output shaft support tool

   The gear train of the THM 440-T4 (4T60) is based on two simple planetary gear sets operating in tandem. The combination of the two planetary units does function much like a compound unit. The two tandem units do not share a common member; rather, certain members are locked together or are integral with each other. The front planetary carrier is locked to the rear ring gear and the front ring gear is locked to the rear planetary carrier. The transaxle houses a third planetary unit, which is used only as the final drive unit and not for overdrive.

   The 440-T4 uses a variable-displacement vane-type oil pump and four multiple-disc clutches, two bands, and two one-way clutches to provide the various gear ranges. One of the one-way clutches is a roller clutch, the other is a sprag. The four multiple-disc clutches are released by several small coil springs when hydraulic pressure is diverted from the clutch's piston. Photo Sequence 15 covers the overhaul procedures for this transaxle.

Photo Sequence 15
Typical Procedure for Overhauling a 4T60 Transaxle

P15-1 Remove the torque converter. Then place the transaxle in a holding fixture. Remove the speedometer sensor and governor assembly.

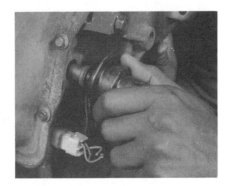

P15-2 Remove the bottom oil pan, oil filter, modulator, and modulator valve.

P15-3 Remove the accumulator cover with governor feed and return pipes and accumulator pistons, gaskets, retainers, and pipes from their bores.

P15-4 Remove the reverse servo cover by applying pressure to it and removing the retaining ring. Then remove the servo assembly from the case.

P15-5 Apply pressure to the 1–2 servo cover and remove the servo cover retaining ring. Then remove the cover and servo assembly.

P15-6 Remove the side cover bolts, nuts, and washers. Then remove the side covers and gaskets. Disconnect and remove the wiring harness to the pressure switches, solenoid, and case connector.

P15-7 Remove the T.V. lever, linkage, and bracket from the valve body assembly. Remove the pump assembly cover bolts, then the pump cover. Remove the servo pipe retainer bolt, retainer plate, mounting bolts, and valve body.

P15-8 Remove the oil reservoir weir. Then mark the location of and remove the check balls between the spacer plate and the valve body and between the channel plate and the spacer plate.

P15-9 Disconnect the manual valve link from the manual valve. Place the detent lever in the park position and remove the retaining clip. Then remove the channel plate with its gaskets.

P15-10 Remove the oil pump drive shaft. Remove the input clutch accumulator and converter clutch piston assemblies. Remove the fourth clutch's plates and the apply plate. Then remove the fourth clutch's thrust bearing, hub, and shaft.

P15-11 Rotate the final drive unit until both ends of the output shaft retaining ring are showing.

P15-12 With a C-ring removal tool, loosen the ring from the output shaft. Rotate the shaft 180 degrees and remove the ring from the shaft. Then remove the output shaft.

P15-13 Remove and discard the O-ring from the input shaft. Then remove the drive sprocket, driven sprocket, and chain as an assembly with the selective fit thrust washers.

P15-14 Remove the driven sprocket support and thrust washer from between the sprockets and the channel plate. Remove the scavenging scoop and driven sprocket support with the second clutch thrust washer.

P15-15 Using the correct tool, remove the second clutch and input shaft clutch housings as an assembly. Remove the reverse band.

P15-16 Remove the thrust washers. Then measure the endplay of the input shaft clutch housing. Select the correct thickness of thrust washer and set it aside for reassembly.

P15-17 Remove the input clutch sprag assembly, third clutch assembly, and the input sun gear. Then remove the reverse band, reverse reaction drum input carrier assembly, and thrust washer.

P15-18 Remove the reaction carrier, reaction sun gear/drum assembly, and forward band.

P15-19 Remove the reaction sun gear thrust bearing and final drive sun gear shaft.

P15-20 Check the final drive endplay and select the correct size thrust washer for the unit.

P15-21 Using the proper tool, remove the final drive assembly and selective thrust washers and bearings.

P15-22 Clean and inspect the transaxle case and all of the transaxle components. Then position the correct-size thrust washers and bearings onto the final drive assembly and install the unit. Petroleum jelly can be used to hold the washers and bearings in place while positioning the unit.

P15-23 Install the final drive sun gear shaft through the final drive ring gear. The splines must engage with the parking gear and the final drive sun gear.

P15-24 Install the forward band into the case, making sure the band is aligned with the anchor pin.

P15-25 Install the reaction sun gear-to-final drive ring gear thrust bearing. Then assemble the reaction sun gear and drum assembly onto the final drive ring gear.

P15-26 Check the endplay of the carriers in the reaction planetary gear set. Then install the thrust washer and carrier assembly into the case. Rotate the carrier until the pinions engage with the reaction sun gear.

P15-27 Install the input carrier with its thrust bearing into the case. Then install the reverse reaction drum, making sure its splines engage with the input carrier.

P15-28 Inspect the roller and sprag clutches. Then put the spacer onto the sun gear, followed by the input sprag retainer, sprag assembly, and roller clutch. Make sure the input sprag and third roller clutch hold and freewheel in opposite directions while holding the input sun gear.

P15-29 Lubricate the inner seal on the input clutch's piston. Then install the seal and assemble the input piston into the input housing. Install a new O-ring onto the input shaft.

P15-30 Install the spring retainer and guide into the piston. Then install the third clutch piston housing into the input housing. Using the proper compressor, install the retaining snap ring.

P15-31 Install the inner seal for the third clutch. Then install the third clutch piston into the housing. Compress the spring retainer and install the retaining snap ring.

P15-32 Install the wave plate. Then install the correct number of clutch plates in the correct sequence. Install the input clutch backing plate and the retaining snap ring. Air check the operation of the clutch.

P15-33 Assemble the second clutch piston in the housing. Install the apply ring, return spring, snap ring, and wave plate. Then assemble the correct number of clutch plates in the correct sequence. Install the backing plate and snap ring. Then air check the clutch's operation.

P15-34 Install the thrust washers. Using the correct tool, install the second clutch and input shaft clutch housings as an assembly. Install the reverse band.

P15-35 Install the scavenging scoop and driven sprocket support with the second clutch thrust washer. Install the driven sprocket support and thrust washer between the sprockets and the channel plate.

P15-36 Install the drive sprocket, driven sprocket, and chain as an assembly.

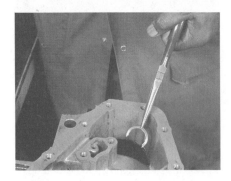

P15-37 Install the output shaft. Start the C-ring onto the output shaft, then rotate the shaft 180 degrees and fully seat the ring onto the shaft.

P15-38 Install the fourth clutch's thrust bearing, hub, and shaft. Then install the fourth clutch's plates and the apply plate. Install the input clutch accumulator and converter clutch piston assemblies. Install the oil pump drive shaft.

P15-39 Install the channel plate with new gaskets. Connect the manual valve link to the manual valve. Place the detent lever in the park position and install the retaining clip.

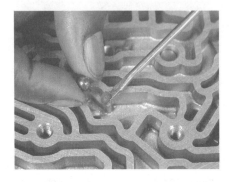

P15-40 Install the oil reservoir weir. Install the check balls in their proper location between the spacer plate and the valve body and between the channel plate and the spacer plate.

P15-41 Install the servo pipe retainer bolt, retainer plate, mounting bolts, and valve body. Install the pump assembly cover bolts, then the pump cover. Tighten the bolts to specifications. Install the T.V. lever, linkage, and bracket onto the valve body assembly..

P15-42 Connect and install the wiring harness to the pressure switches, solenoid, and case connector. Then install the side covers and gaskets. Install the side cover bolts, nuts, and washers. Then tighten them to specifications.

P15-43 Install the 1–2 servo cover, servo assembly, and retaining ring.

P15-44 Install the reverse servo cover, servo assembly, and retaining ring.

P15-45 Install the accumulator cover with governor feed and return pipes and accumulator pistons, gaskets, retainers, and pipes into their bores.

Typical Procedure for Overhauling a 4T60 Transaxle (continued)

P15-46 Install the bottom oil pan, oil filter, modulator, and modulator valve.

P15-47 Install the torque converter, speedometer sensor, and governor assembly.

Honda Transaxles

The Honda CA, F4, and G4 transaxles are used in many Honda and Acura cars. These transmissions do not use a planetary gear set to provide for the different gear ranges. Constant-mesh helical and square-cut gears are used in a manner similar to that of a manual transmission.

These transaxles have a mainshaft and countershaft on which the gears ride. To provide the four forward and one reverse gear, different pairs of gears are locked to the shafts by hydraulically controlled clutches. Reverse gear is obtained through the use of a shift fork, which slides the reverse gear into position. Four multiple-disc clutches, the sliding reverse gear, and a one-way clutch are used to control the gears. Photo Sequence 16 outlines the overhaul procedure for this unique transaxle.

Classroom Manual
Chapter 6, page 211

Special Tools

Dial indicator and holding fixture
Clutch compressor tool
Seal remover/installer
Gear puller
Small chisel
Punch and drift set

Photo Sequence 16
Typical Procedure for Overhauling Honda F4 Automatic Transmission

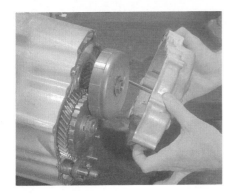

P16-1 With the transaxle on a bench, loosen and remove the bolts that secure the right side cover. Then remove the right side cover.

P16-2 Lock the main shaft using the mainshaft holding tool. Engage the parking pawl and then loosen and remove the mainshaft and countershaft locknuts.

Note: The mainshaft locknut has left-handed threads.

P16-3 Remove the thrust washer, thrust needle bearing, and 1st gear needle bearing.

P16-4 Remove mainshaft 1st gear.

P16-5 Remove the parking pawl, shaft, stop pin, and spring.

P16-6 Install a gear puller and remove the parking gear and countershaft 1st gear as a unit.

P16-7 Remove the reverse idler bearing holder assembly.

P16-8 Remove the parking shift arm and spring.

P16-9 Remove the throttle control lever assembly.

P16-10 Loosen and remove the transmission housing mount bolts.

P16-11 Install the transmission housing puller on the countershaft and screw the puller bolt against the countershaft to separate the housing halves.

P16-12 Remove the transmission housing.

P16-13 Remove the countershaft reverse gear collar and reverse gear.

P16-14 Remove the countershaft 4th gear, needle bearing, distance collar, reverse gear selector hub, needle bearing, reverse gear selector, and reverse shift fork.

P16-15 Remove the mainshaft and countershaft assemblies as a unit.

P16-16 Remove the governor housing, pipe, and separator plate.

P16-17 Remove the governor shaft, holder, and gear assembly.

P16-18 Remove the 2nd/3rd accumulator cover. The cover is spring loaded so press down on the cover while unscrewing the bolts in a star pattern to prevent stripping the threads in the accumulator housing.

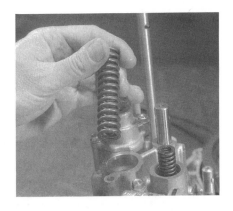

P16-19 Remove the 2nd/3rd accumulator springs.

P16-20 Remove the 4th accumulator cover and spring. Note that this cover is spring loaded so be careful to press down on it as you remove its retaining bolts.

Photo Sequence 16
Typical Procedure for Overhauling Honda F4 Automatic Transmission (continued)

P16-21 Remove the modulator valve body and separator plate.

P16-22 Remove the servo valve body and separator plate.

P16-23 Remove the servo secondary valve body. Be careful not to lose the two steel balls, ball springs, timing accumulator piston and spring, and secondary filter.

P16-24 Remove the LC-shift valve body and separator plate.

P16-25 Remove the regulator valve body.

P16-26 Remove the stator shaft, shaft arm, and stop pin from the main valve body.

P16-27 Remove the cotter pin, washer, rollers, and pin from the manual valve.

P16-28 Remove the main valve body. Be careful not to lose the three steel balls, check ball springs, torque converter check valve, and check valve spring.

P16-29 Remove the pump gears, gear shaft, check valve, check valve spring, and separator plate.

P16-30 Remove the ATF filter screen.

P16-31 Remove the differential assembly from the torque converter housing.

P16-32 Install the differential assembly back into the torque converter housing.

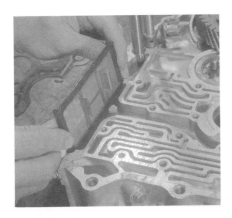

P16-33 Install a new ATF filter screen.

P16-34 Install the separator plate, pump gear shaft, pump gears, check valve, and check valve spring.

P16-35 Install the main valve body and torque to specification. Be careful not to loose the three steel balls, check ball springs, torque converter check valve, and check valve spring.

P16-36 Install the manual valve pin, rollers, washer, and cotter pin.

P16-37 Install the stator shaft, shaft arm, and stop pin in the main valve body.

P16-38 Install the regulator valve body.

P16-39 Install the LC-shift valve separator plate and body. Torque the retaining bolts to specification.

P16-40 Install the servo secondary valve body. Make certain that the two steel balls, ball springs, timing accumulator piston, and spring and secondary filter are in the proper position.

P16-41 Install the servo valve separator plate, valve body, and mount bolts. Torque to specifications.

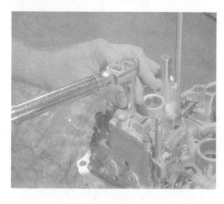

P16-42 Install the modulator valve body and separator plate. Torque its retaining bolts to specification.

P16-43 After replacing the O-ring, install the 4th accumulator spring and cover.

P16-44 Install the 2nd/3rd accumulator springs.

P16-45 Using the handle of a hammer, compress the 2nd/3rd accumulator springs by pressing down on the cover while you tighten the accumulator retaining bolts.

P16-46 Install the servo valve body, and mount bolts. Torque to specification.

P16-47 Install the governor shaft, holder, and gear assembly.

Typical Procedure for Overhauling Honda F4 Automatic Transmission (continued)

P16-48 Install the governor housing, pipe, and separator plate.

P16-49 Install the governor housing, and torque the mount bolts to specification. Bend up the tabs on the lock plates using a steel punch.

P16-50 Install the mainshaft and countershaft assemblies as a unit.

P16-51 Install the countershaft 4th gear, needle bearing, distance collar, reverse gear selector hub, needle bearing, reverse gear selector, and reverse shift fork.

P16-52 Install the countershaft reverse gear collar and reverse gear.

P16-53 Install the reverse shift fork, and torque its mount bolt to specification.

P16-54 Install the reverse idler gear and needle bearing in the transmission housing. Then install the transmission housing.

P16-55 Install the transmission housing mount bolts, and tighten to specification.

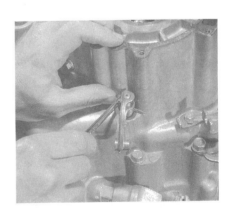

P16-56 Install the throttle control lever assembly, and tighten to specification.

Photo Sequence 16
Typical Procedure for Overhauling Honda F4 Automatic Transmission (continued)

P16-57 Install the parking shift arm and spring, and tighten to specification.

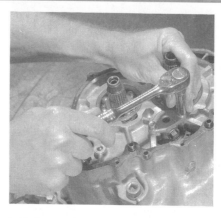

P16-58 Install the reverse idler bearing and holder, and tighten to specification.

P16-59 Install the countershaft 1st gear needle bearing.

P16-60 Install the parking gear and countershaft 1st gear.

P16-61 Engage the parking pawl with the parking gear to ensure proper alignment.

P16-62 Install new countershaft and mainshaft locknuts and, torque to specifications. Stake them to their shafts using a steel punch.

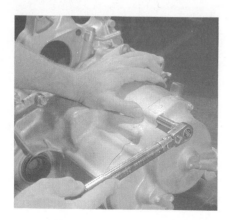

P16-63 Torque end cover mounting bolts to manufacturer's specifications.

CUSTOMER CARE: Because the Honda automatic transaxle uses a sliding reverse gear, damage to the transaxle can occur if the car is towed or moved while the transaxle's gear selector is in neutral but the transaxle is still mechanically in the reverse gear. This can happen because a servo is used to normally delay forward gear engagement after the transaxle has been in reverse. Simply shifting into neutral may not disengage reverse gear. Anytime the vehicle is to be moved, start the engine and place the transaxle into drive, then shift it into neutral. Turn off the engine and the vehicle is ready to be moved.

Special Tools

Dial indicator and
 holding fixture
Clutch compressor
 tool
Oil pump puller
Seal remover/installer

This transaxle is also found in Ford Probes, which are built in the same factory and on the same assembly line as some Mazda cars.

Mazda Motors Transaxles

The Mazda GF4A-EL transaxle is an electronically controlled four-speed unit. There are two different models of this transaxle. The major difference in models is the gear ratio of the final drive unit. This transaxle is used in many FWD vehicles. It uses five multiple-disc clutches, two one-way clutches, and one band to provide the different gear ratios. This type of transaxle is equipped with a rotor-type oil pump driven by the torque converter through a drive shaft. Photo Sequence 17 covers the overhaul procedures for this transaxle.

Photo Sequence 17
Typical Procedure for Overhauling a Mazda GF4A-EL Transaxle

P17-1 Remove the torque converter from the transmission case and pull out the oil pump drive shaft.

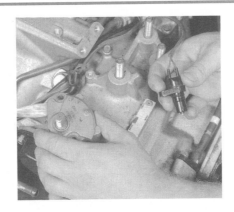

P17-2 Remove the vehicle speed pulse generator and inhibitor switch.

P17-3 Remove the oil pan and gasket. Examine the residue in the pan. Then remove the oil strainer and O-ring.

P17-4 Remove the valve body cover and gasket.

P17-5 Disconnect the solenoid connectors and ATF temperature sensor, then remove the coupler assembly.

P17-6 Remove the valve body assembly.

P17-7 Remove the oil pump and gasket.

P17-8 Remove the clutch assembly by removing the turbine shaft snap ring, pulling out the drum and clutch assembly.

P17-9 Remove the small sun gear and one-way clutch.

P17-10 Remove the band, then its anchor and strut.

P17-11 Remove the piston stem from the band servo.

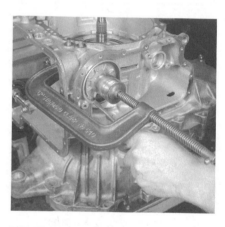

P17-12 Remove the servo.

P17-13 Remove the remaining one-way clutch assembly and the carrier hub assembly.

P17-14 Remove the retaining snap ring. Then remove the internal gear from the output shell.

P17-15 Remove the remaining clutch assembly.

P17-16 Separate the transaxle housing.

P17-17 Remove the output shell from the output shaft.

P17-18 Remove the manual shaft and shift plate.

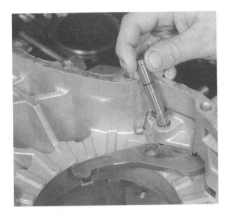

P17-19 Remove the snap ring and the parking assist lever. Then remove the parking pawl assembly.

P17-20 Remove the differential assembly.

P17-21 Remove the bearing housing.

P17-22 Remove the idler gear from the case.

P17-23 Inspect the oil pump and all other parts, and repair or replace as necessary. Make sure to install all new rubber oil seals.

P17-24 Install the idler gear by tapping it into place.

P17-25 Install the selected shim and bearing race into the bearing housing. Then install the bearing housing.

P17-26 Install the differential unit.

P17-27 Install the parking pawl assembly. Then install the parking assist lever and snap ring.

P17-28 Install the manual shaft and shift plate.

P17-29 Install the output shell to the output gear and install the thrust bearing onto the output shell.

P17-30 Fasten the converter housing to the transaxle case. Tighten the bolts gradually and to specifications.

P17-31 Install the turbine shaft and 3–4 clutch assembly.

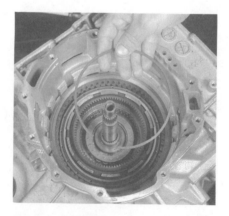

P17-32 Install the internal gear to the output shell. Then install the retaining snap ring.

P17-33 Install the carrier hub assembly.

P17-34 Install the one-way clutch.

P17-35 Install the band servo assembly into the case.

P17-36 Install the band assembly into the case and adjust it so that it is fully expanded.

P17-37 Install the remaining one-way clutch and small sun gear.

P17-38 Install the clutch assembly.

P17-39 Check and correct for total end-play.

P17-40 Adjust the band.

P17-41 Install a new O-ring, oil filter, and pan gasket. Then install the oil pan and tighten the bolts to specifications.

P17-42 Install the valve body. Make sure to tighten the bolts to specifications and according to the order prescribed by the manufacturer.

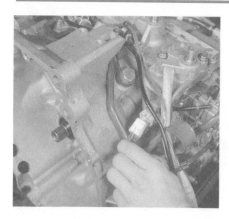

P17-43 Install the solenoid and inhibitor switch with their wiring harness.

P17-44 Install a new gasket. Then install and tighten down the valve body cover.

P17-45 Install the vehicle speed pulse generator.

P17-46 Install the dipstick tube and oil filler tube assemblies. Then install the oil pump drive shaft and the torque converter.

Nissan Motor Company Transmissions

Special Tools

Dial indicator and holding fixture

Clutch compressor tool

Oil pump puller

Seal remover/installer

Widely used Nissan RWD transmissions are the 3N71B, which is a three-speed unit, and the L4N71B/E4N71B series, which provide four forward gears through the use of a Simpson gear set and an additional planetary unit mounted in front of the Simpson gear set. The primary difference between the "L" and the "E" transmissions is that the E-model provides converter lockup in third and fourth gear. Nissan transaxles are similar to other transaxles covered in this chapter.

These transmissions use four multiple-disc clutches, two servos and bands, and a one-way clutch to provide for the different ranges of gears. All of the clutches except the low/reverse unit are released by several small coil springs. The low/reverse unit utilizes a belleville-type spring for greater clamping pressures. Photo Sequence 18 covers the typical overhaul procedures for this model transmission. Always follow the recommended procedure for the transmission or transaxle you are working on.

Photo Sequence 18
Typical Procedure for Overhauling a Nissan L4N71B Automatic Transmission

P18-1 Place the transmission on a suitable workbench. Check the endplay, and record your readings.

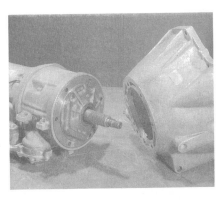

P18-2 Remove the torque converter.

P18-3 Unscrew and remove the electrical kickdown, or downshift solenoid and O-ring.

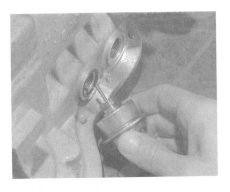

P18-4 Unscrew and remove the throttle modulator valve, its diapragm rod, and O-ring.

P18-5 Remove the speedometer drive assembly with its gear and O-ring.

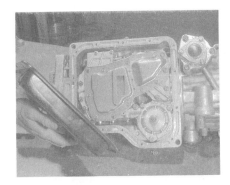

P18-6 Remove the oil pan, and inspect its content. An analysis of foreign matter will provide clues regarding the types of problems to look for during the overhaul procedure. Check the service manual for specific details.

P18-7 Remove the valve body from the case.

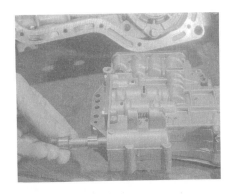

P18-8 Remove the manual valve from the valve body to prevent the valve from dropping out.

P18-9 Back off the servo piston stem locknut, and snuggly tighten the piston stem to prevent the front clutch drum from dropping out when removing the front pump.

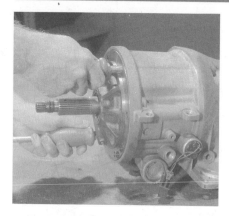

P18-10 Use the correct puller to remove the front pump from the case.

P18-11 Remove the front clutch thrust washer and bearing race.

P18-12 Remove the overdrive servo cover.

P18-13 Loosen the overdrive servo lock-nut, and back off the overdrive band servo piston stem to release the band.

P18-14 Remove the front and rear clutch assemblies. Note the positions of the front pump thrust washers and rear clutch thrust washer.

P18-15 Remove the brake band strut and overdrive brake band.

P18-16 Remove the overdrive housing.

P18-17 Remove the extension. Take care not to lose the parking pawl, spring, pin, and parking actuator.

P18-18 Remove the governor valve assembly.

P18-19 Remove the front drum support and input shaft.

P18-20 Remove the 2nd brake band strut and brake band.

P18-21 Remove the front clutch and planetary gear pack.

P18-22 Remove the output shaft snap ring.

P18-23 Remove the output shaft.

P18-24 Remove the snap ring, connecting drum, and one-way clutch assembly.

P18-25 Use a screwdriver to remove the large retaining snap ring. Then remove the low and reverse brake assembly.

P18-26 Install the low and reverse brake assembly starting with the steel dished plate and then alternating steel and friction plates.

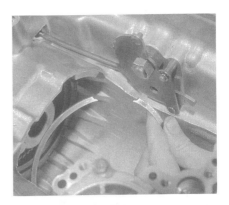

P18-27 Install the retaining plate and snap ring. Check for proper clearance between the snap ring and retaining plate. Select the proper thickness of retaining plate that will give the correct ring to plate clearance if the measurement does not meet the specified limits. Check the low and reverse brake operation using an air gun with a tapered rubber tip.

Photo Sequence 18
Typical Procedure for Overhauling a Nissan L4N71B Automatic Transmission (continued)

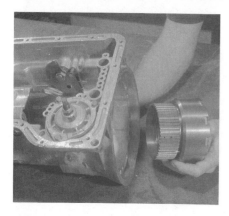

P18-28 Install the one-way clutch assembly, connecting drum, and snap ring.

P18-29 Install the governor needle bearing, thrust washer, output shaft, and oil distributor into the case.

P18-30 Install the planetary gear pack and front clutch assembly.

P18-31 Install the snap ring on the output shaft.

P18-32 Install the 2nd brake band, band strut, and band servo. Lubricate the O-ring seals with ATF.

P18-33 Lubricate the drum support gasket with ATF and install the drum support into the housing.

P18-34 Install the governor valve assembly and torque the retaining bolts to specification.

P18-35 Install the parking actuator and parking pawl assemblies in the extension housing. Then install the extension housing. Torque its mount bolts to specification.

P18-36 Install the front drum support gasket and new O-ring on the overdrive housing. Gently tap it into place using a rubber mallet. Be careful to align the mount bolt holes.

P18-37 Install the needle bearing race and direct clutch thrust washer. Then install the overdrive brake band, strut, and servo assembly. Lubricate the servo O-ring before installing.

P18-38 Install the overdrive pack on the drum support.

P18-39 Adjust the overdrive band. Tighten the piston stem to specification. Then back off two full turns. Secure the servo locknut. Torque the locknut to specification. Test the overdrive servo operation using compressed air.

P18-40 Install the overdrive servo cover, and torque to specification.

P18-41 Adjust the 2nd brake band. Tighten the piston stem to the specified value. Back off three full turns.

P18-42 Secure the piston stem while you tighten the servo locknut to specification.

P18-43 Install the oil pump bearing and thrust washer.

P18-44 Install the oil pump assembly using the special tool. Be sure to lubricate the gasket and O-ring with ATF. Align the mount bolt holes in the pump housing with the mount bolt holes in the overdrive housing.

P18-45 Lubricate and install the manual valve into the control valve body.

P18-46 Install the control valve body and tighten the retaining bolts to the specified value.

P18-47 Install the oil pan gasket, oil pan, and oil pan retaining bolts. Torque the bolts to specified value in a star pattern.

P18-48 Install the speedometer drive assembly with its drive gear and lubricated O-ring.

P18-49 Install the vacuum diaphragm, diaphragm rod, and O-ring.

P18-50 Install the kickdown (or downshift) solenoid and O-ring.

P18-51 Install the torque converter housing. Torque the mount bolts to specification in a star pattern.

A very irate customer had his late-model Honda towed back to the shop. The transmission had failed, again. Just three weeks had passed since the transmission had been rebuilt and now it had failed again.

The technician who did the overhaul was assigned the car. She conducted a visual inspection and found nothing obviously wrong with the electrical system or the transaxle itself. However, the fluid had a burnt smell. She conducted an oil pressure test and found extremely low pressures. She suspected that the pump or pressure regulator had failed.

She pulled the transaxle and checked the pressure regulator and oil pump. She found the pump's gears seized together. This was the first Honda transaxle she had ever rebuilt and she was disappointed that it had failed. Wondering how this could happen, she reviewed the service manual and technical service bulletins. In bold print in the service manual she found the answer, she had failed to correctly align the oil pump shaft during assembly. Based on her experiences with other transmissions she had rebuilt, she didn't know this was critical and assumed that the shaft would only fit one way.

She aligned the shaft, replaced the oil pump gears, and reassembled the transmission. She road tested the car and conducted a pressure test. Everything was fine. She then gave the car back to the customer, after apologizing, knowing that the transaxle was right now. The customer appreciated her honesty and left with confidence that the transmission was okay. The technician learned a lesson: always refer to and follow the directions given in the service manual.

ASE Style Review Questions

1. *Technician A* says some seals must be cut to size before they are installed.
   Technician B says square-cut seals are designed to roll over when a part is fit over them.
   Who is correct?
   A. A only   **C.** Both A and B
   B. B only   **D.** Neither A nor B

2. While assembling a transmission:
   Technician A coats the steel clutch discs with petroleum jelly.
   Technician B soaks the friction discs in clean ATF before installing them.
   Who is correct?
   A. A only   **C.** Both A and B
   B. B only   **D.** Neither A nor B

3. *Technician A* says endplay is often corrected by selective snap rings.
   Technician B says endplay is often corrected by selective thrust washers.
   Who is correct?
   A. A only   **C.** Both A and B
   B. B only   **D.** Neither A nor B

4. *Technician A* says Honda and Acura transaxles use a planetary gear set as a final drive unit.
   Technician B says Honda transaxles use a tandem planetary gear set to provide for the different forward gears.
   Who is correct?
   A. A only   **C.** Both A and B
   B. B only   **D.** Neither A nor B

5. *Technician A* says the oil pumps in most transmissions are press-fit into the case.
   Technician B says the oil pumps in most transmissions are bolted to the transmission case.
   Who is correct?
   A. A only   **C.** Both A and B
   B. B only   **D.** Neither A nor B

6. *Technician A* says a Chrysler 31TH is a transaxle.
   Technician B says a Chrysler 36RH is a transaxle.
   Who is correct?
   A. A only   **C.** Both A and B
   B. B only   **D.** Neither A nor B

7. While assembling a transaxle:
Technician A reuses all seals unless they are damaged.
Technician B lubricates all seals and bearings with clean bearing grease before installing them.
Who is correct?

A. A only   **C.** Both A and B
B. B only   **D.** Neither A nor B

8. *Technician A* says the 4L60 is a RWD transmission.
Technician B says the A4LD is a RWD transmission.
Who is correct?

A. A only   **C.** Both A and B
B. B only   **D.** Neither A nor B

9. *Technician A* says most transmission solenoids are mounted to the clutch housing.
Technician B says most transmission solenoids are removed as an assembly.
Who is correct?

A. A only   **C.** Both A and B
B. B only   **D.** Neither A nor B

10. While checking transmission endplay:
Technician A measures the movement of the shaft with a dial indicator.
Technician B uses a clutch compressor tool to get the maximum movement reading.
Who is correct?

A. A only   **C.** Both A and B
B. B only   **D.** Neither A nor B

Table 6-1 ASE TASK

Diagnose the cause of the abnormal shifting, then determine the needed repairs.

Problem Area	Symptoms	Possible Causes	Classroom Manual	Shop Manual
HARSH ENGAGEMENT	Harsh engagement, initial engagement clunk with warm engine	1. Improper fluid levels	98	33
		2. Throttle linkage damaged or misadjusted	118	49
		3. Engine idle too high	38	33
		4. Valve body bolts loose or too tight	106	99
		5. Oil filter leaking or misassembled	98	33
		6. Sticking valves or dirty valve body	106	99
		7. Looseness in halfshafts or drive shafts	17	43
		8. Excessive transmission endplay	74	82
	Rough engagement in forward and reverse gears	1. Improper fluid levels	98	33
		2. High engine idle	38	33
		3. Looseness in halfshafts or drive shafts	17	43
		4. Incorrect linkage adjustment	118	49
		5. Faulty clutch or band application	147	122
		6. Sticking valves or dirty valve body	106	99
		7. Check ball missing	106	99
		8. Servo apply rod the wrong length	51	55
		9. Throttle linkage out of adjustment	118	49
		10. Malfunctioning accumulator	156	122
		11. Manual shift linkage damaged or misadjusted	118	49
		12. Internal manual shift linkage damaged or broken	118	49
		13. Malfunctioning electronic controls	51	52
SOFT ENGAGEMENT	Engages slowly in D and R	1. Improper fluid levels	98	33
		2. Shift linkage out of adjustment	118	49
		3. Engine idle incorrect	38	33
		4. Incorrect oil pressure	121	99
		5. Dirty or clogged oil filter	99	34

Table 6-1 ASE TASK (continued)

Diagnose the cause of the abnormal shifting, then determine the needed repairs.

Problem Area	Symptoms	Possible Causes	Classroom Manual	Shop Manual
		6. Worn band or clutch assembly	147	122
		7. Damaged accumulator	156	122
		8. Valves sticking or dirty valve body	106	99
		9. Oil pump worn	101	113
		10. Internal leaks	106	99
		11. Contaminated fluid	99	33
		12. Malfunctioning fluid	99	33
POOR QUALITY SHIFTING	Poor quality shifting from 1–2	1. Incorrect oil pressure	121	99
		2. Loose valve body	106	99
		3. Valve body spacer plate-to-cover gasket damaged or mispositioned	106	99
		4. Intermediate clutch ball missing or not sealed	150	126
		5. Intermediate clutch plates defective	150	126
		6. 1–2 servo accumulator faulty	156	122
		7. Improper fluid level	98	33
		8. Poor engine performance	38	33
		9. Throttle linkage damaged, disconnected, or out of adjustment	118	49
		10. Sticking valves or dirty valve body	106	99
		11. Governor valve sticking	54	111
		12. Malfunctioning electronic controls	51	52
	Poor quality shifting from 2–3	1. Excessive or insufficient oil pressure	121	99
		2. Servo or 2–3 accumulator spring missing or broken	156	122
		3. Servo apply rod the wrong length	51	55
		4. Servo regulator valve stuck, damaged, or nicked	147	122
		5. Servo assembly damaged	147	122
		6. Valves sticking or valve body dirty	106	99
		7. Throttle linkage out of adjustment	118	49
		8. Band or clutch not working properly	147	122
		9. Malfunctioning electronic controls	51	52
	Poor quality shifting from 3–4	1. Oil pressure incorrect	121	99
		2. 3–4 accumulator assembly damaged	156	122
		3. Missing check ball	156	122
		4. Malfunctioning electronic controls	51	52
	Incorrect shift points	1. Throttle valve linkage out of adjustment or damaged	118	49
		2. Defective governor	54	111
		3. Defective valve body	106	99
		4. Binding pressure regulator	101	113
		5. Missing oil pump check ball	101	113
		6. Incorrect pressures	121	99
		7. Electrical problems	51	52

Table 6-1 ASE TASK (continued)

Diagnose the cause of the abnormal shifting, then determine the needed repairs.

Problem Area	Symptoms	Possible Causes	Classroom Manual	Shop Manual
POOR QUALITY SHIFTING	Incorrect shift points	8. Internal leaks	106	99
		9. Defective oil pump	101	113
		10. Incorrect vacuum line routing	117	46
		11. Poor engine performance	38	33
		12. Sticking valves or dirty valve body	106	99
		13. Defective vehicle speed or MAP sensor	51	52
		14. Malfunctioning electronic controls	51	52
	Delayed or harsh upshifts	1. High oil levels	98	33
		2. Throttle linkage out of adjustment	118	49
		3. Kickdown band out of adjustment or defective	147	122
		4. Defective governor	54	111
		5. Internal leaks	106	99
		6. Incorrect fluid pressures	121	99
		7. Sticking valves or dirty valve body	106	99
		8. Gearshift out of adjustment	118	49
		9. Defective clutch or band	147	122
		10. Defective oil pump	106	113
		11. Malfunctioning electronic controls	51	52
	All upshifts occur early	1. Improper fluid level	98	33
		2. Throttle linkage out of adjustment or damaged	118	49
		3. Low fluid pressure	98	33
		4. Sticking valves or dirty valve body	106	99
		5. Internal leaks	106	99
		6. Malfunctioning electronic controls	51	52
IMPROPER DOWNSHIFTING	Flare on forced downshifts	1. Low oil pressure	98	33
		2. Band out of adjustment	147	122
		3. Valves sticking or dirty valve body	106	99
		4. Glazed band or drum	75	73
		5. Improper fluid levels	98	33
		6. Throttle linkage out of adjustment	118	49
		7. Defective servo or clutch	147	122
	Abrupt and unwanted downshifts at high speeds	1. Defective valve body	106	99
		2. Band out of adjustment	147	122
		3. Linkage out of adjustment	118	49
	Harsh downshifting	1. Anticlunk spring positioned wrong	156	122
		2. Throttle linkage misadjusted	118	49
		3. Check ball missing	156	122
		4. Improper fluid level	98	33
		5. Incorrect engine idle speed	38	33
		6. Poor engine performance	38	33
		7. Internal leaks	106	99
		8. Sticking valves or dirty valve body	106	99
		9. Servo apply rod length incorrect	51	55

Table 6-1 ASE TASK (continued)

Diagnose the cause of the abnormal shifting, then determine the needed repairs.

Problem Area	Symptoms	Possible Causes	Classroom Manual	Shop Manual
		10. Defective servo	147	122
		11. Malfunctioning electronic controls	51	52
	No part throttle or manual downshifts	1. Binding external or internal linkage	118	49
		2. Defective valve body	106	99
		3. Throttle cable out of adjustment or damaged	118	49
		4. Incorrect oil pressure	121	99
		5. Detent valve/solenoid defective or disconnected	51	52
		6. Defective governor	54	111
		7. Defective clutch or band operation	147	122
		8. Malfunctioning electronic controls	51	52
	No full throttle downshifts	1. Detent cable broken or out of adjustment	118	49
		2. Detent pressure regulator valve stuck	101	113
		3. Faulty servo/accumulator operation	156	122
		4. Internal leaks	106	99
		5. Throttle pedal linkage out of adjustment	118	49
		6. Improper fluid level	98	33
		7. Faulty governor	54	111
		8. Malfunctioning electronic controls	51	52
TRANSMISSION/ TRANSAXLE SLIPS	Slips in low gear	1. Defective forward clutch	150	126
		2. Defective valve body	106	99
		3. Internal linkage binding	118	49
		4. Defective 1–2 accumulator	156	122
		5. Low fluid pressure	121	99
		6. Throttle linkage misadjusted or damaged	118	49
		7. Incorrect fluid level	98	33
		8. Defective one-way/overrun clutch	51	82
	Slips in all ranges	1. Low fluid level	98	33
		2. Low fluid pressure	121	91
		3. Clogged oil filter	99	34
		4. Defective oil pump	101	113
		5. Internal leaks	106	99
		6. Defective clutches or bands	147	122
		7. Malfunctioning electronic controls	51	52
	Slips at forward engagement	1. Improper fluid level	98	33
		2. Throttle linkage misadjusted or damaged	118	49
		3. Internal leaks	106	99
		4. Sticking valves or dirty valve body	106	99
		5. Defective one-way/overrun clutch	51	82
		6. Forward clutch oil feed restricted	150	126
	Slips in 2nd gear	1. Improper fluid level	98	33
		2. Misadjusted or damaged manual linkage	118	49
		3. Worn clutch or band	147	122

Table 6-1 ASE TASK (continued)

Diagnose the cause of the abnormal shifting, then determine the needed repairs.

Problem Area	Symptoms	Possible Causes	Classroom Manual	Shop Manual
TRANSMISSION/ TRANSAXLE SLIPS	Slips in 2nd gear	4. Low fluid pressure	121	99
		5. Faulty servo/accumulator	147	122
		6. Internal leaks	106	99
		7. Sticking valves or dirty valve body	106	99
	Slips in forward gear ranges	1. Improper fluid level	98	33
		2. Misadjusted or damaged manual linkage	118	49
		3. Worn clutch or band	147	122
		4. Low fluid pressure	121	99
		5. Faulty servo/accumulator	147	122
		6. Internal leaks	106	99
		7. Sticking valves or dirty valve body	106	99
		8. Defective oil pump	101	113
		9. Misadjusted throttle linkage	118	49
		10. Defective one-way/overrun clutch	51	82
		11. Forward clutch oil feed restricted	150	126
	Slips in reverse	1. Improper fluid level	98	33
		2. Misadjusted or damaged manual linkage	118	49
		3. Worn clutch or band	147	122
		4. Low fluid pressure	121	99
		5. Faulty servo/accumulator	147	122
		6. Internal leaks	106	99
		7. Sticking valves or dirty valve body	106	99
		8. Defective oil pump	101	113
		9. Reverse clutch oil feed restricted	150	126
POOR PERFORMANCE	Drags in gear	1. Band misadjusted or worn	147	122
		2. Damaged planetary gear set	61	84
	Sluggish acceleration	1. Improper fluid level	98	33
		2. Torque converter one-way clutch locked up	51	82
		3. Worn forward clutch	150	126
		4. Contaminated fluid	98	33
		5. Poor engine performance	38	33
		6. Throttle linkage misadjusted or damaged	118	49
		7. Incorrect fluid pressure	98	33
		8. Malfunctioning electronic controls	51	52
FAILURE TO SHIFT	Low only, no upshift	1. Faulty governor	54	111
		2. Damaged valve body	106	99
		3. Faulty band or clutch	147	122
		4. Faulty accumulator	147	122
		5. Sticking or damaged 1–2 shift valve	106	99
		6. Throttle linkage out of adjustment	118	49
		7. Defective oil pump seals	101	113
		8. Internal fluid leaks	106	99
		9. Missing check ball	156	122
		10. Damaged planetary gear set	61	84

Table 6-1 ASE TASK (continued)

Diagnose the cause of the abnormal shifting, then determine the needed repairs.

Problem Area	Symptoms	Possible Causes	Classroom Manual	Shop Manual
		11. Vacuum leak in modulator circuit	117	46
		12. Defective oil pump	101	113
	1–2 full-throttle shifts only	1. Throttle linkage out of adjustment or damaged	118	49
		2. Oil passages damaged or clogged	106	99
		3. Improper fluid level	98	33
		4. Faulty governor	54	111
		5. Faulty band or clutch	147	122
	No 2–3 shift	1. Faulty 1–2 servo	147	122
		2. 2–3 shift valve sticking	106	99
		3. Improperly torqued valve body	106	99
		4. Internal fluid leaks	106	99
		5. Faulty clutch or band	147	122
		6. Check ball missing	156	122
		7. Malfunctioning electronic controls	51	52
	No 3–4 shift	1. Faulty governor	54	111
		2. 3–4 shift valve sticking	106	99
		3. Faulty clutch or band	147	122
		4. Malfunctioning electronic controls	51	52
	Shifts 1–3; skips 2nd gear	1. Defective servo	147	122
		2. Internal leaks	106	99
		3. Defective valve body	106	99
		4. Defective or damaged band/servo or clutch assembly	147	122
		5. Throttle linkage out of adjustment or damaged	118	49
		6. Sticking valves or dirty valve body	106	99
		7. Improper fluid level	98	33
		8. Glazed band or drum	147	122
	Does not move in any gear	1. Improper fluid level	98	33
		2. Misadjusted or damaged linkage	118	49
		3. Faulty clutch or band application	147	122
		4. Internal leaks	106	99
		5. Sticking valves or dirty valve body	106	99
		6. Low fluid pressure	121	99
		7. Defective oil pump	101	113
		8. Pump feed clogged	101	113
		9. Broken input or output shaft	74	83
		10. Damaged one-way clutch	51	82
		11. Converter-to-flexplate bolts missing	50	113
		12. Damaged planetary gear set	61	84
		13. Damaged drive link	84	94
		14. Pressure regulator valve stuck open	101	113

Table 6-1 ASE TASK (continued)

Diagnose the cause of the abnormal shifting, then determine the needed repairs.

Problem Area	Symptoms	Possible Causes	Classroom Manual	Shop Manual
FAILURE TO SHIFT	Does not move in any gear	15. Stuck parking pawl	38	57
		16. Clogged oil filter	99	34
		17. Damaged flexplate	50	113
		18. Malfunctioning electronic controls	51	52
	No drive in the forward gears, OK in reverse	1. Improper fluid level	98	33
		2. Misadjusted or damaged linkage	118	49
		3. Faulty clutch or band application	147	122
		4. Internal leaks	106	99
		5. Sticking valves or dirty valve body	106	99
		6. Damaged or worn band or clutch assembly	147	122
		7. Low fluid pressure	121	99
		8. Damaged servo/accumulator piston seal	147	122
		9. Worn oil pump	101	113
		10. Damaged one-way clutch	51	82
		11. Damaged or broken halfshafts	17	43
		12. Broken splines on output shaft	74	83
		13. Forward clutch oil feed clogged	150	126
		14. Drive link sprockets or chain damaged	84	94
		15. Torque converter unbolted from flexplate	50	113
		16. Damaged planetary gear set	61	84
		17. Malfunctioning electronic controls	51	52
	Delayed reverse engagement	1. Improper fluid level	98	33
		2. Low mainline pressure	121	99
		3. Internal leaks	106	99
		4. Sticking valves or dirty valve body	106	99
		5. Defective reverse clutch	147	122
		6. Manual linkage misadjusted or damaged	118	49
		7. Worn oil pump or damaged shaft	101	113
		8. Check ball missing	156	122
		9. Damaged one-way/overrun clutch	51	82
		10. Oil passages plugged	106	99
		11. Damaged drive link assembly	84	94
WRONG GEAR STARTING	Second speed start; misses low gear at times	1. Defective governor	54	111
		2. Defective valve body	106	99
		3. Defective one-way clutch	51	82
		4. Throttle linkage misadjusted or damaged	118	49
		5. Improper fluid level	98	33
		6. Internal leaks	106	99
		7. Sticking valves or dirty valve body	106	99
		8. Malfunctioning electronic controls	51	52
	Starts in 3rd in D range	1. Improper fluid levels	98	33
		2. Damaged or improperly adjusted linkage	118	49
		3. Defective governor	54	111
		4. Internal leaks	106	99

Table 6-1 ASE TASK (continued)

Diagnose the cause of the abnormal shifting, then determine the needed repairs.

Problem Area	Symptoms	Possible Causes	Classroom Manual	Shop Manual
		5. Sticking valves or dirty valve body	106	99
		6. Malfunctioning electronic controls	51	52
	Drives in neutral	1. Main linkage out of adjustment	118	49
		2. Defective oil pump	101	113
		3. Faulty clutch/band application	147	122
		4. Cross leakage in case cover	106	99
		5. Damaged planetary gear set	61	84
		6. Insufficient clutch pack clearance	150	126
ENGINE BRAKING	No engine braking in manual 2nd gear	1. Improper fluid level	98	33
		2. Linkage out of adjustment	118	49
		3. Worn or misadjusted band	147	122
		4. Defective servo or clutch	147	122
		5. Internal leaks	106	99
		6. Glazed band or drum	150	126
		7. Clogged oil passages	106	99
	No engine braking in manual low	1. Improper fluid level	98	33
		2. Linkage out of adjustment	118	49
		3. Defective band/servo or clutch	147	122
		4. Faulty pressure regulator valve	101	113
		5. Glazed band or drum	75	73
		6. Sticking valves or dirty valve body	106	99
		7. Low mainline pressure	121	99
		8. Check ball missing	156	122
		9. Internal leaks	106	99
		10. Stuck manual control valve	106	99
	No engine braking in manual 3rd	1. Sticking valves or dirty valve body	106	99
		2. Clogged oil passages	106	99
		3. Faulty overrun clutch	51	82
		4. Internal leaks	106	99
		5. Linkage out of adjustment	118	49
NO PARK	No Park range or will not hold in Park	1. Parking pawl or gear damaged	38	57
		2. Parking pawl spring damaged or broken	38	57
		3. Manual linkage out of adjustment or damaged	118	49
		4. Defective governor hub	54	111

Torque Converter Service

Upon completion and review of this chapter, you should be able to:

❏ Diagnose torque converter problems and determine needed repairs.

❏ Perform a stall test and determine needed repairs.

❏ Perform lockup converter system tests and determine needed repairs.

❏ Inspect converter flexplate, converter attaching bolts, converter pilot, and converter pump drive surfaces.

❏ Inspect, flush, and measure endplay, and test torque converter.

❏ Explain basic electrical terms.

❏ Discuss the basic theories of electricity.

❏ Perform basic electrical checks.

❏ Diagnose hydraulically and electrically controlled torque converter clutches.

Basic Tools

Basic mechanic's tool set

DVOM

Appropriate service manual

Many engine problems can cause a vehicle to behave as if the torque converter is malfunctioning or can actually cause the converter clutch to either lock up early or not at all. Clogged fuel injectors or bad spark-plug wires can seem like torque-converter complaints. Bad vacuum lines, EGR valves, or engine speed sensors can prevent the clutch from locking up at the proper time.

Common exhaust restrictions are a plugged catalytic converter or a collapsed exhaust pipe.

Sources for engine manifold vacuum are found in the intake manifold, below the throttle plate.

General Diagnostics

Many transmission problems are related to the operation of the torque converter. To test the operation of the torque converter, many technicians perform a stall test. The stall test checks the holding capacity of the converter's stator overrunning clutch assembly, as well as the clutches and bands in the transmission.

However, torque converter problems can often be identified by the symptoms; therefore, the need for conducting a stall test is minimized. If the vehicle lacks power when it is pulling away from a stop or when passing, it has a restricted exhaust or the torque converter's one-way stator clutch is slipping. To determine which of these problems is causing the power loss, test for a restricted exhaust first.

The most efficient way to test the exhaust is to use a vacuum gauge. Connect the vacuum gauge to a source of engine manifold vacuum. Observe the vacuum reading with the engine at idle. Quickly open the throttle plates and observe the vacuum reading. Then quickly release the throttle plates to allow them to close. The vacuum reading should show an increase of about 5 in. Hg. upon initial closing of the throttle plates (Figure 7-1). If the vacuum does not increase with the closing of the throttle plates, a restricted exhaust is indicated. If there is no evidence of a restricted exhaust, it can be assumed that the torque converter's stator clutch is slipping and not allowing any torque multiplication to take place in the converter. To repair this problem, the torque converter should be replaced.

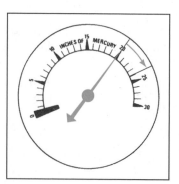

Figure 7-1 To check for an exhaust restriction, connect a vacuum gauge to the engine and observe the readings while the engine speed is raised and quickly lowered. (Courtesy of the Hydra-Matic Division of General Motors Corp.)

A seized one-way clutch will cause the vehicle to have good low-speed operation but to lack power at high speeds. The seized clutch does not allow for the overrunning mode of operation.

If the engine's speed flares up when it is accelerated in Drive and does not have normal acceleration, the clutches or bands in the transmission are slipping. This symptom is similar to the slipping of a clutch in a manual transmission. This problem is often mistakenly blamed on the torque converter.

Technicians often blame the torque converter for problems based on the customer's complaint. Complaints of thumping or grinding noises are often blamed on the converter when they are really caused by bad thrust washers or damaged gears and bearings in the transmission (Figure 7-2). This type of noise can also be caused by nontransmission components, such as bad CV joints and wheel bearings.

Many engine problems can also cause a vehicle to act as if it has a torque converter problem. This is especially true of the lockup converter, which may lockup early or not at all. Ignition and fuel injection problems can behave the same as a malfunctioning converter. Vacuum leaks and bad electrical sensors can prevent the converter from locking up at the correct time.

Two methods are commonly used to check the operation of the stator's one-way clutch: the stall test and bench testing.

Classroom Manual
Chapter 7, page 219

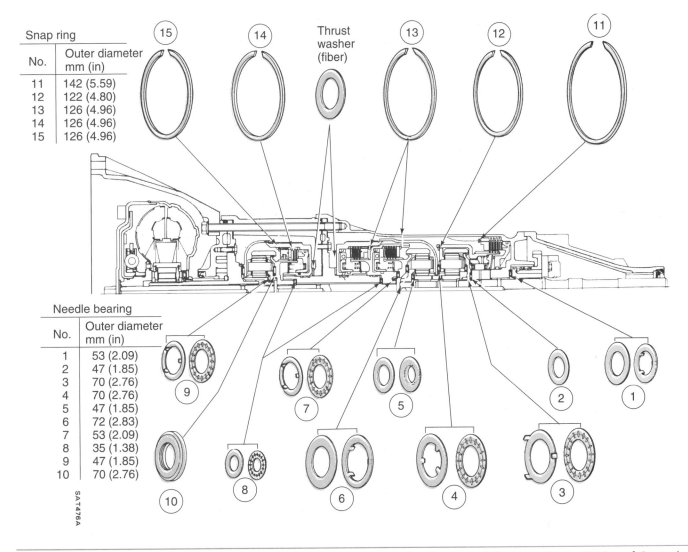

Figure 7-2 Typical locations of various thrust washers in a transmission (Courtesy of the Hydra-Matic Division of General Motors Corp.)

Trouble	Probable cause
Stall rpm high in $\boxed{D_4}$, $\boxed{2}$, $\boxed{1}$ & $\boxed{R}$	• Low fluid level or oil pump output • Clogged oil strainer • Pressure regulator valve stuck closed • Slipping clutch
Stall rpm high in $\boxed{R}$	• Slippage of 4th clutch
Stall rpm high in $\boxed{2}$	• Slippage of 2nd clutch
Stall rpm high in $\boxed{D_4}$	• Slippage of 1st clutch or 1st gear one-way clutch
Stall rpm low in $\boxed{D_4}$, $\boxed{2}$, $\boxed{1}$ & $\boxed{R}$	• Engine output low • Torque converter one-way clutch slipping

Stall rpm high in $\boxed{2}$	• Slippage of 2nd clutch
Stall rpm high in $\boxed{D_4}$	• Slippage of 1st clutch or 1st gear one-way
Stall rpm low in $\boxed{D_4}$, $\boxed{2}$, $\boxed{1}$ & $\boxed{R}$	• Engine output low • Torque converter one-way clutch slipping

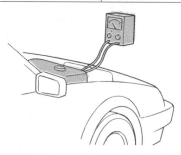

Figure 7-3 Before conducting a stall test, chock the wheels and place the tachometer in a position where it can be easily seen from the driver's seat. (Courtesy of Honda Motor Co.)

Stall Testing

Classroom Manual
Chapter 7, page 230

☑ **SERVICE TIP:** Stall testing is not recommended on many late-model transmissions. This test places extreme stress on the transmission and should only be conducted if recommended by the manufacturer.

CAUTION: Make sure no one is around the engine or the front of the vehicle while a stall test is being conducted.

To conduct a stall test, connect a tachometer to the engine and position it so that it can be easily read from the driver's seat. Set the parking brake and place blocks in front of the vehicle's tires (Figure 7-3). With the engine running, press and hold the brake pedal. Then move the gear selector to the Drive position and press the throttle pedal to the floor. Quickly note the tachometer reading and immediately let off the throttle pedal and allow the engine to idle. Compare the measured stall speed to specifications (Figure 7-4).

Torqueflite transmission stall speed chart

Engine liter	Transaxle type	Converter diameter	Stall rpm
1.7	A–404	9–1/2 inches (241 millimeters)	2300–2500
2.2	A–413	9–1/2 inches (241 millimeters)	2200–2410
2.6	A–470	9–1/2 inches (241 millimeters)	2400–2630

Figure 7-4 A typical torque converter stall speed chart (Courtesy of Chrysler Corp.)

If the torque converter and transmission are functioning properly, the engine will reach a specific speed. If the tachometer indicates a speed above or below specifications, a possible problem exists in the transmission or torque converter. If a torque converter is suspected of being faulty, it should be removed and the one-way clutch should be checked on the bench.

> ⚠️ **WARNING:** To prevent serious damage to the transmission, follow these guidelines while conducting a stall test:

1. Never conduct a stall test if there is an engine problem or engine mount problem.
2. Check the fluid levels in the engine and transmission before conducting the test.
3. The engine should be at normal operating temperature during the test.
4. Never hold the throttle wide open for more than 5 seconds during the test.
5. Do not perform the test in more than two gear ranges without driving the vehicle a few miles to allow the engine and transmission to cool down.
6. After the test, allow the engine to idle for a few minutes to cool the transmission fluid before shutting off the ignition.

If the stall speed is below the specifications, a restricted exhaust or slipping stator clutch is indicated. If the stator's one-way clutch is not holding, ATF leaving the turbine of the converter works against the rotation of the impeller and slows down the engine. With both of these problems, the vehicle will exhibit poor acceleration, either because of a lack of power from the engine or because there is no torque multiplication occurring in the converter. If the stall speed is only slightly below normal, the engine is probably not producing enough power and should be diagnosed and repaired.

If the stall speed is above specifications, the bands or clutches in the transmission may be slipping and not holding properly.

If the vehicle has poor acceleration but had good results from the stall test, suspect a seized one-way clutch. Excessively hot ATF in the transmission is a good indication that the clutch is seized. However, other problems can cause these same symptoms, so be careful during your diagnosis.

A normal stall test will generate a lot of noise, most of which is normal. However, if you hear any metallic noises during the test, diagnose the source of these noises. Operate the vehicle at low speeds on a hoist with the drive wheels free to rotate. If the noises are still present, the source of the noise is probably the torque converter. The converter should be removed and bench tested for internal interference.

Lockup Converter Testing

Nearly all late-model transmissions are equipped with a lockup torque converter. Most of these lockup converters are controlled by the engine control module or computer (Figure 7-5). The

Classroom Manual
Chapter 7, page 232

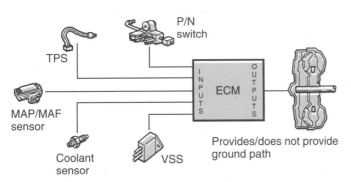

Figure 7-5 Typical electrical control circuitry for a lockup torque converter (Courtesy of the Hydra-Matic Division of General Motors Corp.)

computer turns on the converter clutch solenoid, which opens a valve and allows fluid pressure to engage the clutch. When the computer turns off the solenoid , the clutch disengages.

One of the trickiest parts of diagnosing a converter clutch problem is recognizing a normal-acting converter clutch, as well as an abnormal acting clutch. You should pay attention to the action of all converter clutches, whether or not they are suspected of having a problem. By knowing what a normal clutch feels like, it is easier to feel abnormal clutch activity. A converter clutch can cause a wide variety of driveability problems. The application of the lockup clutch should normally feel like a smooth engagement into another gear. It should not feel harsh, nor should there be any noises related to the application of the clutch.

To properly diagnose lockup converter problems, you must understand their normal operation and the controls involved with the system. Although the actual controls of the lockup clutch assembly vary with the different manufacturers and models of transmissions, they all will have certain operating conditions that must be met before the lockup clutch can be engaged.

Care should be taken during diagnostics because poor lockup clutch action can be caused by engine, electrical, clutch, or torque converter problems.

Before the lockup clutch is applied, the vehicle must be traveling at or above a certain speed. The vehicle speed sensor sends this speed information to the computer. The converter clutch should not be able to engage when the engine is cold; therefore, a coolant temperature sensor provides the computer with information regarding temperature. During sudden deceleration or acceleration, the lockup clutch should be disengaged. One of the sensors used to tell the control computer when these driving modes are present is the TPS. Some transmissions use a third or fourth gear switch to signal to the computer when the transmission is in that gear and to allow for converter lockup. A brake switch is also used in some lockup circuits to disengage the clutch when the brakes are applied (Figure 7-6). These key sensors, the VSS, CTS, TPS, third/fourth gear, and brake switch, should be visually checked as part of your diagnosis of converter problems.

Diagnosis of a lockup converter circuit should be conducted in the same way as any other computer circuit. The computer will recognize problems within the system and store trouble codes that reflect the problem area of the circuit. The codes can be retrieved and displayed by an instrument panel light or a hand-held scanner tool.

The engagement of the lockup clutch should be smooth. If the clutch prematurely engages or is not being applied by full pressure, a shudder or vibration results from the rapid grabbing and slipping of the clutch. The clutch begins to lock up, then slips, because it cannot hold the engine's

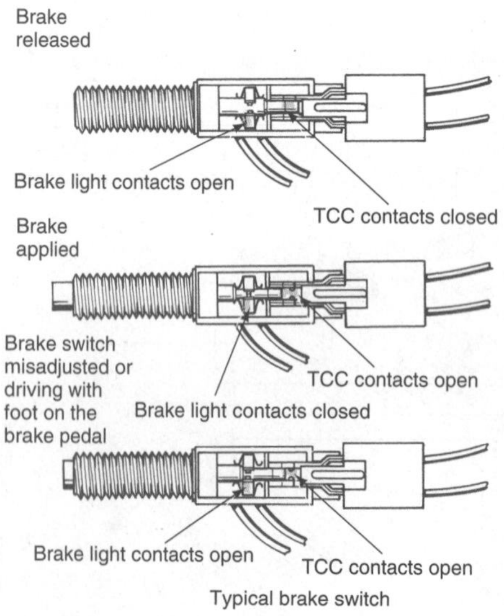

Torque converter clutch shudder is sometimes called chatter or a stick-slip condition.

The ability of a converter clutch to hold torque is its torque capacity.

Figure 7-6 Proper adjustment of the brake light switch is essential for proper operation of the lockup torque converter. (Courtesy of the Hydra-Matic Division of General Motors Corp.)

torque and complete the lockup. The torque capacity of the clutch is determined by the oil pressure applied to the clutch and the condition of the frictional surfaces of the clutch assembly.

⬤ **CUSTOMER CARE:** Torque converter clutch shudder can be both felt and heard through the entire driveline. Although it is a very noticeable and annoying problem, let your customers know that it will not cause immediate damage to the transmission or engine.

If the shudder is only noticeable during the engagement of the clutch, the problem is typically in the converter. When the shudder is only evident after the engagement of the clutch, the cause of the shudder is the engine, transmission, or another component of the driveline. You can identify the source of the shudder by disconnecting the torque converter clutch solenoid or valve. Then road test the vehicle. If the shudder is no longer present, the source of the shudder is the torque converter clutch assembly. If the shudder is caused by the clutch, the converter must be replaced to correct the problem.

When clutch apply pressure is low and the clutch cannot lock firmly, shudder will occur. This may be caused by a faulty clutch solenoid valve or its return spring. The valve is normally held in position by a coil-type return spring. If the spring loses tension, the clutch will be able to prematurely engage. Because insufficient pressure is available to hold the clutch, shudder occurs as the clutch begins to grab and then slips. If the solenoid valve and/or return spring is faulty, it should be replaced, as should the torque converter.

An out-of-round torque converter prevents full clutch engagement, which will also cause shudder, as will contaminated clutch frictional material. This is typically caused by metal particles circulating through the torque converter and collecting on the clutch. Broken or worn dampener springs will also cause shudder.

Besides replacing the torque converter and/or replacing other components of the system, one possible correction for shudder is the use of an ATF with friction modifiers. Some rebuilders may recommend that an oil additive be added to the ATF. The additive is designed to improve or alter the friction capabilities of regular ATF.

⬤ **CUSTOMER CARE:** After replacing a converter because of lockup problems, be sure to explain to the customer that other engine or transmission problems may become more evident with the new lockup converter. Because the old converter didn't lockup, many of the vibrations from the engine and transmission were dampened by the fluid in the converter. The new converter will now lock and the engine and transmission vibrations will now be mechanically transmitted through the drive train.

Vehicles equipped with a converter lockup clutch may stall when the transmission is shifted into reverse gear. The cause of this problem may be plugged transmission cooler lines or cooler itself may be plugged. Fluid normally flows from the torque converter through the transmission cooler. If the cooler passages are blocked, fluid is unable to exhaust from the torque converter and the converter clutch piston remains engaged. When the clutch is engaged, there is no vortex flow in the converter and therefore little torque multiplication is taking place in the converter.

Shop Manual
Chapter 7, page 238

To verify that the transmission cooler is plugged, disconnect the cooler return line from the radiator or cooler (Figure 7-7). Connect a short piece of hose to the outlet of the cooler and allow the other end of the hose to rest inside an empty container. Start the engine and measure the amount of fluid that flows into the container after 20 seconds. One quart of fluid should flow into the container. If less than that filled the container, a plugged cooler is indicated.

To correct a plugged transmission cooler, disconnect the cooler lines at the transmission and the radiator. Blow air through the cooler, one end at a time, then through the cooler lines. The air will clear large pieces of debris from the transmission cooler. Always use low air pressure, no more than 30 psi. Higher pressures may damage the cooler. If there is little air flow through the cooler, the radiator or external cooler must be removed and flushed or replaced.

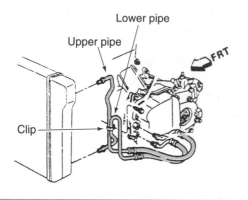

Figure 7-7 Location of cooler lines (Courtesy of Chrysler Corp.)

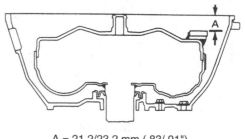

A = 21.2/23.2 mm (.83/.91")

Figure 7-8 Before removing a torque converter, check its installed depth in the bellhousing. (Courtesy of Hydra-Matic Division of General Motors Corp.)

Torque Converter Replacement

Extra care should always be taken when replacing a torque converter. Size and fit are not the only important variables. Nor does size alone determine the stall speed of the converter. Even if the converter has exactly the same stall speed, it may have a different torque ratio and should be used. Always check and double-check the part or model number of the torque converter you are removing and compare it to the one you are going to install. Converters are typically identified by a sticker or a number code stamped into the housing.

When replacing a torque converter, never use an impact wrench on the torque converter bolts. Impact wrenches can drive the bolts through the cover, which will warp the inside surface and prevent proper clutch engagement or it will damage the clutch's pressure plate.

Always perform an endplay check and check the depth of the torque converter in the bellhousing (Figure 7-8) before reinstalling a torque converter or installing a new unit.

Visual Inspection

Nearly all torque converters are welded-together units and therefore cannot be disassembled for inspection or repair. The primary objective when servicing a torque converter should be to check the serviceability of the unit.

Classroom Manual
Chapter 7, page 222

If there was a noise coming from the torque converter during the stall test, visually inspect the converter before pulling the transmission out and replacing the converter. Remove the torque converter access cover on the transmission and rotate the engine. While the torque converter is rotating with the engine, check to make sure the torque converter bolts are not loose and are not contacting the bellhousing. Also observe the action of the converter as it is spinning. If the converter wobbles, it may be due to a damaged flexplate or converter.

CAUTION: Before rotating the engine with a remote starter button, make sure the engine's ignition system is disabled. Also keep all parts of your body away from the rotating flexplate. Serious injury can result from being careless during this check.

Check the converter's balance weights to make sure they are still firmly attached to the unit. Carefully inspect the flexplate for evidence of cracking or other damage. Also check the condition of the starter ring gear. The teeth should not be damaged and the gear should be firmly attached to the flexplate.

Check the torque converter for ballooning. If excessive pressure was able to build up inside the converter, the converter will expand, or balloon. This is typically caused by a stuck converter check valve. If the converter is ballooned, it should be replaced and the cause of the problem also repaired.

A ballooned torque converter will also damage other driveline components. If the converter is ballooned toward the rear of the unit, the transmission's oil pump is most likely damaged. If it is ballooned toward the front, the crankshaft's thrust bearings are undoubtedly damaged.

Further inspection of the torque converter requires that it be removed from the vehicle. These inspections should also be done anytime you have the torque converter out of a vehicle, especially if the oil pump is damaged or if the customer's complaint appears to be related to the torque converter.

The converter's drive lugs or studs should be carefully inspected. These hold the converter firmly to the flexplate and ensure that the converter rotates in line and evenly with the flexplate. The threads on the studs or lugs should be clean and not damaged. They should also be tightly seated into the converter. Also check the shoulder area around the lugs and studs for cracked welds or other damage. If any damage is found, the converter should be replaced. An exception to this is when the internal threads of a drive lug are damaged. These can often be repaired by tapping the threads or by installing a thread insert. Also inspect the converter attaching bolts or nuts and replace them if they are damaged.

Inspect the flexplate for signs of damage, warpage, and cracks. Check the condition of the teeth on the starter ring gear. Replace the flexplate if there is any evidence of damage.

Check the drive hub of the torque converter. It should be smooth and show any signs of wear. If the hub is worn, carefully inspect the oil pump drive and replace the torque converter. Light scratches, burrs, nicks, or scoring marks on the hub surface can be polished with fine crocus cloth. Be careful not to allow dirt to enter into the converter while polishing the hub. If the hub has deep scratches or other major imperfections, the converter should be replaced.

In general, a torque converter should also be replaced if it has fluid leakage from its seams or welds, loose drive studs, worn drive stud shoulders, stripped drive stud threads, a heavily grooved hub, or excessive hub runout.

Transaxles that do not have their oil pump driven directly by the torque converter use a drive shaft that fits into a support bushing inside the converter's hub (Figure 7-9). This bushing should be checked for wear. To do this, measure the inside diameter of the bushing and the outside diameter of the drive shaft. The difference between the two is the amount of clearance. This measurement should be compared to factory specifications. Normally, the maximum allowable clearance is 0.004 in. Excessive clearance can cause oil leaks. If the clearance is excessive, the bushing should be replaced.

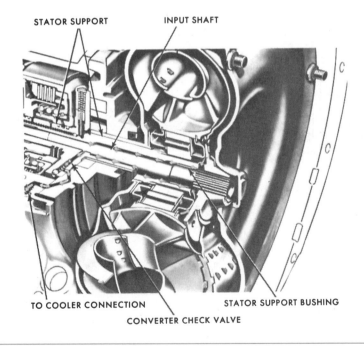

Figure 7-9 Typical stator support and oil flow in a torque converter (Reprinted with the permission of Ford Motor Co.)

Stator One-way Clutch Diagnosis

The operation of the one-way stator clutch inside the torque converter is critical to overall effectiveness of the torque converter. If a problem occurs in this clutch assembly, the clutch will either fail to lock when rotated in either direction or fail to unlock when rotated in either direction. Although these problems are similar, they affect efficiency at opposite ends of the engine's operating speeds. However, in either case, fuel economy will be affected.

When the stator clutch does not lock, there is a disruption in vortex flow and a loss of torque multiplication in the torque converter. A vehicle with this problem will have sluggish low-speed performance, but will perform normally at higher speeds when the stator is supposed to freewheel.

A vehicle with a constantly locked stator will have good low-speed performance and poor high-speed performance. Torque multiplication will always occur, as will speed reduction. A vehicle with a constantly locked stator will show signs of overheating. If you suspect a locked stator, check for a bluish tint on the hub of the converter. This discoloration typically results from overheating.

To check the stator's one-way clutch with the converter on a bench, insert a finger into the splined inner race of the clutch. Attempt to turn the inner race in both directions. You should be able to turn the race freely in a clockwise direction and feel lockup in a counterclockwise direction. If the clutch rotates freely in both directions or if the clutch is locked in both directions, the converter should be replaced.

Because this check does not put a load on the clutch assembly, it does not totally check the unit. Therefore, some manufacturers, such as Ford, recommend the use of a special tool set which holds the inner race and exerts a measurable amount of torque on the outer race, thereby allowing the technician to observe the action of the clutch while under load (Figure 7-10).

Internal Interference Checks

Noises may also be caused by internal converter parts hitting each other or hitting the housing. To check for any interference between the stator and turbine, place the converter face down on a

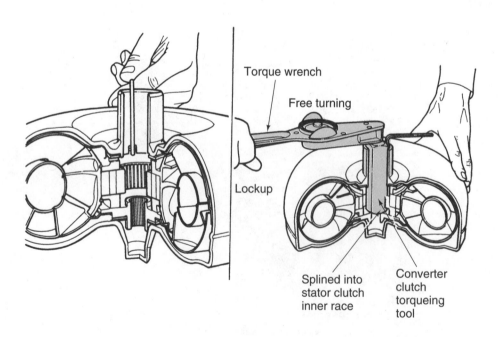

Figure 7-10 Checking one-way clutch with special tool fixture (Reprinted with the permission of Ford Motor Co.)

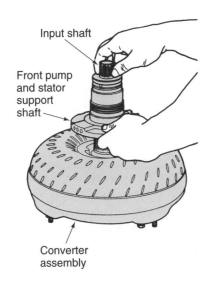

Input shaft

Front pump
and stator
support
shaft

Converter
assembly

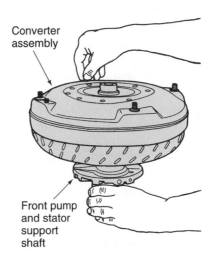

Converter
assembly

Front pump
and stator
support
shaft

Figure 7-11 Checking stator-to-turbine interference
(Reprinted with the permission of Ford Motor Co.)

Figure 7-12 Checking stator-to-impeller interference
(Reprinted with the permission of Ford Motor Co.)

bench (Figure 7-11). Then install the oil pump assembly. Make sure the oil pump drive engages with the oil pump. Insert the input shaft into the hub of the turbine. Hold the oil pump and converter stationary. Then rotate the turbine shaft in both directions. If the shaft does not move freely and/or makes noise, the converter must be replaced.

To check for any interference between the stator and the impeller, place the transmission's oil pump on a bench and fit the converter over the stator support splines (Figure 7-12). Rotate the converter until the hub engages with the oil pump drive. Then hold the pump stationary and rotate the converter in a counterclockwise direction. If the converter does not rotate freely or makes a scraping noise during rotation, the converter must be replaced.

The input shaft may
be referred to as the
turbine shaft on
some transaxles.

Endplay Check

The special tools required to check the internal endplay of a torque converter are typically part of the essential tool kit recommended by each manufacturer. However, these specialty tools can be individually purchased through specialty tool companies. The special tools are a holding tool and a dial indicator with a holding fixture. The holding tool is inserted into the hub of the converter and once bottomed, it is tightened in place (Figure 7-13). This locks the tool into the splines of the

Special Tools

Manufacturer speci-
fied holding tool

Dial indicator

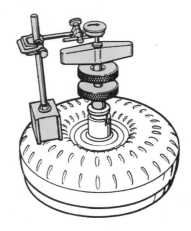

Figure 7-13 Checking converter endplay (Courtesy of the Buick Motor Division of General Motors Corp.)

turbine. The dial indicator is fixed onto the hub. The amount indicated on the dial indicator, as the tool is lifted up, is the amount of endplay inside the converter. If this amount exceeds specifications, replace the converter.

Converter Leakage Tests

If the initial visual inspection suggests that the converter has a leak, special test equipment can be used to determine if the converter is leaking. This equipment uses compressed air to pressurize the converter. Leaks are found in much the same way as tire leaks are; that is, the converter is submerged in water and the trail of air bubbles lead the technician to the source of leakage. This test equipment (Figure 7-14) can only be used to check converters with a drain plug and, therefore, is somewhat limited in its current applications. However, some manufacturers list a procedure for installing a drain plug or draining the converter in their service manuals.

Converter Clutch Friction Material Testing

Classroom Manual
Chapter 7, page 236

To determine if a lockup torque converter should be reused, the following questions should be answered: Is there any obvious damage to the converter? Was there an excessive amount of frictional material in the ATF? Was there any evidence of transmission overheating? If the answer to all of these questions is "no," then the torque converter can be reused, providing the frictional material inside the converter is not severely worn. Unfortunately, the lining cannot be visually inspected or measured. Most often, torque converters are replaced whenever there is any question about the condition of the friction material. To better determine the condition of the material, some specialty tool companies have equipment which uses compressed air to lock the clutch while measuring the clutch's holding power with a torque wrench.

● **CUSTOMER CARE:** Because the converter is an expensive item, it should never be automatically replaced during a transmission overhaul. However, there are certain conditions that mandate converter replacement, such as when the front pump is badly damaged, when the stator's clutch fails, when the converter's hub or housing is damaged, or when the fluid in the transmission has large amounts of aluminum deposits. Failure to replace the converter in these cases will damage a newly overhauled transmission and will turn a customer into an enemy instead of a friend.

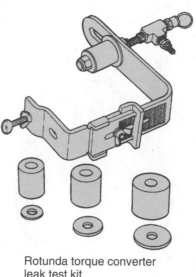

Rotunda torque converter
leak test kit

Figure 7-14 Converter leakage test kit (Reprinted with the permission of Ford Motor Co.)

Starter Ring Gear Replacement

The starter ring gear is most often part of the flexplate. Therefore, whenever the teeth of the ring gear are damaged, the entire flexplate is replaced. Some transmissions are equipped with a torque converter fitted with a ring gear around the outer circumference of the torque converter cover. The ring gear is welded to the front of the converter cover and gear replacement involves breaking or cutting the welds to remove the old gear, then replacing the gear and welding it back onto the converter. This procedure is not typically recommended for torque converters fitted with a lockup clutch, as the heat from welding can destroy the frictional surfaces inside the converter. In these cases, if the ring gear is damaged, the entire converter should be replaced.

Torque converters fitted with a starter ring gear are typically found in Torqueflite and C-4 transmissions.

Cleaning

> ✓ **SERVICE TIP:** A torque converter can collect a lot of debris after a major transmission problem, and no amount of soaking and flushing will remove all of the dirt from the converter. A new converter should be installed.

A torque converter should be flushed anytime there was contaminated fluid in the transmission or when the transmission has been overheated. There are many small corners inside a converter for dirt to become trapped in. Left over debris in the converter can lead to converter and transmission damage. Remember, the fluid that circulates through the converter circulates through the transmission next. Converters are typically cleaned by flushing the inside of the housing with solvent; however, flushing with solvent is not recommended for most lockup converters.

Flushing removes all debris and sludge from the inside of the converter. Flushing is either done by a machine or by hand. Some transmission shops use a torque converter flushing machine which pumps a solvent through the converter as it is rotated by the machine. This method keeps the fluid moving inside the converter. The moving fluid is quite efficient at picking up any dirt present. The dirty solvent is pumped out of the converter and new solvent is added during the flushing procedure.

Shops without a flushing machine clean the inside of converters by hand. The procedure involves pouring about two quarts of clean solvent into the converter, then forcefully rotating and vibrating the converter (Figure 7-15). This action should dislodge any trapped debris. It is also helpful to use the input shaft and spin the turbine while the fluid is inside. The solvent is drained and the process repeated until the drained solvent is clean.

Lockup torque converters can be cleaned by this process except instead of pouring solvent into the converter, you should pour clean ATF. The rest of the procedure is the same.

Some transmission shops may cut the torque converter shell in half, then clean the parts, examine them for wear, and replace any that are worn or broken. The shell is then welded back together. This is a job for only shops that are equipped to do this properly.

After the converter has been flushed, disconnect it from the machine. If the converter has a drain plug, invert and drain the complete assembly. Converters without drain holes or plugs can be

Figure 7-15 When hand flushing a converter, shake it violently. (Courtesy of Mazda Motors)

drilled with a 1/8-inch drill bit between the top end of the impeller fin dimples. This hole will act as an air bleed to maximize flushing. After flushing and draining the converter, the bleeder hole is sealed with a closed-end pop rivet covered with sealant.

General Converter Control Diagnostics

Classroom Manual
Chapter 7, page 237

TCC is a commonly used acronym for torque converter clutch.

All testing of TCC controls should begin with a basic inspection of the engine and transmission. Too often, technicians skip this basic inspection and become frustrated during diagnostics because of conflicting test results. Apparent transmission and torque converter problems are often caused by engine mechanical problems, broken or incorrectly connected vacuum hoses, incorrect engine timing, or incorrect idle speed adjustments. The basic inspection should include:

1. A road test to verify the complaint and further define the problem.
2. Careful inspection of the engine and transmission.
3. A check of the mechanical condition of the engine, the output of the ignition system, and the efficiency of the fuel system.
4. An idle speed and ignition timing check. If the timing is nonadjustable, check the operation of the electronic spark control system.
5. A check of the entire intake system for vacuum leaks.

When inspecting wires and hoses, look for burnt spots, bare wires, and damaged or pinched wires. Make sure the harness to the electronic control unit has a tight and clean connection. Also check the source voltage at the battery before beginning any detailed tests on an electronic control system. If the voltage is too low or too high, the electronic system cannot function properly.

SERVICE TIP: Nearly all electronic converter and transmission controls have a self-diagnostic mode and are capable of displaying trouble codes. However, the basic inspection is important because the computer will not display codes for low compression, open spark plug wires, or problems in the idle air control system.

On early TCC equipped vehicles, lockup was controlled hydraulically. A switch valve was controlled by two other valves, the lockup and fail-safe valves, in the lockup control assembly (Figure 7-16). The lockup valve responds to governor pressure and prevents lockup at speeds below 40 mph. The fail-safe valve responds to throttle pressure and permits lockup in high gear only. Problems with this system are diagnosed in the same way as other hydraulic circuits.

The lockup clutch in most TCC systems is applied when oil flow through the torque converter is reversed. This change can be observed with a pressure gauge. Using the pressure gauge is also a good way to diagnose the clutch's hydraulic system. Connect a pressure gauge to the hydraulic line with a "tee" fitting from the transmission to the cooler. Position the gauge so that is easily seen from the driver's seat. Then raise the vehicle on a hoist with the drive wheels off the ground and able to spin freely. Operate the vehicle until the transmission shifts into high gear. Then maintain a speed of approximately 55 mph. Once the speed is maintained, watch the pressure gauge.

If the pressure decreases 5 to 10 psi, the converter clutch was applied. With this action, you should feel the engagement of the clutch, as well as a change in engine speed. If the pressure changes but the clutch does not engage, the problem may be inside the converter or at the end of the input shaft. If the input shaft end is worn or the O-ring at the end is cut or worn, there will be a pressure loss at the converter clutch. This loss in pressure will prevent full engagement of the clutch. If the pressure does not change and the clutch does not engage, suspect a faulty clutch valve or control solenoid, or a fault in the solenoid control circuit.

Diagnosis of electrical TCC controls isn't as hard as some would lead you to believe. In fact, it becomes easier as you understand how electricity works and how the component or system you are diagnosing works. TCC electrical problems are typically diagnosed through a three-step process: visual inspection, computer code retrieval, and electrical checks of specific circuits. Com-

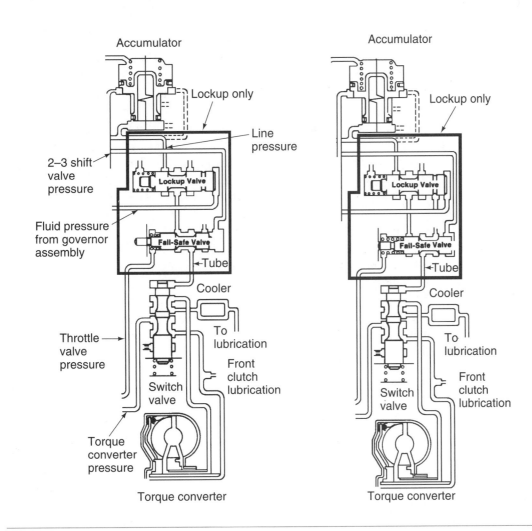

Figure 7-16 Action of the hydraulic fluid from the lockup valve to the torque converter (Courtesy of Chrysler Corp.)

puter trouble code retrieval procedures are rather simple but vary with each manufacturer. Proper inspection procedures and the checking of electrical circuits require an understanding of basic electricity theory, electrical components, and electrical test instruments. These apply to all electrical circuits, no matter who manufactured the vehicle.

Basic Electricity

A good understanding of electricity requires an understanding of the terms commonly used to describe it. Electricity is a form of energy that results from the movement of an electron from one atom to another. Since everything is made up of atoms and all atoms have at least one electron, everything has the potential of having electricity in it. However, in some things, the movement of electrons occurs easier than in others. Materials in which electrons move easily are often referred to as conductors. Materials in which electrons have a very difficult time moving from one atom to another are called insulators.

Wires are the commonly used and most recognizable conductors used on an automobile. These conductors allow electricity to flow to designated parts of the car. The wires, or conductors, are the path for electricity to follow through the car. These wires are covered with an insulator which prevents the electricity from leaking out of the wires. Because insulating materials do not readily allow for the movement of electrons, they are said to have very high electrical resistance.

Electrical resistance is the ability to resist the flow of electricity. The principles of resistance are not only applied to the insulated covering on wires, but are also used within electrical circuits. Using resistance within an electrical circuit controls the flow of electricity within that circuit. Materials used within a circuit do not have extremely high resistance like an insulator; rather, they make it difficult for electricity to flow and cause the flow to lose or give up some of its energy. The amount of resistance in a circuit determines how much flow of electricity there will be in the circuit.

An electrical system can be compared to an automobile's cooling system. Coolant must be able to move throughout the cooling system and return to the radiator. The same is true for electricity. It flows from the battery through its circuit and then back to the battery. A water pump forces the coolant through the system, while valves and the thermostat control the amount of coolant that will flow.

In an electrical circuit, conductors serve the same purpose as the hoses of the cooling system. The work of the water pump is accomplished by the battery—it forces current through the circuit. This force is the result of voltage, which is electrical pressure. Resistors or loads control the amount of current flow in the same way that the thermostat and valves do in the cooling system.

If the pressure is increased in the cooling system, more coolant will flow through it. Likewise, if the valves are nearly closed, little coolant will be able to flow. An electrical circuit behaves in the same way. When resistance is increased, current will decrease and when voltage is increased, current will also increase.

A light bulb is able to shine because the filament inside the bulb resists the flow of electricity. The bulb shines from the heat given off as energy is lost while electricity flows through the filament or resistor. Any device that uses electricity for operation has some resistance. This resistance converts the electrical energy into another form of energy for use by the device. The resistance of a device is measured in ohms with an ohmmeter or with the resistance-measuring function of a multimeter.

The random movement of electrons in a conductor is not current flow. Electrical current is the directed movement of electrons from atom to atom within a conductor. In an automobile, electrons flow from one post of the battery, through a circuit, then back to the other post of the battery. The amount of resistance in the circuit determines how many electrons or how much current will flow through the circuit. If the resistance is high, current will be low. If a wire is cut and the ends placed a distance apart, the air gap between the two ends will create a great enough resistance to stop all current flow. The rate of current is measured in amperes with an ammeter or with the current-measuring function of a multimeter.

The amount of current flow in a circuit is determined by the amount of resistance in the circuit and the amount of pressure pushing the current through. This pressure is set up by the attraction of the negative side of the battery to the positive side. Voltage is the amount of electrical pressure or force present at a particular point within a circuit. This electrical pressure is measured in units of volts. A normal automotive battery has 12 volts present across its positive and negative posts. Voltage pushes current through the circuit's resistors or loads. The amount of voltage in a load decreases as the current passes through the load. By the time the current passes through the last bit of resistance in a circuit, all of the voltage is lost.

In a complete electrical circuit, voltage, current, and resistance are present. The amounts of each depend upon a number of variables. However, most often, the amount of resistance is the factor that determines the amount of current and voltage. Using Ohm's Law, the value of current, voltage, or resistance can be calculated. Ohm's Law states that one volt of electrical pressure is needed to push one ampere of electrical current through one ohm of resistance. Therefore, the amount of voltage (V) needed equals the amount of current (I) multiplied by the amount of resistance (R) or $V = I \times R$.

This law can be used to calculate unknown values in a circuit. To calculate the resistance that would cause a particular amount of current in a fixed voltage circuit, the formula $R = V/I$ is used. To calculate the current of a circuit, $I = V/R$ is used.

Energy cannot be destroyed, nor can it be created. It is merely changed to another form of energy when it is used.

Amperes are commonly referred to as amps.

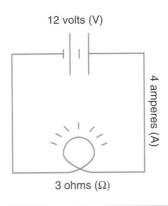

12 volts (V)

4 amperes (A)

3 ohms (Ω)

Figure 7-17 A simple 12-volt series circuit

The most practical use of this law is to explain and predict the effects of a change in a circuit. In a light circuit with 12 volts available from the battery and the resistance of the bulb being 3 ohms, 4 amperes of current will flow through the circuit (Figure 7-17). However, if there is a bad connection at the bulb which creates an additional 1 ohm of resistance, the current will decrease to 3 amperes. As a result of the added resistance and decreased current, the bulb will lose about half of its brightness.

This lack of brightness is also caused by the amount of voltage dropped by the bulb. In the normal circuit, 12 volts is dropped by the bulb. However, in the circuit with the added resistance, 3 volts will be dropped by the bad connection and the remaining 9 volts will be dropped by the bulb.

The amount of power or heat given off by the bulb or any electrical device can be determined by multiplying the current flowing through the load by the amount of voltage dropped by the load. This formula is expressed as $P = I \times E$. The power used to light the bulb in the normal circuit is equal to 4 times 12 or 48 watts. The wattage used by the bulb in the bad circuit is only 27. Electrical power is measured and expressed in watts.

The way resistance affects a circuit depends on the placement of the resistance in the circuit. If the resistance, wanted or unwanted, is placed directly into the circuit, the resistances are said to be in series. If a resistor is placed so that it allows an alternative path for current, it is a parallel resistor.

A series circuit allows current to follow only one path and the amount of current that flows through the circuit depends upon the total resistance of the circuit. To determine the total resistance in a series circuit, all resistance values are added together. At each resistor, voltage will be dropped and the total amount of voltage dropped in a series circuit is equal to the sum of the drops throughout the circuit. Regardless of the possible differences and resistance values of the loads, current in a series circuit is always constant.

Parallel circuits are designed to allow current to flow in more than one path. This allows one power source to power more than one circuit or load. All of a car's accessories and other electrical devices are powered by the battery and can be individually controlled through the use of parallel circuits. Within a parallel circuit, there is a common path from the power source and back to the power source. In an automobile, the path goes from the battery, through the individual legs of the circuit, and back to the battery through a common ground. This common ground is normally the frame of the car. Each branch, or leg, of a parallel circuit behaves as if it were an individual circuit. Current will flow only through the individual circuits when each is closed or completed. All legs of the circuit do not need to be complete in order for current to flow through one of them. In parallel circuits, the total amperage of the circuit is equal to the sum of the amperages in all of the legs of the circuit. No voltage is lost when the circuit splits into its branches; therefore, equal amounts of voltage are applied to each branch of the circuit. The total resistance of a parallel circuit is always less than the resistance of the leg with the smallest amount of resistance. The total resistance of two resistors in parallel can be calculated by dividing the product of the two by their sum.

The legs of a parallel circuit can contain a series circuit. To determine the resistance of that leg, the resistance values are added together. The resistance values of each leg are used to calculate the total resistance of a parallel circuit. Total circuit current flows only through the common power and ground paths; therefore, a change in a branch's resistance will not only affect the current in the branch but will also affect total circuit current.

Electrical Problems

Parallel circuits in an automobile normally have a circuit protection device placed in the common path from the battery. This protection device is usually a fuse or a circuit breaker. Fuses and circuit breakers are designed to protect the wires and components from excessive current. When a great amount of current flows through the fuse or breaker, an element will burn out, causing the circuit to be opened and stopping current flow. This action prevents the high current from burning up the wires or components in the circuit.

High current is caused by low resistance. A decrease in the amount of resistance is typically the result of a short. A short is best defined as an additional and unwanted path to ground. Most shorts, such as a bare wire contacting the frame of the car, create an extremely low resistance parallel branch. Low resistance and high current can also be caused by a slow-turning motor.

A short to ground can be present before the load in the circuit or internally within the load or component. A short can also connect two or more circuits together, causing additional parallel legs and uncontrolled operation of components. An example of a possible result from a wire-to-wire short would be the horn blowing each time the brake pedal is depressed. This could be caused by a wire-to-wire short between the horn and brake light circuits. Shorts are one of the three common types of electrical problems.

Another common electrical fault is the open. An open causes an incomplete circuit and can result from a broken or burnt wire, loose connection, or a faulty component. If a circuit is open, there is no current flow and the component will not operate. If there is an open in one leg of a parallel circuit, the remaining part of that parallel circuit will operate normally.

Excessive resistance at a connector (Figure 7-18), internally in a component, or within a wire is also a common electrical problem. High resistance will cause low current flow and the component will not be able to operate normally, if at all.

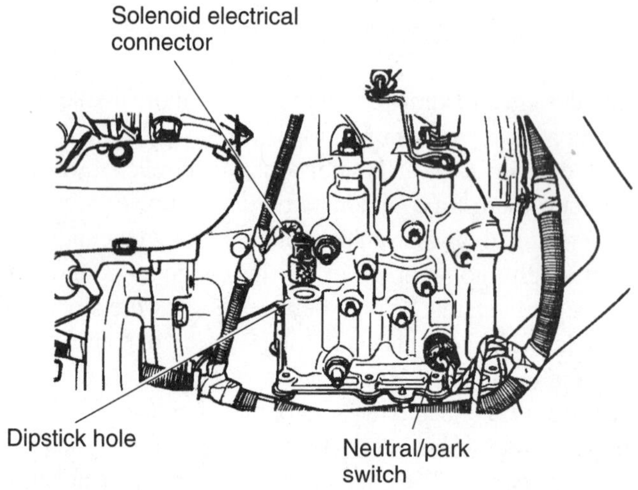

Motors attempting to move a heavier-than-normal load or an immovable object such as a binding window will rotate very slowly or not at all and will draw excessive amounts of current.

Figure 7-18 Typical transmission solenoid wiring harness and connectors (Courtesy of Chrysler Corp.)

Magnetism

Magnetism is a form of energy that can cause current flow in a conductor. It results from the movement of electrons, just like electricity. The movement of electrons cause the atoms of some materials to align and set up magnetic lines of flux. These flux lines establish an attraction of the north pole of the magnet to the south pole.

Current flowing through a conductor normally causes a weak magnetic field to form around the conductor. To strengthen the magnetic field, the conductor can be wound into a coil. This concentrates the flux lines into a smaller area and gives the effect of a large magnet. As long as current is flowing through the wound conductor, a magnetic field will be present. To increase the concentration of the magnetic field, a strip of soft iron can be inserted into the center of the windings. By controlling the current flow through the windings, this magnet can be switched on and off and is referred to as an electromagnet. Many switching devices, such as relays, use this principle.

A relay is both a protection device and a switch (Figure 7-19). A low-current circuit is used to close and open a high-current circuit. Relays use a low current to set up a magnetic field which pulls a set of contacts closed. The closing of the contacts completes the high-current circuit and allows it to operate.

Classroom Manual
Chapter 7, page 237

The most commonly used electromagnetic device used with automatic transmissions is the solenoid (Figure 7-20). A solenoid is a magnetic switch, like a relay. However, rather than moving contacts external to the magnetic field, the soft iron core in the center of the solenoid's windings moves to open or close a circuit. The movement of the core may also be used to mechanically perform a function such as moving a lever or opening and closing a hydraulic valve.

Soft iron can also become a permanent magnet. A permanent magnet is one that retains its magnetic field after current has ceased to flow. Once the electrons of the material are aligned, they tend to stay in their location and retain their magnetic qualities.

When a conductor is passed through lines of magnetic flux, a voltage is induced in the conductor. The amount of induced voltage depends upon the strength of the magnetic field and how quickly the conductor is moved through the field. This principle is used in alternators that recharge the battery and in many engine control devices.

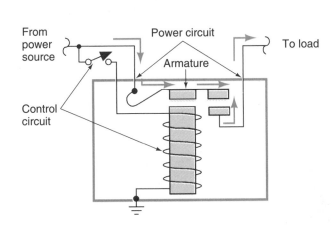

Figure 7-19 A relay uses electrical current to create a magnetic field to draw the contacts closed.

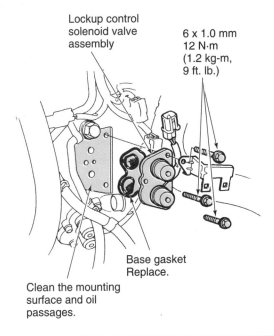

Figure 7-20 TCC solenoid assembly (Courtesy of Honda Motor Co.)

As the conductor moves into the magnetic field, a positive voltage is produced. When the conductor moves out of the field, it induces a negative voltage. Equal amounts of movement in and out of the field will produce equal amounts of positive and negative voltages. Many computer inputs or sensors rely on this change of voltage to inform the computer of the location of a moving component.

In an alternator, a magnetic field is rotated past a number of conductors. As the field rotates, the stationary conductors continuously pass through the field and voltage is induced. The use of multiple conductors increases the amount of induced voltage.

Principles of magnetism are also used in motors. Magnetic poles of the same polarity oppose each other, while opposite poles are attracted to one another. Motors are constructed of two magnets, one inside the other. Rotation is caused by the constant attraction and opposition of the poles. The speed of the rotation is controlled by the strength of the magnetic fields, which can be controlled by the amount of current flow through the electromagnets. Some motors use two electromagnets, while others use a permanent magnet and an electromagnet.

Circuit Controls

Relays are often used to control a motor circuit. The use of a relay allows other circuits to be activated along with the motor. An example of this is the starter motor relay. This relay not only completes the circuit from the battery to the starter motor but also completes the ignition circuit. Both of these actions allow the engine to start.

Circuits are controlled by switches, which open and close the circuit, and by variable resistors, which control the amount of current flow in the circuit. These switches and controls can be manually operated or controlled by electricity and magnetism. Manual switches and variable resistors are either controlled by the driver or through mechanical linkages that move in response to operating conditions.

The different models of vehicles have many different automatic switches. These switches open or close a circuit depending upon the conditions set up for the switches. The driver is usually unaware of the action of the switches unless one of them fails. These switches can be controlled by temperature (Figure 7-21) or pressure and will activate or deactivate a circuit whenever the conditions cause them to close or open. Automobiles equipped with electronic engine controls rely heavily on this type of switch for the activation of circuits, as well as for sending information to the computer. Many manual switches in a TCC circuit are controlled by fluid pressure. When pressure is present at the diaphragm of the switch, the diaphragm moves and either opens or closes the switch.

Semiconductors

Computerized engine and transmission controls rely heavily on semiconductors, both in the computer and in the sensing devices. A semiconductor is simply a material that conducts electricity only when the conditions are right. Two types of semiconductors are used in automobiles: diodes and transistors.

A diode is the most basic semiconductor in that it allows current to flow through it in one direction only. A diode is placed in series within a circuit. As current attempts to flow through one side of

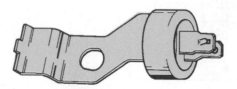

Figure 7-21 A TCC temperature switch (Courtesy of the Hydra-Matic Division of General Motors Corp.)

a diode, it is met with a great amount of resistance and little current, if any, flows through the circuit. When current flows toward the opposite side, it is met with very little resistance and the current flows through it.

An alternator produces an alternating current (ac), as a result of the conductors passing in and out of the magnetic field. Alternating current is a type of electricity in which current flows both from positive to negative and negative to positive within a circuit. The normal current flow in an automobile is direct current (dc), which flows in one direction only. In order to recharge a car's battery, the ac from the alternator must be changed to dc. This is accomplished by diodes.

Diodes are also used in a number of other circuits. They are used in some warning circuits to light an indicator lamp, activate a buzzer or chimes, or activate a light and buzzer. Specially designed diodes are used to control more than the direction of current flow. Zener diodes will allow current to flow in both directions when the voltage is above a particular level. These diodes are often used to protect systems from voltage surges or to precisely control the voltage in a circuit. They are commonly found in electronic control modules, voltage regulators, and instrumentation displays.

A light emitting diode is often used in ignition systems, instrumentation displays, and in many measuring instruments. When the proper amount of current is introduced to these diodes, they release some electrical energy in the form of light. Through careful arrangement of these diodes, it is possible to light a display of numbers and letters.

Transistors are semiconductors that are used as amplifiers and switches. Activated by low amounts of voltage and current, they can amplify current and voltage and serve as a relay to switch a high-current circuit on or off. When used in combination with diodes, transistors can quickly and precisely open and close a high-current circuit. The advantage of using a transistor to serve as a relay is that it has no moving parts to wear or break.

Amplifying transistors are commonly used in radios and electronic ignitions to strengthen weak signals. They provide for a high current in response to low voltage signals.

The main switching component of a transistor is its base. When the emitter and collector are connected into a circuit, the transistor will not function until the correct amount of voltage is applied to the base. At that time, the transistor will either serve as a switch or will amplify the signal it receives.

Combinations of diodes and transistors are the basis of solid-state electronics used in today's electronic control modules and computers. They often are compactly constructed to form an integrated circuit. A single IC chip can perform the functions of thousands of semiconductors at one-millionth of their size.

Semiconductor materials are also used in many sensors to monitor the speed of shafts. The most common of these sensors are called Hall Effect sensors. The action of a semiconductor when it is moved through a magnetic field is similar to that of a conductor. However, the semiconductor must have current flowing through it before the passing of the magnetic field will induce a voltage in it. As the magnetic field moves toward the semiconductor material, the voltage potential at the semiconductor begins to change. Throughout the movement of the field, the voltage will respond to the position of the field. This change in voltage serves as an information signal to the computer or module, indicating the exact position and speed of the rotating object.

In automotive systems, individual semiconductors are not normally replaced. When a component containing semiconductors is found to be defective, it is replaced as a unit. Proper diagnostics and testing are the keys to replacing the correct component. Accurate testing begins with using the proper procedure for measuring current flow, voltage, and resistance.

Basic Electrical Testing

Ammeters

Measuring current flow can provide a summary of the activity in a circuit. If the current flow is lower than expected, there is some extra resistance in the circuit. This resistance can be the result of corrosion, frayed wires, or a loose connection. When current is higher than expected, a short is usually indicated. A circuit operating normally will have the expected amount of current flow.

An ammeter is used to measure current flow. There are two types of ammeters and both require a different hookup into the circuit. Prior to using an ammeter in a circuit, make sure the meter is capable of handling the current in the circuit. Circuits like the starting motor circuit have very high amounts of current and only specially designed meters are able to withstand the current flow of these circuits. Ammeters often have ranges of measurement. Always select the proper range prior to connecting the meter.

One type of ammeter has two leads, a positive and a negative lead. In order for this type of ammeter to accurately read current flow, the meter must be connected in series to the circuit. The circuit should be disconnected from its power source, then opened by disconnecting a connector in the circuit and inserting the red lead in the connector half coming from the positive side of the battery and inserting the black lead into the other half. The circuit is now completed through the meter. When activated, current will flow through the circuit and the meter, which will indicate the amount of current flow.

Inductive ammeters are not connected into series with the circuit. This type of meter utilizes an inductive pickup that monitors the magnetic field formed by current flowing through a conductor. These meters normally have three leads: a positive, negative, and an inductive lead. The positive and negative leads are connected to their appropriate posts of the battery. The inductive lead is clamped around a conductor in the circuit. When the circuit is activated, current flow will be read on the meter. Meters used to measure current flow in the starting and charging systems are normally the inductive type.

The positive lead is colored red and the negative lead black.

Voltmeters

Electrical circuits and components cannot operate properly if the proper amount of voltage is not available. Not only must the battery be able to deliver the proper amount of voltage but the proper amount of voltage must be available to operate the intended component. As current flows through a resistance, voltage is lost or dropped. In circuits where undesired resistance is present, the voltage available for the circuit is decreased. Because of these conditions, voltage is measured at the source, at the electrical loads, and across the loads and circuits.

Battery voltage is typically called source voltage.

A voltmeter is used to measure voltage and normally has two leads: a positive and a negative lead. Connecting the positive lead to any point within a circuit and connecting the negative lead to a ground will measure the voltage at the point where the positive lead is connected. If battery voltage is being measured, the positive lead is connected to the positive terminal of the battery and the negative lead to the negative post or to a clean spot on the common ground. The voltmeter will read the potential difference between the two points and display that difference on its scale. In the case of a 12-volt battery, the potential at the positive post is 12 volts and the potential at the negative post is zero; therefore, the meter will read 12 volts.

The positive lead can be connected anywhere within the circuit and it will measure the voltage present at that point. To measure the amount of available voltage, the negative lead can be connected to any point in the common ground circuit. To measure the amount of voltage drop across any wire and component of the circuit, the positive lead is connected to the battery side of the component and the negative lead is placed directly to the other side of the component. To measure the voltage drop of a light bulb in a simple circuit, connect the positive lead to the battery or power lead at the bulb and the other lead to the ground side of the bulb. The meter will read the amount of voltage drop across the bulb. If no other resistances are present in the circuit, the amount of voltage drop will equal the amount of source voltage. However, if there is a resistance present in the ground connection, the bulb will have a voltage drop that is less than battery voltage. If the voltage drop across the ground is measured, it will equal the source voltage minus the amount of voltage dropped by the light. In order to measure voltage drop and available voltage at various points within a circuit, the circuit must be activated (Figure 7-22).

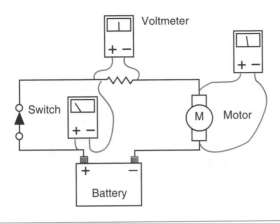

Figure 7-22 Using a voltmeter to measure voltage drops

Ohmmeters

Ohmmeters measure resistance and should never be connected into an activated circuit. The meter uses its own power to determine the amount of resistance present between two points. If the circuit has its own power source and an ohmmeter is connected to it, the meter will easily burn out from the excessive power. An ohmmeter is frequently used to test for circuit completeness or continuity. If a circuit is complete, the meter will display a low resistance. If a circuit is open, a large amount of resistance will be displayed.

Most ohmmeters have a variety of scales to measure within. A technician chooses between scales of 1, 100, or 10,000 ohms. The anticipated resistance determines the scale to be selected. After the scale has been selected, the meter must be zeroed in that scale. To do this, the leads of the meter are held together and the needle or number display is adjusted until it reads zero. An ohmmeter is usually equipped with an adjustment control to zero the meter.

To measure the resistance of a circuit or component, disconnect the power from the circuit or remove the component from the circuit. With the meter zeroed in the appropriate scale, connect one of the meter's leads to the power side of the component and the other lead to the ground side. The meter will display the amount of resistance present between the two points. If the resistance is greater than the selected scale, the meter will show a reading of "infinity." When this reading results, the technician should move up a scale on the meter. Whenever the scales have been changed, the meter must be zeroed for that range prior to measuring the resistance. If subsequent scale changes result in continued readings of infinity, it can be assumed that there is no continuity between the two measured points.

An infinite reading simply means that the resistance is greater than what can be measured in that scale on the ohmmeter.

Ohmmeters can also be used to compare the resistance of a component to the value it should have. Many electrical components have a specified resistance value, which is listed in the service manuals. This resistance value is important because it controls the amount of voltage dropped and the amount of current that will flow in the circuit. If a component does not have the proper amount of resistance, the circuit will not operate properly.

Prior to testing a component or circuit with an ohmmeter, the service manual should be checked for precautions regarding the impedance of the meter. Ohmmeters supply their own power to measure resistance and often electronic components can be damaged by causing excessive current to flow in them. Using the proper impedance ohmmeter will avoid damage to the components. If an impedance is not given in the manual, it can be safely assumed that any ohmmeter is appropriate for that circuit.

Test Lights

A test light is often used to verify the presence of voltage at a particular point. Although the test light cannot give a reading of the amount of voltage present, it can clearly display the presence of

Figure 7-23 A typical analog multimeter

voltage. A test light normally has one wire lead. This is connected to ground. The positive lead is actually a piercing probe which is slipped into the circuit at the desired point to verify voltage. If voltage is present, the bulb in the test light will shine.

A self-powered test light is often used to test for continuity instead of an ohmmeter. A self-powered test light does not rely on the power of the circuit to light its bulb; rather, it contains a battery for a power supply and when connected across a completed circuit the bulb will light. This type of test light is connected to the circuit in the same manner as an ohmmeter.

Multimeters

A multimeter is often referred to as a VOM (volt-ohm-milliammeter).

DVOM is a common abbreviation for digital volt-ohm-milliammeter.

Impedance is best defined as the operating resistance of an electrical device.

Classroom Manual
Chapter 7, page 237

Ohmmeters, ammeters, and voltmeters can be separate meters or they can be combined in one meter called a multimeter. Most multimeters can measure dc volts, dc current, ac volts, and ohms. Multimeters have either needle type, analog (Figure 7-23), or numeric, digital, displays. To use this type of meter, the desired function and scale are selected prior to connecting the meter. Once the meter is set in a function, the VOM should be used as an individual function-type meter.

A DVOM with high impedance is normally required for testing electronic components and circuits. The high impedance prevents a surge of high current through the sensitive semiconductors when the tester is connected into a circuit. High current can destroy electronic circuits and components.

Testing Components

Meters can be used to test electrical components such as switches, loads, and semiconductors. Switches can be tested for operation and for excessive resistance with a voltmeter, test light, or ohmmeter. To check the operation of a switch with a voltmeter or a test light, connect the meter's positive lead to the battery side of the switch. With the negative lead attached to a good ground, voltage should be measured at this point. Without closing the switch, move the positive lead to the other side of the switch. If the switch is open, no voltage will be present at that point. The amount of voltage present at this side of the switch should equal the amount on the other side when the switch is closed. If the voltage decreases, the switch is causing a voltage drop due to excessive resistance. If no voltage is present on the ground side of the switch with it closed, the switch is not functioning properly and should be replaced.

If a switch has been removed from the circuit, it can be tested with an ohmmeter or a self-powered test light. By connecting the leads across the switch connections, the action of the switch should open and close the circuit.

Ohmmeters are used to test semiconductors. Because semiconductors allow current flow only in certain directions, they can be tested by connecting an ohmmeter to its connections. By placing the leads of the meter on the semiconductor connections, continuity should be observed.

Reversing the meter leads to the same connections should result in different readings of continuity. For example, a diode should show good continuity when the leads are connected to it and poor continuity when the leads are reversed.

Troubleshooting electrical problems involves the use of meters, test lights, jumper wires, circuit breakers, and/or short detection devices. A jumper wire is simply a length of wire normally equipped with alligator clips on either end. These clips allow the wire to be easily and safely connected to various points within a circuit. Jumper wires are primarily used to bypass a particular point within a circuit.

A circuit breaker is a protection device that resets itself after it has been tripped by high current. Circuit breakers open due to the heat of high current. While open, the breaker cools and closes again to complete the circuit. When a fuse is blown or burned out by high current, it must be replaced to reactivate the circuit. To diagnose a circuit with a short, a completed circuit makes locating the fault easier. By inserting a circuit breaker in place of the fuse, the circuit can be activated and still protected.

A number of techniques and test instruments are used to locate a short, the most common of which is called a short detector. This device utilizes the magnetic field formed by current flow to indicate the location of a short. The cycling of the circuit breaker will be indicated by the sweeps of the detector's needle as it is moved along the length of the circuit. When the needle no longer sweeps in response to the breaker, the location of the short can be assumed to be before that point in the circuit. Carefully moving the detector back through the circuit should locate the exact place of the short.

To determine the location of an open, a test light or voltmeter is used. By probing along the circuit, the completed path can be traced. The point where voltage is no longer present is the point where the open is located. If the circuit is open, there will be no current flow. This means that if the open is after the load, source voltage will be measured after the load because there is no voltage drop in the circuit. Often a jumper wire is used to verify or isolate the problem area.

Lockup Converter Control Diagnostics

The method used to diagnose computer-controlled TCC systems varies with each manufacturer (Figure 7-24). Although many of the procedural steps are quite different, all follow basically the same scheme. To illustrate a typical TCC diagnostic procedure, a detailed look at diagnosing a TCC system fitted to a GM transmission and a GM computerized engine control system follows. Photo Sequence 19 shows this procedure using the dash lights and a hand-held scanner.

General Motors TCC Diagnostic Routine

GM PFI and TBI engines use an ECM to perform their self-diagnostic feature. The ECM will detect failures in many of its systems or circuits and the failed system can be identified through displayed trouble codes.

To retrieve trouble codes, turn the ignition on with the engine off. Verify that the "Check Engine" or "Service Engine Soon" lamp is lit. This lamp should go off when the ignition switch is off. If the lamp fails to light, check that circuit before proceeding.

Then ground the diagnostic test terminal by connecting a jumper wire from terminal A to terminal B at the ALDL connector located below the instrument panel on most GM vehicles. The CEL/SES light should flash a code 12. The code 12 is indicated by a flash, a pause, and two flashes. A code 12 indicates that the system is now in its self-diagnostic mode. Code 12 will flash three times and will be followed by any trouble codes stored in the memory of the ECM. Each of these trouble codes will flash three times, starting with the lowest numbered code. When all of the trouble codes have been displayed, a code 12 will be repeated. If no codes are present, the system will continue to flash a code 12 until the jumper wire at the ALDL is removed or the engine is started.

Because electronic torque converter controls are integrated with other electronic transmission controls on late-model vehicles, TCC trouble code retrieval for specific models of transmissions are included in the next chapter.

Special Tools

Jumper wire
Paper clip
Diagnostic key
Scan tool

PFI is an abbreviation used for port fuel injection.

TBI is an abbreviation used for throttle body injection.

The "Check Engine" lamp will be referred to as the CEL and the "Service Engine Soon" lamp will be referred to as the SES.

The ALDL is the assembly line data link.

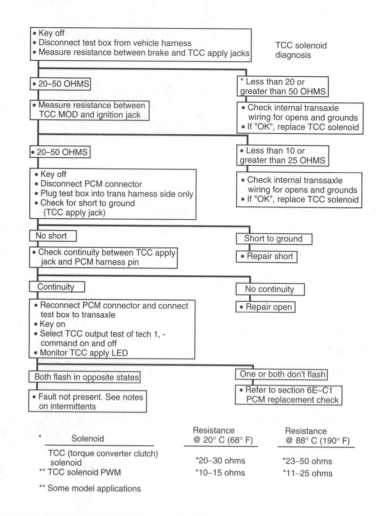

```
┌─────────────────────────────────────────┐
│ • Key off                               │     TCC solenoid
│ • Disconnect test box from vehicle harness│     diagnosis
│ • Measure resistance between brake and TCC apply jacks│
└─────────────────────────────────────────┘

┌──────────────────┐              ┌──────────────────────────┐
│ • 20–50 OHMS     │              │ * Less than 20 or        │
└──────────────────┘              │   greater than 50 OHMS   │
┌──────────────────────────┐      └──────────────────────────┘
│ • Measure resistance between│    ┌──────────────────────────┐
│   TCC MOD and ignition jack │    │ • Check internal transaxle│
└──────────────────────────┘      │   wiring for opens and grounds│
                                  │ • If "OK", replace TCC solenoid│
                                  └──────────────────────────┘

┌──────────────────┐              ┌──────────────────────────┐
│ • 20–50 OHMS     │              │ • Less than 10 or        │
└──────────────────┘              │   greater than 25 OHMS   │
┌──────────────────────────┐      └──────────────────────────┘
│ • Key off                │      ┌──────────────────────────┐
│ • Disconnect PCM connector│     │ • Check internal transsaxle│
│ • Plug test box into trans│     │   wiring for opens and grounds│
│   harness side only      │      │ • If "OK", replace TCC solenoid│
│ • Check for short to ground│    └──────────────────────────┘
│   (TCC apply jack)       │
└──────────────────────────┘

┌──────────────────┐              ┌──────────────────┐
│ No short         │              │ Short to ground  │
└──────────────────┘              └──────────────────┘
┌──────────────────────────┐      ┌──────────────────┐
│ • Check continuity between TCC apply│ • Repair short │
│   jack and PCM harness pin │      └──────────────────┘
└──────────────────────────┘

┌──────────────────┐              ┌──────────────────┐
│ Continuity       │              │ No continuity    │
└──────────────────┘              └──────────────────┘
┌──────────────────────────┐      ┌──────────────────┐
│ • Reconnect PCM connector and connect│ • Repair open  │
│   test box to transaxle  │      └──────────────────┘
│ • Key on                 │
│ • Select TCC output test of tech 1, -│
│   command on and off     │
│ • Monitor TCC apply LED  │
└──────────────────────────┘

┌──────────────────────────┐      ┌──────────────────────────┐
│ Both flash in opposite states│  │ One or both don't flash  │
└──────────────────────────┘      └──────────────────────────┘
┌──────────────────────────┐      ┌──────────────────────────┐
│ • Fault not present. See notes│ │ • Refer to section 6E–C1 │
│   on intermittents       │      │   PCM replacement check  │
└──────────────────────────┘      └──────────────────────────┘
```

* Solenoid	Resistance @ 20° C (68° F)	Resistance @ 88° C (190° F)
TCC (torque converter clutch) solenoid	*20–30 ohms	*23–50 ohms
** TCC solenoid PWM	*10–15 ohms	*11–25 ohms

** Some model applications

Figure 7-24 A typical TCC solenoid diagnostic chart (Courtesy of the Chevrolet Motor Division of General Motors Corp.)

Photo Sequence 19
Typical Procedure for Diagnosing a Computer-controlled TCC System

P19-1 Turn the ignition on with the engine off. Verify that the "Check Engine" or "Service Engine Soon" lamp is lit.

P19-2 Locate the ALDL.

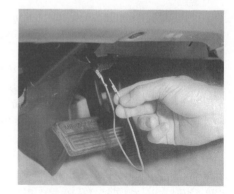

P19-3 Ground the diagnostic test terminal by connecting a jumper wire from terminal A to terminal B at the ALDL connector.

P19-4 The CEL/SES light should flash a code 12. The code 12 is indicated by a flash, a pause, and two flashes.

P19-5 Connect a scan tool to the ALDL.

P19-6 Connect the scan tool power lead to the cigar lighter receptacle.

P19-7 The scan tool must be programmed with the correct identification of the vehicle.

P19-8 The scan tool will display the trouble codes numerically.

P19-9 A scan tool will not only display the trouble codes held in the memory of the ECM, but it can also be used to display the data or signals that the ECM is receiving from inputs or is sending out to outputs.

P19-10 Compare the trouble codes that are retrieved to the appropriate trouble code chart.

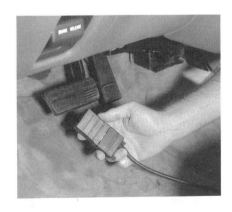

P19-11 After the codes have been retrieved, remove the jumper wire or scan tool.

P19-12 Then remove the ECM fuse to erase the codes from the memory.

A scan tool can also be used to retrieve codes and gather other information. The scan tool is connected to the ALDL and after it is properly set up, it will display the codes numerically. A scan tool normally requires only two connections to the vehicle, one at the ALDL and the other to the cigar lighter where it picks up battery voltage. A scan tool not only will display the trouble codes held in the memory of the ECM, but it can also be used to display the data or signals that the ECM is receiving from inputs or is sending out to outputs. Most scan tools are capable of displaying engine and vehicle speed, intake air and coolant temperature, and oxygen and throttle position sensor voltages. In all of these cases, changes in voltage will relate to a particular condition or set of conditions.

The displayed codes should be recorded. By referring to the appropriate trouble code chart, you will be able to identify the circuit or system that has the fault.

After the codes are retrieved, erase the codes from the memory of the ECM. To do this, locate and remove the ECM fuse. Keep the fuse out for at least ten seconds. This simple act stops power to the memory of the computer and erases the codes.

The codes are erased before diagnosis continues because you need to know if the new codes recorded during your road test are soft or hard codes. A hard code is a code that reappears after you clear the codes and perform a retest of the system. A soft code usually does not reappear after you clear the codes and perform a routine retest. Trouble code charts are designed to repair hard codes, not soft codes.

If no hard codes are present after you have started the engine, use the field service mode or a scan tool to verify that the ECM has control over the fuel delivery and other systems. With the engine running, connect the jumper wire across terminals A and B. The CEL/SES should respond to inputs from the oxygen sensor. If the lamp flashes two times per second, the system is in open loop. Open loop indicates that the voltage from the oxygen sensor is not being recognized by the ECM. The CEL/SES lamp will display open loop for 30 seconds to 2 minutes after the engine has been started or until the oxygen sensor reaches normal operating temperature. If the system never goes into closed loop, you should check the oxygen sensor circuit.

Closed loop is indicated by the CEL/SES lamp flashing one time per second. Closed loop confirms that the oxygen sensor signal is recognized by the ECM to control fuel delivery and that the system is working normally.

When you have an intermittent problem, use any soft codes from previous self-diagnostics as a troubleshooting aid. Soft codes may allow you to isolate the faulty circuit. The customer's explanation of the problem is also of great help when diagnosing these problems. Finding out when and where the problem occurs can lead you to properly identifying the problem area. This information will also give you the operating conditions in which the problem exists, allowing you to duplicate the conditions to verify the complaint.

The ECM controls the TCC solenoid, which in turn controls the flow of ATF to the torque converter clutch. The clutch mechanically connects the engine to the transmission. This action increases fuel economy and reduces heat in the transmission. The TCC solenoid receives power from the ignition switch through the brake light switch. The ECM provides the ground for the solenoid. The ECM completes the TCC solenoid circuit based on inputs from the CTS, VSS, MAF, MAP, and TPS sensors. The presence of trouble codes 14, 15, 21, 24, 32, or 34 can be a clue as to the cause of faulty TCC operation.

The ALDL contains a test terminal for the TCC circuit. This lead, terminal F, is to be used during pinpoint testing of the TCC circuit. By activating the TCC and watching the engine's speed, you can quickly determine if the TCC is actually being fully engaged. Firm and total engagement of the clutch should cause the engine's rpm to drop approximately 300 rpm. If the engine speed doesn't change, you know that the TCC has not been applied. If the change in engine speed is slight, the clutch may not be fully applied. Some transmissions have sensors that inform the ECM what gear the transmission is operating in. They allow the ECM to apply the TCC in second, third, and fourth gears, depending on engine load and system design.

If the clutch does not engage when the ALDL converter terminal is jumped to ground, check the clutch control solenoid and its circuit. Check for battery power to the solenoid. To identify the power feed wire, look at the appropriate wiring diagram in the service manual. If power is being supplied to the solenoid, ground the other end of the solenoid. This should activate the solenoid. If the solenoid does activate, check the ground circuit. If the solenoid does not respond, check the solenoid. Most solenoids have a resistance value specification. Measure the resistance across the solenoid and compare the readings to specifications.

All faulty wires and connectors should be repaired or replaced. TCC solenoids are not rebuildable and are replaced when they are faulty.

CASE STUDY

A customer brought her early-model Ford Escort into the shop with a complaint of the transmission not shifting into the higher gears. The car had over 100,000 miles on it. The technician assumed that after this many miles the transaxle needed to be rebuilt and assured the customer that this would take care of the problem. This was not an easy decision for the customer as the car was worth only slightly more than the cost of the repair. She decided, however, to go ahead with the repair since she really liked the car.

While the technician was rebuilding the transaxle, he found nothing major wrong with it, just normal wear, especially considering the mileage on the car. When things wear, oil pressure goes down. He assumes this was the cause of the no-shift problem. After rebuilding the transaxle and installing it into the car, he road tested it. It still wouldn't shift.

He was baffled and frustrated. He knew he had the transmission together right and he had used good parts. He called a friend, who also did transmission work, for advice. The first question his friend asked him was "Did you check the converter?" He had not!

What his friend had told him was that early style ATXs used the splitter gear-type torque converter and if the planetary gear inside the converter was damaged, there would be no high gears. With this information, he knew what the problem was and proceeded to repair the transaxle.

Terms to Know

ALDL	Impedance	Series circuit
Ammeter	Insulator	Shudder
Amperes	Magnetism	Stall speed
Ballooning	Ohmmeter	Stick-slip
Conductor	Ohm's Law	Voltage
Current	Parallel circuit	Voltage drop
DVOM	Power transistor	VOM
Electricity	Resistance	

ASE Style Review Questions

1. *Technician A* says a ballooned torque converter can cause damage to the oil pump.
   Technician B says a ballooned torque converter is caused by excessive pressure in the torque converter.
   Who is correct?
   - **A.** A only
   - **B.** B only
   - **C.** Both A and B
   - **D.** Neither A nor B

2. While checking torque converter endplay:
   Technician A says the torque converter must be installed in the transmission.
   Technician B says the torque converter endplay is corrected by installing a thrust washer between the oil pump and the direct or front clutch.
   Who is correct?
   - **A.** A only
   - **B.** B only
   - **C.** Both A and B
   - **D.** Neither A nor B

3. *Technician A* says an ammeter is always connected in series with the circuit it is testing.
   Technician B says an ammeter is only used when the power for the circuit is disconnected.
   Who is correct?
   - **A.** A only
   - **B.** B only
   - **C.** Both A and B
   - **D.** Neither A nor B

4. *Technician A* says an ohmmeter can be used to test semiconductors.
   Technician B says low-impedance ohmmeters should be used on electronic circuits.
   Who is correct?
   - **A.** A only
   - **B.** B only
   - **C.** Both A and B
   - **D.** Neither A nor B

5. While checking a switch:
   Technician A uses a voltmeter at the output of the switch in the various switch positions.
   Technician B uses a test light at the output of the switch in the various switch positions.
   Who is correct?
   - **A.** A only
   - **B.** B only
   - **C.** Both A and B
   - **D.** Neither A nor B

6. *Technician A* says if the stall speed of a torque converter is below specifications, a restricted exhaust or slipping stator clutch is indicated.
   Technician B says if the stall speed is above specifications, the bands or clutches in the transmission may not be holding properly.
   Who is correct?
   - **A.** A only
   - **B.** B only
   - **C.** Both A and B
   - **D.** Neither A nor B

7. *Technician A* says torque converter clutch control problems are always caused by electrical malfunctions.
   Technician B says nearly all converter clutches are engaged through the application of hydraulic pressure on the clutch.
   Who is correct?
   - **A.** A only
   - **B.** B only
   - **C.** Both A and B
   - **D.** Neither A nor B

8. While checking the stall speed of a torque converter:
   Technician A maintains full-throttle power until the engine stalls.
   Technician B manually engages the TCC before running the test.
   Who is correct?
   - **A.** A only
   - **B.** B only
   - **C.** Both A and B
   - **D.** Neither A nor B

9. While checking voltage drops:
   Technician A connects the voltmeter across the load with the circuit activated.
   Technician B connects the negative meter lead to ground and the positive lead to the component being tested.
   Who is correct?
   - **A.** A only
   - **B.** B only
   - **C.** Both A and B
   - **D.** Neither A nor B

10. *Technician A* says a seized one-way stator clutch will cause the vehicle to have good low-speed operation but poor high-speed performance.
    Technician B says a free-wheeling or nonlocking one-way stator clutch will cause the vehicle to have poor acceleration
    Who is correct?
    - **A.** A only
    - **B.** B only
    - **C.** Both A and B
    - **D.** Neither A nor B

Table 5-1 ASE TASK

Diagnose lockup converter system faults and determine needed repairs.

Problem Area	Symptoms	Possible Causes	Classroom Manual	Shop Manual
LOCKUP CONVERTER CLUTCH NOT WORKING PROPERLY	Clutch does not apply	1. Electrical system has fault	237	256
		2. Clutch check ball missing, damaged, or not seating	242	252
		3. Input shaft seals damaged	235	86
		4. Defective valve body	238	105
		5. Clutch bypass valve stuck, nicked, or damaged	235	252
		6. Clutch shift, apply, or throttle valve stuck	238	105
		7. Defective oil pump	234	246
		8. Malfunctioning electronic controls	237	258
	Clutch does not release	1. Malfunctioning electronic controls	237	258
		2. Clutch bypass valve stuck, nicked, or damaged	235	252
		3. Internal leaks	225	246
		4. Damaged torque converter	223	245
		5. Clutch apply valve stuck	235	252
	Rough application of clutch	1. Damaged clutch pressure plate	236	250
		2. Missing check ball	242	252
		3. Converter clutch apply valve feed clogged	235	252
		4. Damaged input shaft seals	235	86
		5. Poor engine performance	234	240
		6. Sticking valves or dirty valve body	238	105
		7. Low oil pressure	234	245

Servicing Electronic Transmission Controls

Upon completion and review of this chapter, you should be able to:

❏ Diagnose electronic control systems and determine needed repairs.

❏ Conduct a road test to determine if the fault is electrical or hydraulic.

❏ Inspect, test, and replace electrical/electronic switches.

❏ Inspect, test, and replace electrical/electronic sensors.

❏ Inspect, test, bypass, and replace electrical/electronic solenoids.

❏ Inspect, test, adjust and/or replace electrical/electronic relays, connectors, and harnesses.

❏ Retrieve trouble codes from common electronically controlled automatic transmissions.

Basic Tools

Basic mechanic's tool set

DVOM

Jumper wires

Appropriate service manual

The use of electronics in transmission control systems has blossomed in the last few years. A transmission technician now must not only be able to diagnose and repair the mechanical and hydraulic components of automatic transmissions, but must also be able to work with advanced electronic systems.

EAT is a commonly used acronym for electronic automatic transmission.

Classroom Manual
Chapter 8, page 259

General Diagnosis

Some electronic transmissions are only partially controlled; that is, only the engagement of the converter clutch and third-to-fourth shifting is electronically controlled. Other models feature electronic shifting into all gears plus electronic control of the TCC.

The controls of an EAT direct hydraulic flow through the use of solenoid valves. When it is used to control TCC operation, the solenoid opens a hydraulic circuit to the TCC spool valve, causing the spool valve to move and direct mainline pressure to apply the clutch. Electronically controlled shifting is accomplished in much the same way. Shifting occurs when a solenoid is either turned on or turned off (Figure 8-1). At least two shift solenoids are incorporated into the system

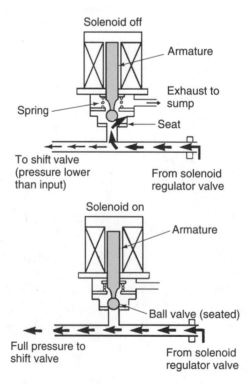

Figure 8-1 Shift solenoid action (Reprinted with the permission of Ford Motor Co.)

and shifting takes place by controlling the solenoids. The desired gear is put into operation through a combination of on and off solenoids.

Several sensors and switches are used to inform the control computer of the current operating conditions. Most of these sensors are also used to calibrate engine performance. The computer then determines the appropriate shift time for maximum efficiency and best feel. The shift solenoids are controlled by the computer, which either supplies power to the solenoids or supplies a ground circuit. The techniques for diagnosing electronic transmissions are basically the same techniques used to diagnose TCC systems.

Although EATs are rather reliable, they have introduced new problems for the automatic transmission technician. Some of the common problems that affect the shift timing and quality, as well as the timing and quality of TCC engagement are incorrect battery voltage, a blown fuse, poor connections, a defective TPS or VSS, defective solenoids, crossed wires to the solenoids or sensors, corrosion at an electrical terminal, or faulty installation of some accessory, such as a cellular telephone.

Improper shift points can be caused by electrical circuit problems, faulty electrical components, or bad connectors, as well as a defective governor or governor drive gear assembly. Some EATs do not rely on the hydraulic signals from a governor, rather they rely on the electrical signals from electrical sensors to determine shift timing.

Computer-controlled transmissions often start off in the wrong gear. This can happen for several reasons, either internal transmission problems or external control system problems. Internal transmission problems can be faulty solenoids or stuck valves. External problems can be the result of a complete loss of power or ground to the control circuit or a fail-safe protection strategy initiated by the computer to protect itself or the transmission from an observed problem. Typically, the default gear is simply the gear that is applied when the shift solenoids are off.

A visual inspection of the transmission and the electrical system should include a careful check of all electrical wires and connectors for damage, looseness, and corrosion (Figure 8-2). Loose connections, even when clean, usually only make intermittent contact. They will also cor-

Shop Manual
Chapter 3, page 40

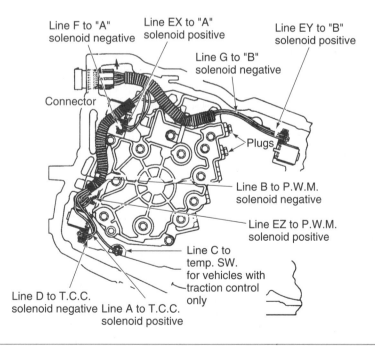

Figure 8-2 The position and condition of the transmission wiring harness and its connectors should be carefully checked. (Courtesy of the Chevrolet Motor Division of General Motors Corp.)

rode and collect foreign material, which can prevent contact altogether. Use an ohmmeter to check the continuity through a connector suspected of being faulty. Check all ground cables and connections. Corroded battery terminals and/or broken or loose ground straps to the frame or engine block will cause problems. This part of your inspection is especially important for electronically controlled transmissions and for transmissions that have a lockup torque converter. All faulty electrical components found during the inspection should be repaired or replaced.

Continue your basic inspection with checking the fuse or fuses to the control module. To accurately check a fuse, either test it for continuity with an ohmmeter or check each side of the fuse for power when the circuit is activated.

⚠️ **WARNING:** Remember to disconnect power to the component before checking it with an ohmmeter. Failure to do so can destroy the meter.

Make sure the battery's voltage and the alternator's output voltage is at least 12.6 volts. If a problem is found here, correct it before continuing to diagnose the system. With the system on, measure the voltage dropped across the connectors and circuits. If there is an excessive amount of voltage drop, the circuit cannot function properly because of the higher-than-normal resistance in the circuit.

⬤ **CUSTOMER CARE:** Vehicles equipped with engine or transmission computers may require a relearn procedure after the battery is disconnected. Many vehicle computers memorize and store vehicle operation patterns for optimum driveability and performance. When the vehicle battery is disconnected, this memory is lost. The computer will use default data until new data from each key start is stored. Customers often complain of driveability problems during the relearn stage because the vehicle acts differently than before it was serviced. To reduce the possibility of complaints, after any service that requires that the battery be disconnected, the vehicle should be road tested and the correct relearn procedure followed.

Road Test

Shop Manual
Chapter 3, page 43

───────────────

Manifold vacuum is often referred to as a negative pressure because it is lower than atmospheric pressure.

Critical to proper diagnosis of EAT and TCC control systems is a road test. The road test should be conducted in the same way as one for a nonelectronic transmission except that a scanner tool should also be connected to the circuit to monitor engine and transmission operation.

During the road test, the vehicle should be driven in the normal manner. All pressure and gear changes should be noted. The various computer inputs should also be monitored and the readings recorded for future reference. Some scan tools are capable of printing out a report of the test drive. Critical information from the inputs includes engine speed, vehicle speed, manifold vacuum, operating gear, and the time it took to shift gears (Figure 8-3). If the scanner does not have the ability to give a summary of the road test, you should record this same information after each gear or operating condition change.

✓ **SERVICE TIP:** If the transmission will not upshift after you have serviced a Hyundai with a Mitsubishi KM-series transmission, check the connectors. It is very easy to install them backwards by installing the output connector to the input sensor harness connector and vice versa. The connectors are identical. Only the wire colors are different and they should be used to verify a correct connection.

Basic Electrical Circuit

───────────────

ECA is the abbreviation for electronic control assembly.

To summarize how the circuitry of an electronically controlled transmission interrelates to a computerized engine control system, a summary of the transmission controls for a Ford E4OD (Figure 8-4) and AXOD-E follows. This system is similar to the other systems used by Ford, as well as the other manufacturers. The transmission is controlled by the ECA of the EEC-IV system. The computer

```
                TECH 1 DATA LIST

 1. INPUT SPEED                    Ø RPM
 2. OUTPUT SPEED                   Ø RPM
 3. ENGINE SPEED                   Ø RPM
 4. INPUT SPEED                      RPM
 5.   1  /  2  /  3  RNG      ON    ON    ON
 6. PRNDL SWITCH              DRIVE 2
 7. CURRENT GEAR                   2
 8.   SOL A   SOL B           OFF        OFF
 9. DES.  FORCE MOTOR         0.00 AMPS
10. ACT.  FORCE MOTOR         0.00 AMPS
11. BRAKE SWITCH              OFF
12. SYSTEM VOLTAGE            12.2 V
13. COOLANT TEMP                 28°C     82°F
14. TRANS TEMP                   25°C     77°F
15. THROTTLE ANGLE            Ø %
16. THROT POSITION            0.66 V
17. TCC DUTY CYCLE            Ø %
18. TCC SLIP                  Ø RPM
19. 1–2 SHIFT TIME            0.00  SEC
20. 2–3 SHIFT TIME            0.00  SEC
21. TRANS GEAR RATIO          0.00
22. TURBINE SPEED             Ø RPM
```

Figure 8-3 A typical printout from a Tech 1 scan tool (Courtesy of the Hydra-Matic Division of General Motors Corp.)

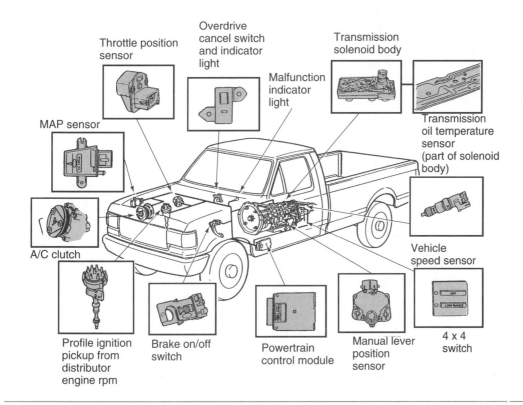

Figure 8-4 Typical location of input sensors in an EAT system (Reprinted with the permission of Ford Motor Co.)

273

As a comparison of cold-start shifting and schedule shifting, if the engine is cold, the transmission will shift later and converter clutch engagement will also be later. When the engine warms up, the transmission will shift more quickly and have an earlier converter clutch engagement.

The electronic controls used with Ford's E4OD are similar to the AXOD-E, the exception being that the E4OD doesn't use a turbine speed sensor; therefore, the E4OD doesn't have the same shift feel as the AXOD-E.

PIP is the acronym used for the profile ignition pickup, which sends an engine speed signal to the ECA.

The static electricity that can be generated by a vinyl seat is 20,000 volts. It only takes 35 volts to destroy any electronic automotive device.

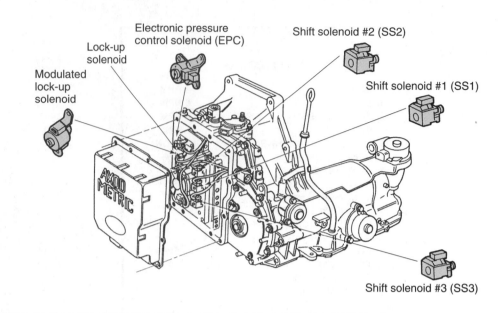

Figure 8-5 Location of output solenoids on an AXOD-E transaxle (Reprinted with the permission of Ford Motor Co.)

receives inputs from both the engine and the transmission. Based on this information, the ECA can send signals to operate components in both the engine and transmission. There are many inputs to the ECA from the transmission. The manual lever position sensor informs the ECA what gear has been selected. The vehicle speed sensor, located in the same place as the governor would be, inputs the mph at which the vehicle is traveling. The turbine speed sensor sends signals to inform the ECA of the transaxle's speed. The TSS is not used on all models. The transmission oil temperature sensor monitors the temperature of the ATF. This signal determines whether the ECA should go to a cold-start shift cycle or shift according to its schedule.

There are four nontransmission inputs: the TPS, MAF, PIP, and "brake on" switch. The system uses five solenoids, which are located in the transmission: the electronic pressure control solenoid, which regulates transmission operating pressure; the modulated lockup solenoid, which controls the converter clutch; and three shift solenoids, which control fluid flow to the various holding devices (Figure 8-5).

As you can readily tell in this system, as well as the others, engine performance problems can cause the transmission to perform unsatisfactorily. Since the computer controls the outputs to the engine and the transmission based on input sensors common to both, it is very important that you perform basic engine checks before deeply diagnosing the transmission. This is important when diagnosing nonelectronically controlled transmissions and it is even more important when diagnosing an EAT.

✓ **SERVICE TIP:** Static electricity can damage electronic components. To prevent damage from static electricity while working inside the car, make a copper bracelet for your wrist and attach a wire to it. Clip the other end of the wire to a good ground. This will relieve any static electricity generated by your movement inside the vehicle.

Because the engine and transmission control systems share information, a common component may affect both systems. Nearly all computer control systems have some form of self-diagnostics. During this mode of operation, the computer will display a trouble code that represents a circuit or component it determines is faulty. Part of the basic procedure for diagnosing an EAT is to retrieve the trouble codes from the computer. This information will determine what step you should take next (Figure 8-6).

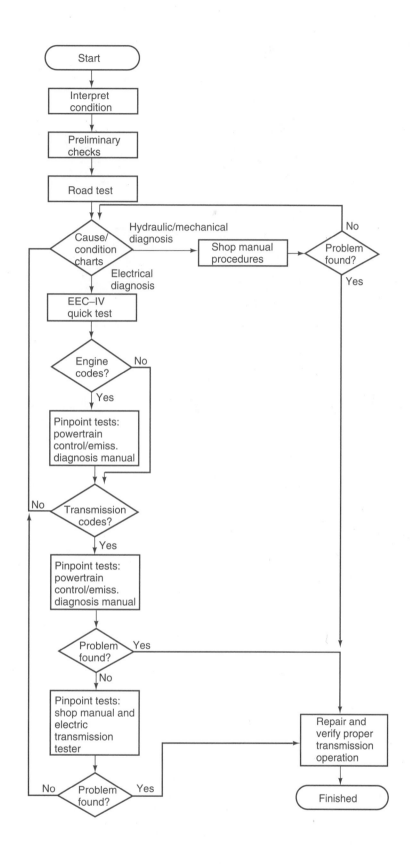

Figure 8-6 A flow chart showing the major steps that should be followed while diagnosing an EAT (Reprinted with the permission of Ford Motor Co.)

Figure 8-7 A Tech 1 scan tool designed for GM products (Courtesy of Kent-Moore Tools)

Basic Diagnostics of Specific Systems

Electronic testing on vehicles equipped with a computer can be accomplished by diagnostic procedures that can be entered into the computer's program, self-diagnostics, and/or by plugging in a manufacturer-provided or aftermarket-designed scan tool (Figure 8-7). Use the following discussions for learning purposes only. They are intended to expose you to the world of electronic controls, not to guide you totally through a diagnostic procedure on a particular model transmission. Much of the information listed under one manufacturer will also be true for others.

Always refer to the correct service manual when diagnosing electronic control systems. The manufacturers list the procedures for self-diagnostics in their service manuals. Also listed are the interpretation tables for the trouble codes retrieved from the computer.

Chrysler

Special Tools

DRB-II

Assortment of test
cartridges

Classroom Manual
Chapter 8, page 264

Chrysler has two electronic control systems. One has been used since 1988 in a variety of models. The other system is a multiplexed system that is used on current mini-vans and late-model cars. The original system is quite basic and similar to those used by other manufacturers. The control computer has self-diagnostic capabilities. The recommended way to retrieve trouble codes is through the use of a DRB-II.

The DRB-II is connected to the serial communication interface, which is located by the left front shock tower. The red power supply lead is connected to the positive post of the battery. The scan tool has a variety of cartridges, which contain the test parameters for the vehicle being tested. After the correct cartridge has been installed in the tester and the scan tool is connected to the vehicle, the tester's display will show a test pattern. After a few seconds, the display will change to read out the copyright date and revision level of the cartridge. After a few seconds, the display will ask for the vehicle's model year. The model year is selected by pressing either right or left arrow key. Once the display shows the correct year, press the down arrow key.

The display will then ask for the system that will be tested. The system is selected by pressing either left or right arrow key. Once the display indicates power train, press the down arrow key. Other typical system choices are engine or EFI/Turbo.

The display will then ask for the desired test selection. The desired test is selected by pressing either the left or right arrow key. The diagnostic test mode is used to see if there are any fault codes stored in the on-board diagnostic system memory of the computer. The circuit actuation test mode (ATM) is used to make the computer cycle a solenoid on and off. The sensor test mode is used to see the output signals of certain sensors as they are received by the computer when the engine is not running. The engine running test mode is used to see sensor output signals as they are received by the computer while the engine is running.

To enter into the diagnostic mode, set the DRB-II to the "engine off" test. Turn the ignition switch to "on-off, on-off, on" within five seconds. Then record all codes displayed on the tester. By depressing the "read/hold" key until "hold" is displayed, you can stop the display of codes. To continue the display, press the "read/hold" key.

Following are typical codes and the conditions that would cause the computer to display them:

> Code 15—Vehicle Speed Sensor: This vehicle speed sensor code will light the "check engine" lamp on California vehicles only. The code will be set if a speed sensor signal is not detected for over an 11–13 second period.
>
> Code 22—Engine Coolant Sensor: This code will cause the "check engine" lamp to light. The code will be set if the coolant sensor voltage is above 4.96 volts when the engine is cold or below 0.51 volt when the engine is warm.
>
> Code 24—Throttle Position Sensor: This code will light the "check engine" lamp. This code will be set if the signal from the TPS is below 0.16 volt or above 4.7 volts.
>
> Code 37—Transmission Lockup Solenoid: This code will not illuminate the "check engine" lamp. The code will be set if the lock-up solenoid does not turn on and off when it should.
>
> Code 55—End of Diagnostic Mode: This code will not illuminate the check engine lamp and it appears at the end of the diagnostic series.
>
> Code 88—Start of Diagnostic Mode: This code will not illuminate the "check engine" lamp. The code must appear at the start of the diagnostic series. If it does not, the fault codes will be inaccurate.

To erase the fault codes, put the system into ATM test mode 10. Then press the "read/hold" key until "hold" is displayed. The display will then flash alternating "0s." When the flashing 0s stop and the display shows "00," all fault codes have been erased.

The other transmission control system is the Chrysler collision detection system, which is a multiplex serial data bus or network. The network consists of a pair of twisted wires, which interconnect the CCD modules. Each CCD module uses this network to communicate and exchange data with other modules on the network. The number of modules varies with the model of vehicle.

Diagnosis of this transaxle system is relatively simple. Nearly all of the important information is available through the use of a scan tool (Figure 8-8) and the transaxle is equipped with many pressure taps. Both of these features give you an accurate look at the operation of the computer.

When the transaxle controller detects a problem, it will automatically go into the limp-in mode of operation. This mode defaults the transaxle to second gear only with second gear starts, reverse, and park operations. Limp-in mode is selected by the computer because it received a signal from an input device that it interpreted as an out-of-range signal. The computer will go into this mode if there is an electrical control system problem, a hydraulic apply circuit problem, an internal mechanical problem, or a sticking valve problem.

To determine the reason why the computer has defaulted to this mode, connect a scan tool to the blue six-pin CCD connector under the dash. Retrieve and record the trouble codes that are displayed. Then clear the memory and drive the vehicle until the complaint occurs again. Retrieve and write down the codes again. These codes are the most current and should be compared to the original codes.

If the scan tool is unable to display a code or is unable to recognize the computer or its circuits, the problem may be a faulty or dead bus. A dead bus is commonly caused by a dead body

The Chrysler collision detection system is referred to as the CCD system.

Classroom Manual
Chapter 8, page 265

A bus is a common connector and is used as an information source for the vehicle's various control units.

Each module is connected to the bus through separate 60-pin connectors.

Figure 8-8 A table for Chrysler 42LE transaxles, showing trouble codes, the condition that caused the code, and possible causes of the problem (Courtesy of Chrysler Corp.)

NOTE:
Code 36 is not stored alone. It is stored if a speed error (codes 50 through 58) is detected immediately after a shift. Look at the possible causes associated with the speed error code.

In the table below, the columns under "Damaged or failed clutches:" are L/R clutch, 2/4 clutch, Reverse clutch, OD clutch, and UD clutch.

Fault Code Number	Condition	Planetary gear sets broken or seized	Faulty cooling system	Torque converter clutch failure	Internal solenoid leak	Pressures too high	Valve body leakage	Regulator valve	Torque converter control valve	Torque converter clutch switch valve	Solenoid switch valve	Stuck/sticky valves	Plugged filter	Worn or damaged accumulator seal rings	Damaged clutch seals	L/R clutch	2/4 clutch	Reverse clutch	OD clutch	UD clutch	Worn pump	Worn or damaged input shaft seal rings	Worn or damaged reaction shaft support seal rings	Aerated fluid (high fluid level)	Low fluid level
21	OD clutch—pressure too low			X	X	X	X	X	X			X	X	X	X				X		X		X	X	X
22	2/4 clutch—pressure too low			X	X	X	X	X	X			X	X	X	X		X				X			X	X
23	2/4 clutch and OD clutch—pressures too low			X	X	X	X	X	X			X	X	X							X			X	X
24	L/R clutch—pressure too low			X	X	X	X	X	X			X	X	X	X	X					X			X	X
25	L/R clutch and OD clutches—pressures too low			X	X	X	X	X	X			X	X	X							X			X	X
26	L/R clutch and 2/4 clutches—pressures too low			X	X	X	X	X	X			X	X	X							X			X	X
27	OD, 2/4, and L/R clutches—pressures too low			X	X	X	X	X	X			X	X	X							X			X	X
31	OD clutch pressure switch response failure				X	X		X										X	X					X	X
32	2/4 pressure switch response failure				X			X									X		X					X	X
33	2/4 and O/D clutch pressure response failures				X			X					X						X					X	X
37	Solenoid switch valve stuck in the LO position				X			X			X	X													
38	Partial torque converter clutch out of range			X	X	X		X		X	X	X								X	X				
47	Solenoid switch valve stuck in the LR position				X			X			X	X													
50	Speed ratio default in reverse	X	X		X	X	X	X			X	X	X	X	X			X			X	X	X		X
51	Speed ratio default in 1st	X	X		X	X	X	X			X	X	X	X	X				X		X	X	X		X
52	Speed ratio default in 2nd	X	X		X	X	X	X			X		X	X			X		X		X	X	X		X
53	Speed ratio default in 3rd	X	X		X	X	X	X			X		X	X				X	X		X	X	X		X
54	Speed ratio default in 4th	X	X		X	X	X	X			X		X	X			X		X		X	X	X		X
60	Inadequate LR element volume	X					X					X	X	X											
61	Inadequate 2/4 element volume	X					X					X	X				X								
62	Inadequate OD element volume	X					X					X	X						X						

computer, transaxle controller, or any other module on the CCD bus. The exact bad module can be identified by disconnecting each module, one at a time, from the bus.

After you have one module disconnected, rerun the start-up of the scan tool. If the bus becomes active, the disconnected control module was the cause of the previous dead bus. Repeat this process by reconnecting the disconnected module and unplugging another. The module that is disconnected when the bus becomes active is the dead module.

If the cause of the dead bus was not identified, voltage checks should be made. Measure the voltage drop across the ground side of the diagnostic connector by inserting the positive voltmeter lead to the ground lead of the connector and the negative lead to the negative terminal of the battery. The maximum allowable voltage drop across the ground is 0.1 volt. The source voltage to the computer should also be checked. If battery voltage is not present at the computer, check the voltage of the battery. If it is okay, check the voltage drop in the supply circuit. If there is more than 0.1 volt dropped, identify the source of the resistance and repair the problem. Also check the bus bias voltage by connecting the voltmeter's positive lead to either bias terminal in the diagnostic connector and the negative lead to the ground terminal in the connector. The bias voltage should be between 2.1 to 2.6 volts. Any incorrect voltages could be the cause for a dead bus and the cause of the problems should be corrected. After the electrical problems are repaired, the scan tool test should be conducted.

Bias voltage is voltage applied across a diode.

> **SERVICE TIP:** On some A-604-equipped vehicles built before 1/16/90, there was a problem of surging above 45 mph. This was caused by a headlamp/dash wiring harness that generated a radio frequency. The computer misinterpreted the radio frequency and started to cycle the torque converter clutch on and off, creating the surge. The repair procedure involves some alterations to the wiring harness. Check the appropriate service manual or technical service bulletin for the proper repair.

Using the scan tool, the trouble codes are retrieved and compared to the trouble code chart given in the service manual. These codes will identify problem circuits. The service manual also gives step-by-step procedures for pinpoint diagnostics of each trouble code. Always follow these as they are efficient and exact.

Ford

Most Ford EAT systems are based on their EEC IV or EEC V systems. Although the exact make-up of the systems varies according to the transmission used and the model vehicle being tested, all follow similar procedures. The following equipment is recommended to diagnose and test an EEC system.

Classroom Manual
Chapter 8, page 270

Self-Test Automatic Read-out (STAR) Diagnostic Tester—This tester is recommended, but not required. The tester was designed for EEC systems and is used to display the service codes. There are also many aftermarket testers available for testing these systems. An analog volt/ohmmeter with a 0–20 Vdc range can be used as an alternate to a diagnostic tester.

DVOM—This multimeter must have a minimum impedance of 10 megohms.

Breakout Box—This is a jumper wire assembly that connects between the vehicle harness and the ECA. The breakout box is required to perform certain tests on the system. During individual circuit tests, the procedures will call for probing particular pin numbers. These pin numbers relate to the pins of the breakout box.

Vacuum Gauge and Vacuum Pump—This can be one assembly or separate units.

Tachometer—Must have a 0-6000 rpm range defined in 20-rpm increments.

Spark Tester—A modified side electrode spark plug with the electrode removed and an alligator clip attached may be used in place of the spark tester.

MAP/BP Tester—This tester plugs into the MAP/BP sensor circuit and the DVOM. It is used to check input and output signals from the sensor, which produces a frequency signal. This signal may also be monitored on a scope set to a millivolt scale.

Other Equipment:—Timing light, fuel injection pressure gauge, nonpowered 12-volt test light, and a jumper wire, about 15 inches long.

> **WARNING:** Some of this test equipment is required to perform diagnostics on an EEC system. NEVER attempt to test this system without the proper equipment. Damage to vehicle components will result if improper test equipment is used.

The EEC system offers a variety of test sequences. These steps are called the Quicktest and should be carefully followed in sequence to get accurate readings. Failure to do so may result in misdiagnosis or the replacement of nonfaulty components. The steps are test preparation and equipment hook-up, key on engine off (KOEO) self-test, key on engine running (KOER) self-test, and continuous self-test. Photo Sequence 20 shows the basic procedure for diagnosing EAT problems.

After all tests, servicing, or repairs have been completed, the Quicktest should be repeated to ensure that all systems are working properly. The KOEO and KOER self-tests are designed to identify faults present at the time of the testing, not intermittent faults. Intermittent faults are detected by the Continuous self-test mode.

Before hooking up any equipment to diagnose the EEC system, the following checks should be made in preparation for further testing:

1. Verify the condition of the air cleaner and air ducting.
2. Check all vacuum hoses for leaks, restrictions, and proper routing.
3. Check the electrical connectors of the EEC system for corrosion, bent or broken pins, loose wires or terminals, and proper routing.
4. Check the ECA, sensors, and actuators for physical damage.
5. Set the parking brake and place the shift lever in "P." DO NOT move the shift lever during the test unless the procedure calls for it.
6. Turn off all lights and accessories. Make sure the vehicle's doors are closed when making voltage or resistance readings.
7. Check the level of the engine's coolant.
8. Start the engine and allow it to idle until the upper radiator hose is hot and pressurized.
9. Check for leaks around the exhaust manifold, exhaust gas oxygen sensor, and vacuum hose connections.

To retrieve codes with the analog VOM, turn the ignition key off. Then connect the jumper wire from the self-test input (STI) pigtail to pin #2 at the self-test connector (Figure 8-9). Set the VOM

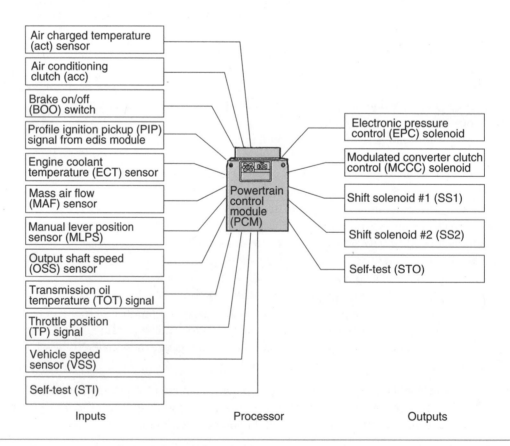

Figure 8-9 The electronic system for a Ford EAT. Note that both the inputs and outputs have a self-test sequence. (Reprinted with the permission of Ford Motor Co.)

Photo Sequence 20
Typical Procedure for Diagnosing EAT Problems

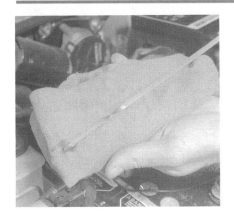

P20-1 Check fluid level and condition.

P20-2 Check all connectors.

P20-3 Road test the vehicle to verify the complaint.

P20-4 Connect the STAR tester.

P20-5 Observe the codes.

P20-6 Connect the breakout box.

P20-7 Install the template over the breakout box.

P20-8 Perform pinpoint tests according to the service manual.

P20-9 After the fault has been corrected, rerun the Quicktest.

to the correct dc voltage scale. Connect the positive lead of the VOM to the positive battery terminal. The negative lead of the meter is connected to the #4 pin of the self-test connector. Connect the timing light and proceed to the KOEO self-test. Codes are shown as voltage pulses (needle sweeps). Pay attention to the length of the pauses in order to read the codes correctly.

To retrieve codes with the STAR tester, turn off the ignition. Then connect the color-coded adapter cable leads to the diagnostic tester. Connect the adapter cable's two service connectors to the vehicle's self-test connectors. Then connect the timing light and proceed to the KOEO self-test.

> ✓ **SERVICE TIP:** The self-test connector is found in different locations, depending upon the model of the vehicle. On most FWD models, the self-test connector is located in the right rear of the engine compartment, near the firewall.

All service codes are two-digit numbers, which are generated one-digit at a time. There will be a 2-second pause between each digit in a code. There will be a 4-second pause between each code. The continuous memory codes are separated from the functional test service codes by a 6-second delay, a single 1/2-second sweep, and another 6-second delay. Always record the codes in the order received. If a diagnostic tester is used, it will count the pulses and display them as a digital code. If the "check engine" light is used, numeric service codes will be displayed at the light.

Following is the proper procedure for conducting the KOEO self-test:

1. Turn the ignition switch to the "off" position.
2. Connect the tester.
3. Turn on the ignition.
4. The ECA will run through a test cycle, opening and closing various solenoids and switches. After the completion of the test cycle, the codes will appear. Record all displayed codes.

The first set of codes that will be displayed are the KOEO self-test codes. These codes will be repeated twice. They are followed by the separator pulse signal and the continuous memory codes. These codes will also be repeated twice. A KOEO code 11 indicates the system has passed the self-test. Whenever a repair is made, repeat the Quicktest.

General Motors

Special Tools

Scan tool (Tech 1)
Jumper wire
Diagnostic key

Classroom Manual
Chapter 8, page 274

Shop Manual
Chapter 7, page 263

The 4T60-E and other GM transaxles use two electric solenoids to control transaxle upshifts and downshifts. Each solenoid is turned on and off by the PCM. The 4L80-E transmission functions in much the same way; however, it is also fitted with a force motor, which controls hydraulic line pressure, and a TCC solenoid. Some models have a transmission control module in addition to the PCM. In all systems, the PCM has self-diagnostic capabilities, which help identify which parts or circuits may need further testing.

The PCM constantly monitors all electrical circuits. If the PCM detects a problem or an out-of-range sensor input, it records a trouble code in its memory. If the problem continues for some time, the CEL or SES light will glow. It is possible for the PCM to have detected a trouble and not light the CEL/SES lamp. However, the appropriate code may be stored in its memory.

Many methods are commonly used to retrieve the trouble codes from the PCM's memory. The most simple method is to use the CEL/SES light. Other methods include the use of special scan tools.

To display the trouble codes with the CEL/SES lamp, locate the ALDL connector (Figure 8-10). Then, with the ignition on and the engine off, jump across the A and B terminals of the connector. The CEL/SES light should begin to flash codes. Each code will be repeated three times. The first series of flashes is the first digit of the code and the second series is the second digit.

Trouble codes are displayed starting with the lowest numbered code. Codes will continue until the jumper is disconnected from the ALDL.

A scan tool is a specialized tester, which, when connected into the ALDL, can be used to diagnose the control system (Figure 8-11). It provides access to circuit voltage information, as well as trouble codes. By observing the various input voltages, you can identify out-of-specification input

Terminal identification

A	Ground	F	T.C.C. (if used)
B	Diagnostic terminal	G	Fuel pump (CK)
C	A.I.R. (if used)	H	Brake sense speed input (CK)
		M	Serial data (see special tools)

Figure 8-10 An ALDL connector with the functions of critical pins noted (Courtesy of the Hydra-Matic Division of General Motors Corp.)

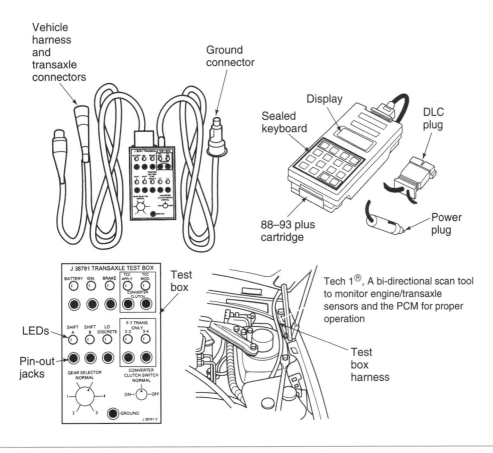

Figure 8-11 A transmission test box and Tech 1 scan tool used to diagnose GM EATs (Courtesy of the Chevrolet Motor Division of General Motors Corp.)

signals. Out-of-specification signals will not set a code unless they are out of the normal operating range. However, these signals will cause a driveability problem. To retrieve trouble codes with a scan tool, simply plug the tester lead into the ALDL. Then program the scan tool for that particular vehicle according to the scan tool's instructions. Trouble codes will appear digitally on the tester.

To erase the codes after repairs have been made, turn off the ignition. With the jumper wire still connected, remove the control module fuse for 10 seconds. If the fuse cannot be located, disconnect

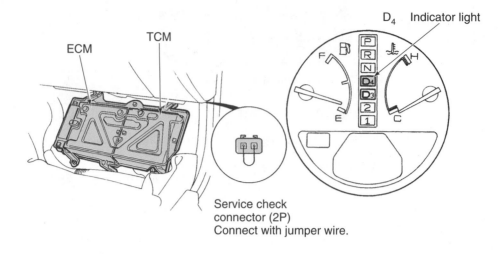

Figure 8-12 To retrieve trouble codes on a Honda, connect two terminals of the service check connector and observe the D4 indicator. (Courtesy of Honda Motor Co.)

the negative battery lead for 10 seconds. After the power has been disconnected from the PCM, the codes, and the operating instructions on some models, will be removed. For this reason it is important that you follow the relearn procedures for the transmission.

Honda

Classroom Manual
Chapter 8, page 280

Honda and Acura transaxle trouble codes are communicated by either flashing an LED on a side of the transaxle controller or by flashing the S or D4 shifter status light on the instrument panel after connecting a service connector (Figure 8-12). Always refer to the correct service manual when attempting to retrieve codes.

> **SERVICE TIP:** All Honda/Acura systems flash a shift status light on the instrument panel when a problem is first noticed by the computer. This initial flashing is not a trouble code, rather it is an indicator that a code is stored.

The connector is a two-pin, two-wire connector that is not connected to anything. Using a jumper wire, hook the two pins together, turn the ignition switch on and watch either the S or D4 light on the instrument panel. The three possible locations for the connector are under the right side of the dash, behind the right side of the center console, or behind the left front edge of the passenger's carpet up against the firewall. If a trouble code is present, the light will either give short flashes or long and short flashes. Short flashes are a ones digit of either a one- or two-digit trouble code (Figure 8-13). Long flashes are the tens digit of a two-digit code. Short flashes are

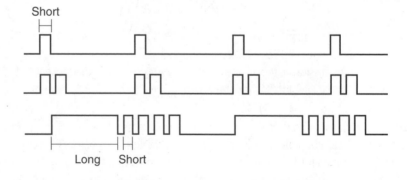

Figure 8-13 Honda trouble codes are displayed in a series of short and long flashes. (Courtesy of Honda Motor Co.)

Number of D4 indicator light blinks while Service Check Connector is jumped.	D4 indicator light	Possible Cause	Symptom	Refer to page
1	Blinks	• Disconnected lock-up control solenoid valve A connector • Short or open in lock-up control solenoid valve A wire • Faulty lock-up control solenoid valve A	• Lock-up clutch does not engage. • Lock-up clutch does not dis-engage. • Unstable idle speed.	14-38
2	Blinks	• Disconnected lock-up clontrol solenoid valve B connector • Short or open in lock-up control solenoid valve B wire • Faulty lock-up control solenoid valve B	• Lock-up clutch not engage.	14-39
3	Blinks or OFF	• Disconnected throttle position (TP) sensor connector • Short or open in TP sensor wire • Faulty TP sensor	• Lock-up clutch does not engage.	14-40
4	Blinks	• Disconnected vehicle speed sensor (VSS) connector • Short or open in VSS wire • Faulty VSS	• Lock-up clutch does not engage.	14-41
5	Blinks	• Short in A/T gear position switch wire • Faulty A/T gear position switch	• Fails to shift other than 2nd ↔ 4th gears. • Lock-up clutch does not engage.	14-42
6	OFF	• Disconnected A/T gear position switch connector • Open in A/T gear position switch wire • Faulty A/T gear position switch	• Fails to shift other than 2nd ↔ 4th gears. • Lock-up clutch does not engage. • Lock-up clutch engages and disengages alternately.	14-44
7	Blinks	• Disconnected shift control solenoid valve A connector • Short or open in shift control solenoid valve A wire • Faulty shift control solenoid vavle A	• Fails to shift (between 1st ↔ 4th, 2nd ↔ 4th or 2nd ↔ 3rd gears only). • Fails to shift (stuck in 4th gear)	14-46
8	Blinks	• Disconnected shift control solenoid valve B connector • Short or open in shift control solenoid valve B wire • Faulty shift control solenoid valve B	• Fails to shift (stuck in 1st or 4th gears).	14-47
9	Blinks	• Disconnected countershaft speed sensor connector • Short or open in the counter-shaft speed sensor wire • Faulty countershaft speed sensor	• Lock-up clutch does not engage.	14-48

Figure 8-14 A Honda trouble code diagnostic chart (Courtesy of Honda Motor Co.)

about a half-second long and long flashes are about a second and a half long. There is about a one-second pause between each code. All codes are shown in sequence, from the lowest number to the highest code (Figure 8-14). The sequence will repeat as long as the key is on and the connector is jumped across.

Some models have the diagnostic LED on the side of the transaxle controller (Figure 8-15). Read the flashes of the LED to determine the trouble code. The LED blinks all short flashes. Whenever the key is on and a problem exists, the LED will blink a series of short flashes, pauses, and then either repeat the same number of flashes (if there is only one code), or advance to the next number of blinks. Just count each series of flashes, and you've got the code. The computer for

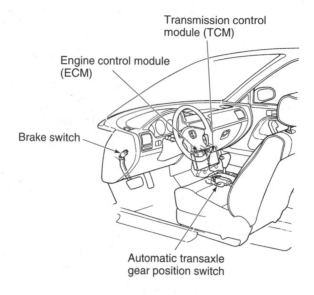

Transmission control
module (TCM)

Engine control module
(ECM)

Brake switch

Automatic transaxle
gear position switch

Figure 8-15 Location of the transmission control module and other EAT components that are in the car's interior (Courtesy of Honda Motor Co.)

Honda cars is located either under the front edge of the front passenger's carpet, behind the center console, or under the driver's seat. In Acuras, the computer is located behind the left side of the instrument panel, next to the inner fender panel, or under the right or left front seat with the LED facing the rear of the car.

The single- and double-digit codes should be compared to the appropriate trouble code chart provided in the service manual. Like other computer self-diagnostic systems, these systems do not determine the exact fault. They identify the problem area. After repairs have been made, the codes should be erased.

To erase the codes, remove the appropriate fuse for at least one minute or pull the negative cable of the battery for a minute. Because Honda and Acura systems use different fuses for different models to keep the memory alive, make sure you check the service manual to identify the proper fuse.

⬤ **CUSTOMER CARE:** Before disconnecting the battery, record the preset stations on the radio. Then reset the stations and the clock before delivering the vehicle back to the customer.

Mazda

Trouble codes in Mazda EAT systems are displayed by either the Hold light or the Check Engine light. If no problem exists, the Hold or Check Engine light will come on for three seconds after turning on the key, then it will go out. If a trouble does exist, the light will flash in a regular pattern until the trouble is no longer there. This is not a trouble code; it is an indication that a problem exists.

To retrieve the codes, locate the diagnostic request lead (Figure 8-16). This lead may be a single pin and wire blue connector, a single pin and wire green connector, or one of the pins in an integrated diagnostic connector. Connect the lead to a good ground with a jumper wire. When the diagnostic pin is grounded, the Hold or Check Engine light will display the code in long and short flashes. Long flashes, which are 1.2 seconds long, are the tens digit of a two-digit trouble code. Short flashes, which are 0.4 second long, are the ones digit of a one- or two-digit code. The lowest number codes will be shown first, then the higher number codes. There is a 4-second pause between codes. The series of codes will repeat as long as the diagnostic request pin is grounded (Figure 8-17). If no codes are present when you ground this pin, the indicator light will remain off.

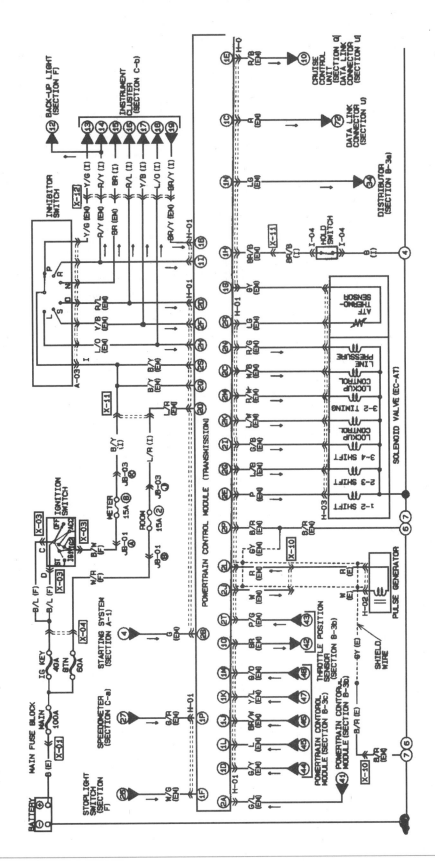

Figure 8-16 Wiring diagram for a Mazda EAT system (Courtesy of Mazda Motors)

Table A

Code No.	Buzzer Pattern	Diagnosed circuit	Condition	Point	Memorized
01		Engine rpm signal (Ne1 signal)	No input signal from distributor Ne1 signal while driving at drum speed above 600 rpm in D, S, or L ranges	• Distributor connector • Wiring from distributor to powertrain control module (transmission) (PCMT)	Yes
06		Vehicle speed sensor	No input signal from vehicle speed sensor while driving at drum speed above 600 rpm in D, S, or L ranges	• Vehicle speed sensor connector • Wiring from vhicle speed sensor to instrument cluster • Wiring from instrument cluster to powertrain control module (transmission) (PCMT) • Vehiche speed sensor resistance	Yes
12		Throttle position sensor	Open or short circuit	• Throttle position sensor connector • Wiring from throttle position sensor to powertrain control module (transmission) (PCMT) • Throttle position sensor resistance	Yes
14		Barometric absolute pressure sensor	Open or short circuit	• Barometric absolute pressure sensor connector • Wiring from atmospheric pressure sensor to powertrain control module (PCME)	Yes
55		Vehicle speed pulse generator	No input signal from vehicle speed pulse generator while driving at vehicle speed 40 km/h (25 mph) or higher in D, S or L range	• Vehicle speed pulse generator connector • Wiring from vehicle speed pulse generator to powertrain control module (transmission) (PCMT) • Vehicle speed pulse generator resistance	Yes
56		ATF thermosensor	Open or short circuit	• ATF thermosensor connector • Wiring from ATF thermosensor to powertrain control module (transmission) (PCMT) • ATF thermosensor resistance	Yes
57		Reduce torque signal 1	Open or short circuit of reduce torque signal 1 wire harness	• Wiring from powertrain control module (PCME) to powertrain control module (transmission) (PCMT)	Yes
58		Reduce torque signal 2	Open or short circuit of reduce torque signal 2 wire harness	• Wiring from powertrain control module (PCME) to powertrain control module (transminnion) (PCMT)	Yes
59		Torque reduced signal/engine coolant temperature signal	Open or short circuit of torque reduced signal/engine coolant temperature signal wire harness	• Wiring from powertrain control module (PCME) to powertrain control module (transmission) (PCMT)	Yes
60		Solenoid valve (1-2 shift)	Open or short circuit of solenoid valve and/or wiring	• Solenoid valve connector • Wiring from solenoid valve to powertrain control module (transmission) (PCMT) • Solenoid valve resistance	Yes

A

Table B

Code No.	Buzzer Pattern	Diagnosed circuit	Condition	Point	Memorized
61		Solenoid valve (2-3 shift)	Open or short circuit of solenoid valve and/or wiring	• Solenoid valve connector • Wiring from solenoid valve to powertrain control module (transmission) (PCMT) • Solenoid valve resistance	Yes
62		Solenoid valve (3-4 shift)	Open or short circuit of solenoid valve and/or wiring	• Solenoid valve connector • Wiring from solenoid valve to powertrain control module (transmission) (PCMT) • Solenoid valve resistance	Yes
63		Solenoid valve (lockup control)	Open or short circuit of solenoid valve and/or wiring	• Solenoid valve connector • Wiring from solenoid valve to powertrain control module (transmission) (PCMT) • Solenoid valve resistance	Yes
64		Solenoid valve (3-2 timing)	Open or short circuit of solenoid valve and/or wiring	• Solenoid valve connector • Wiring from solenoid valve to powertrain control module (transmission) (PCMT) • Solenoid valve resistance	Yes
65		Solenoid valve (lockup)	Open or short circuit of solenoid valve and/or wiring	• Solenoid valve connector • Wiring from solenoid valve to powertrain control module (transmission) (PCMT) • Solenoid valve resistance	Yes
66		Solenoid valve (line pressure)	Open or short circuit of solenoid valve and/or wiring	• Solenoid valve connector • Wiring from solenoid valve to powertrain control module (transmission) (PCMT) • Solenoid valve resistance	Yes

B

Figure 8-17 Trouble code chart for Mazda vehicles (Courtesy of Mazda Motors)

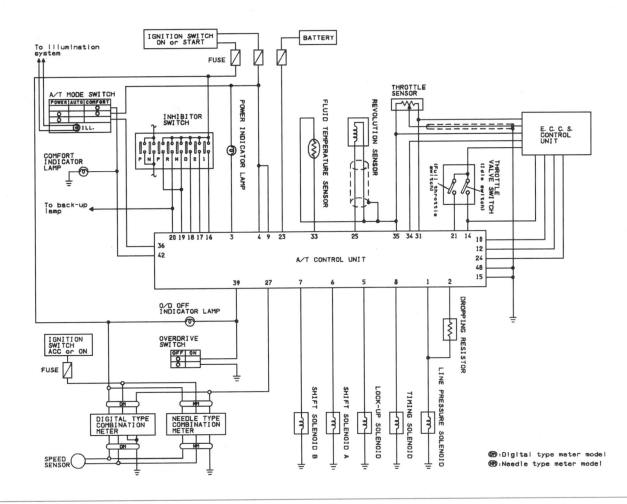

Figure 8-18 A typical Nissan EAT wiring diagram (Courtesy of Nissan Motor Co., Ltd.)

Nissan

Transmission trouble code retrieval on Nissan computer-shifted transmissions and transaxles is easy and uniform throughout the different models of vehicles (Figure 8-18). Simply turn the ignition switch to the on position and move the shifter and the O/D off switch. Then an instrument panel light will flash codes. The codes are displayed by the O/D off light, power (Figure 8-19), or A/T check light.

Classroom Manual
Chapter 8, page 284

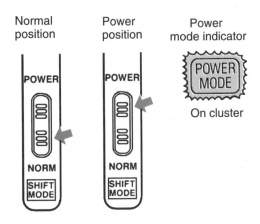

Figure 8-19 A typical shift mode switch (Reprinted with the permission of Ford Motor Co.)

To begin the procedure, make sure the Mode switch is in the Auto position. Then run the engine until it has reached normal operating temperature. Turn the key off, set the parking brake, and do not touch the brake or throttle pedals until the procedure calls for you to do so. Move the shifter to D and turn off the O/D switch. Turn the ignition switch back to the on position. After waiting at least 2 seconds, move the shifter to "2" and turn on the O/D switch. Then move the shifter to "1" and turn off the O/D switch. Now slowly push the throttle pedal to the floor and slowly release it. The codes will now appear on the instrument panel.

The codes will appear in a sequence of eleven flashes. This sequence always begins with a long, 2-second starter flash, followed by ten shorter flashes. If a problem does not exist, the ten following flashes will be very short flashes of 0.2 second. If a problem exists, one of the ten short flashes will be a little longer, 0.8 second, than the rest in that sequence. To identify the code, count the short flashes and determine which one was the longer one. Begin counting after the longer initial flash. If the first flash was the longest short flash, the code is 1. After each sequence of 11 flashes, there is a 2.5-second pause, then either the code or codes will repeat. All codes will automatically clear if the problem is fixed and the engine has been started two times after the system has been repaired.

Saturn

Classroom Manual
Chapter 8, page 283

Saturn EATs use five electronic valves, which are controlled by the PCM to control shift feel and timing. The PCM has a self-diagnostic function. If an electrical problem occurs, the SES or "Shift to D2" lamp will come on or will start flashing. The PCM stores trouble codes whenever it has detected a fault in the engine or transaxle circuit. Problems are stored as hard or intermittent problems. The PCM will also store other problems, but it may not turn on the SES light in response to the problem. These codes are stored in the memory to aid in diagnostics.

The PCM recognizes three different types of faults: hard, intermittent, and malfunction history/information flags. Hard codes cause the SES light to glow and remain on until the problem is repaired. These codes can be interpreted by looking up the code in the appropriate trouble code chart. Intermittent codes cause the SES to flicker or glow and go out approximately 10 seconds after the fault disappears. The corresponding trouble code will remain, however, in memory to be retrieved by a technician. If the intermittent fault does not occur for 50 engine starts, the code will be erased from the computer's memory. Engine information flags will not cause the SES light to glow. Unlike the other types of codes, information flags and malfunction history will not be erased from the PCM memory. These codes can only be retrieved and erased from memory by using a Saturn portable diagnostic tool (PDT).

Trouble codes are read by counting the flashes of the SES lamp. The first series of flashes represents the first digit of the trouble code and the second series represents the second digit. The first code will always be code 12, followed by all other stored codes. Each code is repeated three times. When code 11 appears, that signals the end of the engine's self-diagnostic test and the beginning of the transaxle self-test. The "Shift to D2" lamp will start flashing any transaxle-related codes stored in the transaxle controller. If the SES doesn't flash a code 11, there are no transaxle codes stored in memory.

To set the PCM into the self-diagnostic mode, turn the ignition on with the engine off. Connect a jumper wire from terminal B to terminal A at the PCM diagnostic connector (Figure 8-20). The SES

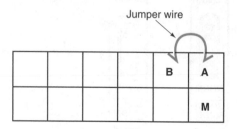

Figure 8-20 A Saturn diagnostic connector with critical terminals shown and a jumper wire inserted to activate the self-test mode

should begin to flash codes. To end the self-diagnostic mode, turn the ignition off and remove the jumper wire.

To erase the trouble codes, turn the ignition on and make contact three times within 5 seconds with a jumper wire between from terminals A and B at the diagnostic connector.

The PCM has a relearn function and the proper relearn process should be followed anytime the battery has been disconnected.

Toyota

A self-diagnostic feature is built into all Toyota EATs. A warning of a fault is indicated by the over-drive off indicator lamp. If a malfunction occurs in the speed sensor or solenoid circuits, the over-drive off lamp will blink to warn the driver of the fault. On some models, the diagnostic codes can be read by the number of blinks on the overdrive off lamp when terminals ECT and E2 in the diagnostic connector are shorted together. Other models require that terminals TE and E be shorted together (Figure 8-21). Make sure you refer to the appropriate service manual to determine which terminals should be shorted. Once the computer is set into its diagnostic mode, codes will begin to be displayed.

If a malfunction is in memory, the overdrive off lamp will blink once every 0.5 second. The first number of blinks equals the first digit of a two-digit code. After a 1.5-second pause, the second number of blinks will equal the second digit. If there are two or more codes, there will be a 2.5-second pause between each code. All codes should be interpreted by using the trouble code chart provided in the service manual. The diagnostic codes can be erased from the computer's memory by removing the DOME, ECU +B, or EFI fuse for at least 10 seconds with the key off.

 SERVICE TIP: Warning and diagnostic codes can only be read the overdrive switch is on. When the switch is off, the lamp is on continuously and will not flash.

Classroom Manual
Chapter 8, page 283

Input Testing

If the self-test sequence points to a problem in an input circuit, the input should be tested to determine the exact malfunction. The manufacturers often list specific procedures for specific sensors. Always follow them.

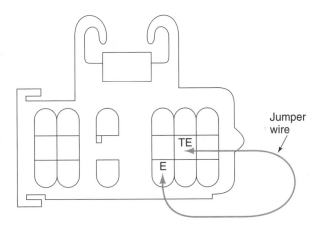

Figure 8-21 A Toyota Previa diagnostic connector with a jumper wire inserted to activate the self-test mode

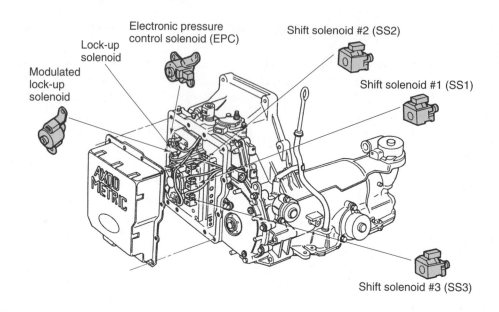

Figure 8-22 Typical location of the various transaxle switches used in an EAT system (Reprinted with the permission of Ford Motor Co.)

Switches

Special Tools

High-impedance
   DVOM

Jumper wire

20-A fuse or circuit
   breaker

Classroom Manual
Chapter 8, page 260

Many different switches are used as inputs (Figure 8-22) or control devices for EATs. Most of the switches are either mechanically or hydraulically controlled. The operation of these switches can be easily checked with an ohmmeter (Figure 8-23). With the meter connected across the switch's leads, there should be continuity or low resistance when the switch is closed and there should be infinite resistance across the switch when it is open. A test light can also be used but requires that the switch be hot-wired with source voltage. When the switch is closed, power should be present at both sides of the switch. When the switch is open, power should be present at only one side.

The manual lever position sensor is a switch that provides information to the computer as to what operating range has been selected by the driver. Based on that information, the computer determines the proper shift strategy for the transmission. Following is a list of problems that may result from a faulty manual lever position sensor:

- No upshifting
- Slipping out of gear

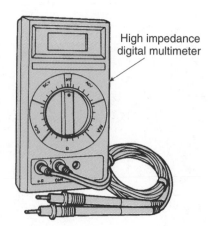

Figure 8-23 A high-impedance DVOM (Courtesy of the Chevrolet Motor Division of General Motors Corp.)

- High line pressure in transmissions equipped with a pressure control solenoid
- Delayed gear engagements
- Engine starts in other lever positions besides park and neutral

Because this switch is open or closed depending on position, it can be checked with an ohmmeter. By referring to a wiring diagram, you should be able to determine when the switch should be open. Then connect the meter across the input and out of the switch. Move the lever into the desired position and measure the resistance. An infinite reading is expected when the switch is open. If there is any resistance, the switch should be replaced.

The pressure switches used in today's transmissions either complete or open an electrical circuit. These switches are either grounding switches or they connect or disconnect two wires. In early TCC controls, a grounding-type switch was tapped into governor pressure to ground the TCC solenoid. Different spring tensions were used in these switches to control the speed at which the TCC could be applied. Other grounding-type switches are used to tell the computer or back-up lights when a gear is operating or not operating.

The other type of switch can be normally open or normally closed. Normally open switches will have no continuity across the terminals until oil pressure is applied to them. Normally closed switches will have continuity across the terminals until oil pressure is applied to them. Refer to the wiring diagram to determine the type of switch and test the switch with an ohmmeter. Base your expected results on the type of switch you are testing. By using air pressure you can easily see if the switch works properly or if it has a leak.

When possible, you should check switches when they are installed and controlled by the vehicle. By watching an ohmmeter connected to a governor switch, you can check its spring rating. Connect the ohmmeter to the "A" terminal and ground. Bring the engine's speed up. When the ohmmeter reads about 25 ohms, the governor switch is closed. The speed at which this occurs is the speed required to close the grounding switch. Other grounding-type switches can be checked in the same way. However, it is important that you identify if it is a normally open or closed switch before testing.

Throttle Position Sensor

Another type of switch is a potentiometer (Figure 8-24). Rather than opening and closing a circuit, a potentiometer controls the circuit by varying its resistance in response to something. A TPS is a potentiometer (Figure 8-25). It sends very low voltage back to the computer when the throttle plates are closed and increases the voltage as the throttle is opened.

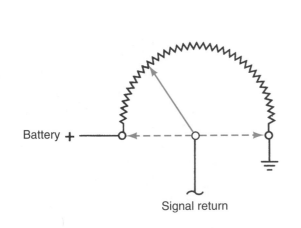

Figure 8-24 A potentiometer is used to send a signal voltage from the wiper of the switch.

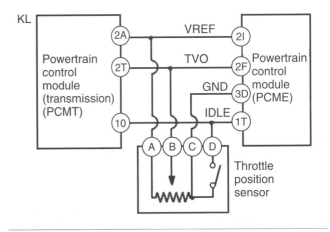

Figure 8-25 A TPS is a potentiometer. The movement of the throttle changes the voltage on circuit TVO. (Courtesy of Mazda Motors)

A TPS sends information to the control computer as to what position the throttle is in. If this sensor fails, the following problems can result:

- No upshifts
- Quick upshifts
- Delayed shifts
- Line-pressure problems with transmissions that have a line pressure control solenoid
- Erratic converter clutch engagement

Most TPSs receive a reference voltage of 5 volts. What the TPS sends back to the computer is determined by the position of the throttle. When the throttle is closed, it sends approximately 0.5 volt back to the computer. When the throttle begins to open, the voltage begins to increase. When the throttle reaches the wide-open position, approximately 4.5 volts are sent back to the computer. This increase in voltage should rise and fall smoothly with a change in throttle position. All changes in throttle position should result in a change in voltage.

A TPS can be checked with an ohmmeter or a voltmeter. If it is checked with an ohmmeter, you should be able to watch the resistance across the TPS change as the throttle is opened and closed. A resistance specification is usually given in the service manual. Compare your reading to this. With a voltmeter, you will be able to measure the reference voltage and the output voltage. Both of these should be within specified amounts. If the reference voltage is lower than normal, check the voltage drop across the reference voltage circuit from the computer to the TPS. If the TPS is found to be defective, it should be replaced.

Speed Sensors

Vehicle speed sensors provide road speed in formation to the computer. When this sensor fails or sends faulty readings, it can cause complaints that are similar to those caused by a bad TPS. The most common complaints are no overdrive, no converter-clutch engagement, and no upshifts.

There are two types of speed sensors: ac voltage generator sensors and reed-style sensors. Both of these rely on magnetic principles. The reed style is simply a switch which closes every time a magnet passes by. The magnet is attached to a rotating shaft, typically the output shaft. The activity of the sensor can be checked with an ohmmeter. If the switch is disconnected and the output shaft rotated one complete revolution, the ohmmeter should show an open and closed circuit within the one revolution of the shaft.

Ac voltage generators rely on a stationary magnet and a rotating shaft fitted with iron teeth (Figure 8-26). Each time a tooth passes through the magnetic field, an electrical pulse is present. By counting the number of teeth on the output shaft, you can determine how many pulses per one revolution you will measure with a voltmeter set to ac volts (Figure 8-27).

Mass Airflow Sensor

A mass airflow sensor is used to determine engine load by measuring the mass of the air being taken into the throttle body. The computer uses this information for fuel and air control. When this sensor fails or sends faulty signals, the engine runs roughly and tends to stall as soon as you put the transmission into gear. The mass airflow sensor is a wire, located in the intake air stream, that receives a fixed voltage. The wire is designed so that it changes resistance in response to temperature changes. When the wire is hot, the resistance is high and less current flows through the wire. When the wire is cold, the resistance is low and larger amounts of current pass through it. The amount of air passing over the wire determines the amount of resistance the wire has. The computer monitors the amount of current flow and interprets the flow as the mass of the air.

This sensor can be measured with a multimeter set to the Hz frequency range. Check the service manual for specific values. At idle, 30 Hz is normally measured, with the frequency increasing as the throttle opens. A scan tool can also be used to test this sensor. Most scanners have a test mode that monitors mass airflow sensors. While diagnosing these systems, keep in mind that cold air is denser than warm air.

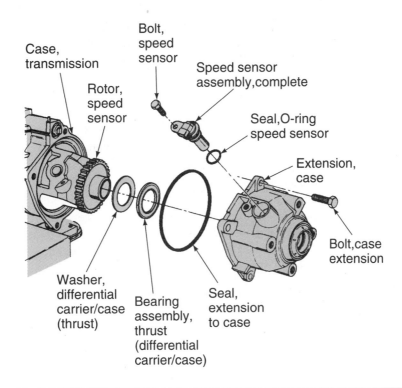

Figure 8-26 Ac generators rely on a stationary magnet, the sensor assembly, and a rotor fitted with teeth. (Courtesy of the Chevrolet Motor Division of General Motors Corp.)

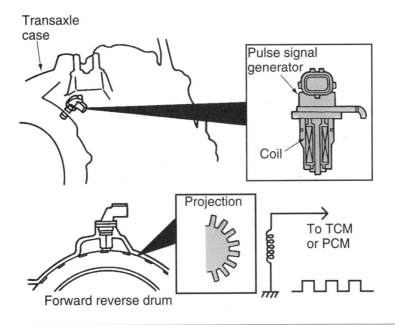

Figure 8-27 The vehicle speed sensor will display a pulse each time a tooth of the rotor passes by the sensor. (Reprinted with the permission of Ford Motor Co.)

Temperature Sensors

These sensors are designed to change resistance with changes in temperature. A temperature sensor is based on a thermistor. Some thermistors increase resistance with an increase of temperature. Others decrease the resistance as temperature increases. Obviously, these sensors can be checked with an ohmmeter. To do so, remove the sensor. Then determine the temperature of the sensor and measure the resistance across it. Compare your reading to the chart of normal resistances given in the service manual (Figure 8-28).

Solenoid Testing

Special Tools

Jumper wire

20-A fuse or circuit breaker

Classroom Manual
Chapter 8, page 264

If you were unable to identify the cause of a transmission problem through the previous checks, you should continue your diagnostics with testing the solenoids. This will allow you to determine if the shifting problem is the solenoids or their control circuit, or if it is a hydraulic or mechanical problem in the transmission.

Before continuing, however, you must first determine if the solenoids are case-grounded and fed voltage by the computer or if they always have power applied to them and the computer merely supplies the ground. While looking in the service manual to find this, also find the section that tells you which solenoids are on and which are off for each of the different gears.

CAUTION: When jumping solenoids with a hot wire, be sure to have a 20-ampere fuse in line to prevent damage to the solenoids and to your hand. A shorted solenoid will draw very high current, which will melt the wire.

To begin this test, you should secure the tools and/or equipment necessary to manually activate the solenoids. Switch panels that connect into the solenoid assembly and allow the technician to switch gears by depressing or flicking a switch are available.

DIAGNOSTIC AID
TRANSMISSION SENSOR—TEMP TO RESISTANCE (APPROXIMATE)

°C	°F	Ω MINIMUM RESISTANCE	Ω NORMAL RESISTANCE	Ω MAXIMUM RESISTANCE
-40°	-40°	80965	100544	120123
-30°	-20°	42701	52426	62151
-20°	-04°	23458	28491	33524
-10°	14°	13366	16068	18770
0°	32°	7871	9379	10869
10°	50°	4771	5640	6508
20°	68°	2981	3500	4018
30°	86°	1915	2232	2550
40°	104°	1260	1460	1660
50°	122°	848.8	977.1	1105
60°	140°	584.1	668.1	753.4
70°	158°	410.3	467.2	524.2
80°	176°	293.7	332.7	371.7
90°	194°	213.9	241.0	268.2
100°	212°	158.1	177.4	196.8
110°	230°	118.8	132.6	146.5
120°	248°	90.40	100.6	110.8
130°	266°	69.48	77.29	85.11
140°	284°	53.96	60.13	66.29
150°	302°	42.43	47.31	52.20
160°	320°	32.51	36.13	39.73
170°	338°	25.13	27.92	30.71
180°	356°	19.42	21.58	23.74
190°	374°	15.01	16.68	18.35
200°	392°	11.60	12.89	14.18

Figure 8-28 A chart showing the relationship between temperature and resistance for a temperature sensor (Courtesy of the Hydra-Matic Division of General Motors Corp.)

SERVICE TIP: This type of tester is easily made. Simply secure a harness for the transmission you want to test. Connect the leads from the harness to simple switches. Follow the solenoid/gear pattern when doing this. To change gears, all you will need to do is turn off one switch and turn on the next.

The solenoids will be energized in the correct pattern. Observe the action of the solenoids. A Toyota A140E is used as an example of this procedure. This transmission has two shift solenoids and the computer sends voltage to the solenoids to activate them. In first gear, solenoid #1 is on. In second gear, both solenoids are on. In third gear, only solenoid #2 is on and neither of the solenoids are on in fourth gear. A white wire in the transmission harness goes to solenoid #1 and a black wire goes to solenoid #2.

To begin this test sequence, disconnect the wiring harness that leads to the solenoids. Then connect 12 volts to the switch panel. Start the engine and move the shift lever into Drive. Now turn on solenoid #1. The transmission should be in first gear. Increase your speed, then turn on solenoid #2. The transmission should immediately shift into second gear. When you want to shift into third, turn off solenoid #1. Likewise, when you want fourth gear, turn off solenoid #2.

To totally test the transmission, you should shift gears under light, half, and full throttle. If the transmission shifts fine with the movement of the switches, you know that the transmission is fine. Any shifting problem has to be caused by something electrical. If the transmission doesn't respond to the switch movements, the problem is probably in the transmission.

This same technique can be used to check solenoids with a controlled ground. Instead of running a 12-volt hot wire, you run a good ground wire. This check can be conducted on all transmission solenoids except those that have a duty cycle. Check the service manual to identify this type of solenoid.

At times, a solenoid valve will work fine during light-throttle operation, but will leak when pressure increases. To verify that the valve is leaking, activate the solenoid, then increase engine speed while pulling on the throttle cable. If the solenoid valve leaks, the transmission will downshift. Leaking solenoids should be suspected whenever the transmission shifts roughly under heavy loads or full throttle but shifts fine under light throttle.

Solenoids can be checked for circuit resistance and shorts to ground. This can typically be done without removing the oil pan. The test can be conducted at the transmission case connector. By identifying the proper pins in the connector, individual solenoids can be checked with an ohmmeter. Remember, lower-than-normal resistance indicates a short, whereas higher-than-normal resistance indicates a problem of high resistance. If you get an infinite reading across the solenoid, the solenoid windings are open. The ohmmeter can also be used to check for shorts to ground. Simply connect one lead of the ohmmeter to one end of the solenoid windings and the other lead to a good ground. The reading should be infinite. If there is any measurable resistance, the winding is shorted to ground.

Solenoids can also be tested on a bench. Resistance values are typically given in service manuals for each application (Figure 8-29). A solenoid may be functioning well electrically, but it may still fail mechanically or hydraulically. A solenoid's check valve may fail to seat or the porting can be plugged. This is not an electrical problem, rather it could be caused by the magnetic field collecting metal particles in the ATF and clogging the port or check valve. This would cause erratic shifting, no shift conditions, or binding shifts. When a solenoid affected in this way is activated, it will make a slow, dull thud. A good solenoid tends to snap when activated.

Chrysler Solenoids

The solenoid assembly of a Chrysler A-604 consists of four solenoids, four ball valves, three pressure switches, and three resistors. The ball valves are operated by the solenoids and together they cause the transaxle to shift. The three pressure switches and resistors are used to signal the computer that a shift has occurred. The solenoids should be checked for proper operation. This can be done through the eight-pin connector at the assembly (Figure 8-30).

Terminal	Solenoid valve	Resistance (Ω)
A	1-2 shift	11—27
B	2-3 shift	11—27
C	3-4 shift	11—27
D	Lockup control	11—27
E	3-2 timing	11—27
F	Lockup	9—18
G	Line pressure	9—18

ATF temperature: –40—160°C (–40—320°F)

Figure 8-29 Service manuals show the location of the terminals and the resistance values needed to test transmission solenoids. (Courtesy of Mazda Motors)

Pin #		Function
1	·	Overdrive pressure SW
2	·	Low reverse pressure SW
3	·	2–4 pressure SW
4	·	12-volt input
5	·	Underdrive solenoid
6	·	Overdrive solenoid
7	·	Low reverse/lockup solenoid
8	·	2–4 low reverse solenoid

Figure 8-30 The eight-pin transaxle connector for a Chrysler A-604 transaxle and the function of each pin

Using an ohmmeter with the harness disconnected from the assembly, check the resistance across pins 1 and 4, 2 and 4, and 3 and 4. The resistance should be between 270 and 330 ohms. With this test, you have checked the resistor for each of the switches. To check the action of these switches, apply air pressure to the feed holes for the overdrive, low-reverse, and overdrive pressure switches. With air applied, the resistance reading across each of the pin combinations should now be zero ohms.

Each of the four solenoids can also be checked with an ohmmeter. Connect the meter across pins 4 and 5, 4 and 6, 4 and 7, and 4 and 8. The resistance across any of these combinations should be 1.5 ohms.

Ford Solenoids

Ford's E4OD can be shifted without the computer by energizing the #1 and #2 shift solenoids in the correct sequence (Figure 8-31). Refer to the illustration (Figure 8-32) for the location of the pins in the E4OD harness connector. Connect a 12-volt power source to pin #1. Then connect a ground lead to the following pins to shift the transmission: first gear—pin #3; second gear—pins #3 and #2; third gear—pin #2, and fourth gear—no pins grounded. If the transmission doesn't shift, check the solenoids.

Using an ohmmeter, check the resistance of solenoid #1 by measuring the resistance across pins #1 and #3. To check solenoid #2, measure across pins #1 and #2. And to check solenoid #3, measure the resistance across pins #1 and #4. The resistance across any of these should be 20 to 30 ohms.

To check the solenoids on a AXOD-E with an ohmmeter, disconnect the harness to the solenoid assembly. Measure the resistance across the appropriate pins and compare your readings to specifications. The shift solenoids should have a reading of 12–30 ohms; the EPC solenoid should have a reading of 2.5–6.5 ohms; the modulated lockup solenoid should have a reading of 0.75–22.0 ohms; and the regular lockup solenoid should have 16–40 ohms across its terminals.

GM Solenoids

The solenoids of 4T60-E transaxles can be checked with an ohmmeter. Both shift solenoids and the TCC solenoid should have a resistance of 20–30 ohms at 68°F and 23–50 ohms at 190°F. The

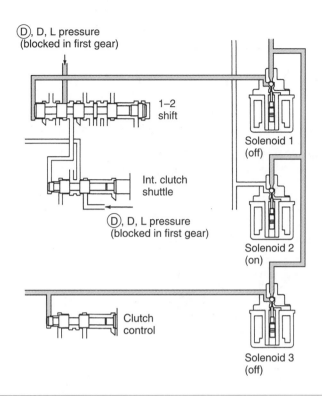

Figure 8-31 The operation of the shift solenoids while an E4OD is in first gear (Reprinted with the permission of Ford Motor Co.)

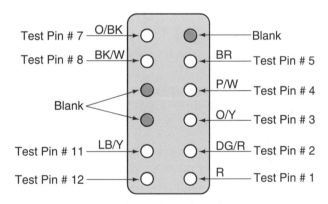

Figure 8-32 The transmission harness connector for an E4OD. The pin numbers are needed for testing of the solenoids.

TCC PWM solenoid should have a value of 10–15 ohms at 68°F and 11–25 ohms at 190°F. The terminals for each of these solenoids is accessible at the transmission harness (Figure 8-33).

Like most solenoids, these are simple on/off devices. Because of this, they can be checked through normal diagnostic methods. However, some transmissions, such as the GM 4L80-E, control line pressure according to engine load signals. A simple on/off shift solenoid cannot regulate the flow of oil in a metered manner. Therefore, the 4L80-E is equipped with a solenoid called a force motor, which controls line pressure. This solenoid is a three-port, spool valve, electronic pressure regulator that controls pressure based on current flow through its coil winding. The amount of current flow through the solenoid is controlled by the PCM. As VFS current increases, line pressure decreases.

A force motor is sometimes referred to as a variable force solenoid or VFS.

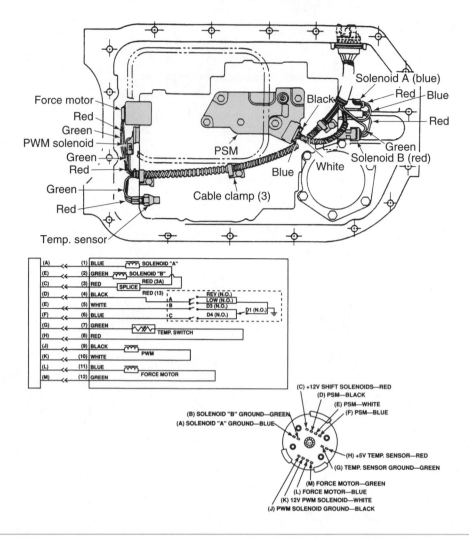

Figure 8-33 Testing of the electrical components in a 4L80-E can be done through the transmission harness connector. (Courtesy of the Hydra-Matic Division of General Motors Corp.)

DSO is a common abbreviation for digital storage oscilloscope.

A PWM is a pulse width modulated solenoid.

Solenoids that are controlled by duty cycle or a constantly changing signal are more active than the simple on/off solenoids; therefore, they are more likely to wear. Many transmission rebuilders recommend that this type of solenoid be automatically replaced during a transmission overhaul. Duty-cycle solenoids are less prone to clogging because they are cycled full on every 10 seconds. This full-on time helps to keep the regulator valve clean. The action of a force motor can be monitored on a digital storage oscilloscope. Connect the DSO directly to the force motor circuit. The pattern should be compared to those given in the service manual (Figure 8-34). A PWM solenoid is used to apply and release the TCC in such a manner that the TCC will apply and release smoothly under all conditions. The PWM solenoid in the 4L80-E uses a negative duty cycle. Its activity can also be monitored on a scope (Figure 8-35).

Honda Solenoids

Honda's shift solenoids are normally closed. When the computer sends battery voltage to them, they open and allow fluid to flow through them. When the fluid flows through the solenoid, it releases pressure that was preventing a shift valve from moving. When the pressure is exhausted, the valve is able to move and the transaxle shifts.

A Honda transmission can be shifted without the computer by jumping 12 volts to the shift solenoids. Apply voltage to solenoid B to operate in first gear, to both shift solenoids for second gear, to

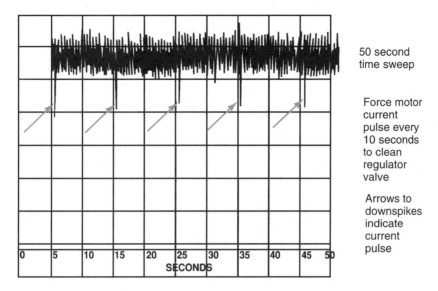

50 second
time sweep

Force motor
current
pulse every
10 seconds
to clean
regulator
valve

Arrows to
downspikes
indicate
current
pulse

Figure 8-34 A normal scope pattern for a force motor (Courtesy of the Hydra-Matic Division of General Motors Corp.)

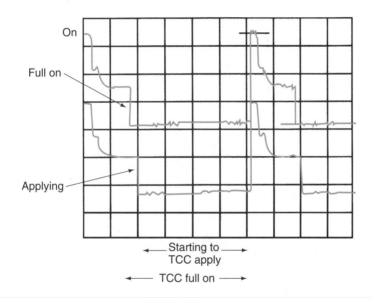

Figure 8-35 A normal scope pattern for a pulse width modulated solenoid (Courtesy of Hydra-Matic Division of General Motors Corp.)

solenoid A for third, and to none of them for fourth gear. If the transaxle shifts normally when the solenoids are energized, the solenoids or the controller is at fault and the mechanical and hydraulic portions of the transaxle are normal.

To observe the action of the solenoids, connect a pressure gauge to the transaxle taps for the solenoids. Observe the pressure readings when the transaxle upshifts. If the pressure is greater than 3 psi when the solenoid valve is open, the valve has a restriction or the solenoid is not opening fully and not allowing a full release of pressure. If the valving is restricted, the solenoids should be replaced.

SERVICE TIP: There are small filters located in the valve body to protect the solenoids. If these filters are plugged, they will cause a restriction to oil flow. The filters are mounted in a preformed rubber gasket and can be cleaned; however, the rubber gaskets should be replaced on every rebuild.

Use an ohmmeter to electrically check the solenoids. The resistance across a shift solenoid should be 12–24 ohms and 14–30 ohms across the lockup solenoid. If the resistance of the solenoids is not within these figures, they should be replaced.

Guidelines for Diagnosing EATs

1. Make sure the battery has at least 11.5 volts before troubleshooting the transmission.
2. Check all fuses and identify the cause of any blown fuses.
3. Compare the wiring to all suspected components against the colors given in the service manual.
4. When testing electronic circuits, always use a high-impedance test light or DVOM.
5. If an output device is not working properly, check the power circuit to it.
6. If an input device is not sending the correct signal back to the computer, check the reference voltage it is receiving and the voltage it is sending back to the computer.
7. Compare the voltages in and out of a sensor with the voltages the computer is sending out and receiving.
8. Before replacing a computer, check the solenoid isolation diodes according to the procedures outlined in the service manual.
9. Make sure computer wiring harnesses do not run parallel with any high-current wires or harnesses. The magnetic field created by the high current may induce a voltage in the computer harness.
10. Take necessary precautions to prevent the possibility of static discharge while working with electronic systems.
11. While checking individual components, always check the voltage drop of the ground circuits. This becomes more and more important as cars are made of less and less material that conduct electricity well.
12. Make sure the ignition is off whenever you disconnect or connect an electronic device in a circuit.
13. All sensors should be checked in cold and hot conditions.
14. All wire terminals and connections should be checked for tightness and cleanliness.
15. Use T.V.-tuner cleaning spray to clean all connectors and terminals.
16. Use a dielectric grease at all connections to prevent future corrosion.
17. If you must break through the insulation of a wire to take an electrical measurement, make sure you tightly tape over the area after you are finished testing.

CASE STUDY

A customer came into the shop with a high-mileage Hyundai Excel with a Mitsubishi KM-175 transmission. The car was delivered by a tow truck because the transmission wouldn't shift gears. The technician assigned to the car pulled the transmission and overhauled it. After installing an overhaul kit and inspecting all other transmission parts, the transmission was reassembled and installed back into the car. The transmission still didn't shift.

Next the technician checked the electronics involved with the transmission. After a little research in the service manual, she found that the transmission was in the fail-safe mode. With a multimeter, the technician then checked the solenoids, connectors, and wiring only to find nothing wrong. Then she checked complete circuits and found that the computer was doing nothing.

She located the computer under the passenger's seat and found heavy corrosion on the computer and the wiring harness terminals. The technician then cleaned the area and replaced the computer. The transmission now shifted well.

Terms to Know

Bias voltage	DSO	Negative pressure
Bus	EAT	PWM
CCD	ECA	SES
CEL	KOEO	STAR tester
Digital storage oscilloscope	KOER	VFS
DRB-II	Manifold vacuum	

ASE Style Review Questions

1. *Technician A* says delayed shifts can be caused by a faulty TPS.
   Technician B says delayed shifts can be caused by an open shift solenoid.
   Who is correct?
   A. A only **C.** Both A and B
   B. B only **D.** Neither A nor B

2. While checking a manual lever position sensor:
   Technician A says the switch should be open in all positions except Park and Neutral.
   Technician B says a faulty manual lever position sensor will allow the engine to start in other gear positions besides Park and Neutral.
   Who is correct?
   A. A only **C.** Both A and B
   B. B only **D.** Neither A nor B

3. *Technician A* says scan tools are needed to retrieve codes on all models of cars.
   Technician B says all scan tools provide a historical report of the computer system.
   Who is correct?
   A. A only **C.** Both A and B
   B. B only **D.** Neither A nor B

4. *Technician A* says Chrysler systems are best diagnosed with a DRB-II.
   Technician B says General Motors systems are best diagnosed with a Tech 1.
   Who is correct?
   A. A only **C.** Both A and B
   B. B only **D.** Neither A nor B

5. While checking the codes on a Ford system:
   Technician A uses a STAR tester.
   Technician B uses an analog VOM.
   Who is correct?
   A. A only **C.** Both A and B
   B. B only **D.** Neither A nor B

6. *Technician A* says some shift solenoids can be activated by providing a ground for the solenoid.
   Technician B says some shift solenoids can be activated by applying hydraulic pressure to their valve.
   Who is correct?
   A. A only **C.** Both A and B
   B. B only **D.** Neither A nor B

7. While checking the trouble codes on a Honda:
   Technician A says the short flashes should be counted as this is the trouble code.
   Technician B says there is about a one-second delay in the flashing between codes.
   Who is correct?
   A. A only **C.** Both A and B
   B. B only **D.** Neither A nor B

8. While checking codes on a Nissan:
   Technician A says the short flashes should be counted as this is the trouble code.
   Technician B says the codes will appear in a sequence of eleven flashes, beginning with a long flash.
   Who is correct?
   A. A only **C.** Both A and B
   B. B only **D.** Neither A nor B

9. *Technician A* says the PCM of a Saturn system recognizes three different types of faults.
Technician B says the PCM of a Saturn will not display information flags unless their specific scan tool is used.
Who is correct?
A. A only
B. B only
C. Both A and B
D. Neither A nor B

10. While checking the codes on a Ford AXOD-E:
Technician A says during the KOEO self-test, the codes are repeated only twice.
Technician B says the continuous memory codes appear after the separator pulse signal.
Who is correct?
A. A only
B. B only
C. Both A and B
D. Neither A nor B

Table 8-1 ASE TASK

Diagnose electrical systems to determine the cause of engine no-start conditions and determine needed repairs.

Problem Area	Symptoms	Possible Causes	Classroom Manual	Shop Manual
ENGINE WILL NOT CRANK	Engine will not start in Park or Neutral	1. Misadjusted manual linkage	46	47
		2. Bad connection to the neutral safety switch	48	51
		3. Wiring to neutral/safety switch broken, disconnected, or corroded	48	51
		4. Malfunctioning electronic controls	259	270

APPENDIX A

Automatic Transmission Special Tool Suppliers

Automatic Transmission Rebuilders
Association
Ventura, CA

Baum Tools Unlimited, Inc.
Longboat Key, FL

Big A Auto Parts, APC Inc.
Houston, TX

Carquest Corp.
Tarrytown, NY

Hastings Manufacturing
Hastings, MI

KD Tools, Danaher Tool Group
Lancaster, PA

Kent-Moore, Dov. SPX Corp.
Warren, Mi

Lisle Corp.
Clarinda, IA

MacTools
Washington Courthouse, OH

MATCO Tool Co.
Stow, OH

NAPA Hand/Service Tools
Lancaster, PA

OTC, Div. SPX Corp.
Owatonna, MN

Parts Plus
Memphis, TN

Snap-on Tools Corp.
Kenosha, WI

Transtar Industries, Inc.
Cleveland, OH

Metric Conversions

	to convert these	to these,	multiply by:
TEMPERATURE	Centigrade Degrees	Fahrenheit Degrees	1.8 then + 32
	Fahrenheit Degrees	Centigrade Degrees	0.556 after − 32
LENGTH	Millimeters	Inches	0.03937
	Inches	Millimeters	25.4
	Meters	Feet	3.28084
	Feet	Meters	0.3048
	Kilometers	Miles	0.62137
	Miles	Kilometers	1.60935
AREA	Square Centimeters	Square Inches	0.155
	Square Inches	Square Centimeters	6.45159
VOLUME	Cubic Centimeters	Cubic Inches	0.06103
	Cubic Inches	Cubic Centimeters	16.38703
	Cubic Centimeters	Liters	0.001
	Liters	Cubic Centimeters	1000
	Liters	Cubic Inches	61.025
	Cubic Inches	Liters	0.01639
	Liters	Quarts	1.05672
	Quarts	Liters	0.94633
	Liters	Pints	2.11344
	Pints	Liters	0.47317
	Liters	Ounces	33.81497
	Ounces	Liters	0.02957
WEIGHT	Grams	Ounces	0.03527
	Ounces	Grams	28.34953
	Kilograms	Pounds	2.20462
	Pounds	Kilograms	0.45359
WORK	Centimeter Kilograms	Inch-Pounds	0.8676
	Inch-Pounds	Centimeter-Kilograms	1.15262
	Meter Kilograms	Foot-Pounds	7.23301
	Foot-Pounds	Newton-Meters	1.3558
PRESSURE	Kilograms/Square Centimeter	Pounds/Square Inch	14.22334
	Pounds/Square Inch	Kilograms/Square Centimeter	0.07031
	Bar	Pounds/Square Inch	14.504
	Pounds/Square Inch	Bar	0.06895

GLOSSARY

Abrasion Wearing or rubbing away of a part.
Abrasión El desgaste o consumo por rozamiento de una parte.

Acceleration An increase in velocity or speed.
Aceleración Un incremento en la velocidad.

Accumulator A device used in automatic transmissions to cushion the shock of shifting between gears, providing a smoother feel inside the vehicle.
Acumulador Un dispositivo que se usa en las transmisiones automáticas para suavizar el choque de cambios entre las velocidades, así proporcionando una sensación más uniforme en el interior del vehículo.

Adhesives Chemicals used to hold gaskets in place during the assembly of an engine. They also aid the gasket in maintaining a tight seal by filling in the small irregularities on the surfaces and by preventing the gasket from shifting due to engine vibration.
Adhesivo Los productos químicos que se usan para sujetar a los empaques en una posición correcta mientras que se efectúa la asamblea de un motor. También ayuden para que los empaques mantengan un sello impermeable, rellenando a las irregularidades pequeñas en las superficies y previniendo que se mueva el empaque debido a las vibraciones del motor.

Alignment An adjustment to a line or to bring into a line.
Alineación Un ajuste que se efectúa en una linea o alinear.

Antifriction bearing A bearing designed to reduce friction. This type of bearing normally uses ball or roller inserts to reduce the friction.
Cojinetes de antifricción Un cojinete diseñado con el fin de disminuir la fricción. Este tipo de cojinete suele incorporar una pieza inserta esférica o de rodillos para disminuir la fricción.

Antiseize Thread compound designed to keep threaded connections from damage due to rust or corrosion.
Antiagarrotamiento Un compuesto para filetes diseñado para proteger a las conecciones fileteados de los daños de la oxidación o la corrosión.

Apply devices Devices that hold or drive members of a planetary gear set. They may be hydraulically or mechanically applied.
Dispositivos de aplicación Los dispositivos que sujeten o manejan los miembros de un engranaje planetario. Se pueden aplicar mecánicamente o hidráulicamente.

Arbor press A small, hand-operated shop press used when only a light force is required against a bearing, shaft, or other part.
Prensa para calar Una prensa de mano pequeña del taller que se puede usar en casos que requieren una fuerza ligera contra un cojinete, una flecha u otra parte.

Asbestos A material that was commonly used as a gasket material in places where temperatures are extreme. This material is being used less frequently today because of health hazards that are inherent to the material.
Amiante Una materia que se usaba frecuentemente como materia de empaques en sitios en los cuales las temperaturas son extremas. Esta materia se usa menos actualmente debido a los peligros al salud que se atribuyan a esta materia.

ATF Automatic Transmission Fluid.
ATF Fluido de Transmisión Automática

Automatic transmission A transmission in which gear or ratio changes are self-activated, eliminating the necessity of hand-shifting gears.
Transmisión automática Una transmisión en la cual un cambio deengranajes o los cambios en relación son por mando automático, así eliminando la necesidad de cambios de velocidades manual.

Axial Parallel to a shaft or bearing bore.
Axial Paralelo a una flecha o al taladro del cojinete.

Axis The centerline of a rotating part, a symmetrical part, or a circular bore.
Eje La linea de quilla de una parte giratoria, una parte simétrica, o un taladro circular.

Axle The shaft or shafts of a machine upon which the wheels are mounted.
Semieje El eje o los ejes de una máquina sobre los cuales se montan las ruedas.

Axle ratio The ratio between the rotational speed (rpm) of the driveshaft and that of the driven wheel; gear reduction through the differential, determined by dividing the number of teeth on the ring gear by the number of teeth on the drive pinion.
Relación del eje La relación entre la velocidad giratorio (rpm) del árbol propulsor y la de la rueda arrastrada; reducción de los engranajes por medio del diferencial, que se determina por dividir el número de dientes de la corona por el número de los dientes en el piñón de ataque.

Axle shaft A shaft on which the road wheels are mounted.
Flecha del semieje Una flecha en la cual se monta las ruedas.

Axle shaft end thrust A force exerted on the end of an axle shaft that is most pronounced when the vehicle turns corners and curves.
Golpe en la flecha del semieje Una fuerza que se aplica en el extremo de la flecha del semieje que se pronuncia más cuando un vehículo da la vuelta.

Backlash The amount of clearance or play between two meshed gears.
Juego La cantidad de holgura o juego entre dos engranajes endentados.

Balance Having equal weight distribution. The term is usually used to describe the weight distribution around the circumference and between the front and back sides of a wheel.

Equilibrio Lo que tiene una distribución igual de peso. El término suele usarse para describir la distribución del peso alrededor de la circunferencia y entre los lados delanteros y traseros de una rueda.

Balance valve A regulating valve that controls a pressure of just the right value to balance other forces acting on the valve.

Válvula niveladora Una válvula de reglaje que controla a la presión del valor correcto para mantener el equilibrio contra las otras fuerzas que afectan a la válvula.

Ball bearing An antifriction bearing consisting of a hardened inner and outer race with hardened steel balls which roll between the two races, and supports the load of the shaft.

Rodamiento de bolas Un cojinete de antifricción que consiste de una pista endurecida interior e exterior y contiene bolas de acero endurecidos que ruedan entre las dos pistas, y sostiene la carga de la flecha.

Ball joint A suspension component that attaches the control arm to the steering knuckle and serves as the lower pivot point for the steering knuckle. The ball joint gets its name from its ball-and-socket design. It allows both up-and-down motion as well as rotation. In a MacPherson strut FWD suspension system, the two lower ball joints are nonload carrying.

Articulación esférica Un componente de la suspensión que une el brazo de mando a la articulación de la dirección y sirve como un punto pivote inferior de la articulación de la dirección. La articulación esférica derive su nombre de su diseño de bola y casquillo. Permite no sólo el movimiento de arriba y abajo sino también el de rotación. En un sistema de suspensión tipo FWD con poste de MacPherson, las articulaciones esféricas inferiores no soportan el peso.

Band A steel band with an inner lining of friction material. Device used to hold a clutch drum at certain times during transmission operation.

Banda Una banda de acero que tiene un forro interior de una materia de fricción. Un dispositivo que retiene al tambor del embrague en algunos momentos durante la operación de la transmisión.

Bearing The supporting part that reduces friction between a stationary and rotating part or between two moving parts.

Cojinete La parte portadora que reduce la fricción entre una parte fija y una parte giratoria o entre dos partes que muevan.

Bearing cage A spacer that keeps the balls or rollers in a bearing in proper position between the inner and outer races.

Jaula del cojinete Un espaciador que mantiene a las bolas o a los rodillos del cojinete en la posición correcta entre las pistas interiores e exteriores.

Bearing caps In the differential, caps held in place by bolts or nuts which, in turn, hold bearings in place.

Tapones del cojinete En un diferencial, las tapas que se sujeten en su lugar por pernos o tuercas, los cuales en su turno, retienen y posicionan a los cojinetes.

Bearing cone The inner race, rollers, and cage assembly of a tapered roller bearing. Cones and cups must always be replaced in matched sets.

Cono del cojinete La asamblea de la pista interior, los rodillos, y el jaula de un cojinete de rodillos cónico. Se debe siempre reemplazar a ambos partes de un par de conos del cojinete y los anillos exteriores a la vez.

Bearing cup The outer race of a tapered roller bearing or ball bearing.

Anillo exterior La pista exterior de un cojinete cónico de rodillas o de bolas.

Bearing race The surface upon which the rollers or balls of a bearing rotate. The outer race is the same thing as the cup, and the inner race is the one closest to the axle shaft.

Pista del cojinete La superficie sobre la cual ruedan los rodillos o las bolas de un cojinete. La pista exterior es lo mismo que un anillo exterior, y la pista interior es la más cercana a la flecha del eje.

Belleville spring A tempered spring steel cone-shaped plate used to aid the mechanical force in a pressure plate assembly.

Resorte de tensión Belleville Un plato de resorte del acero revenido en forma cónica que aumenta a la fuerza mecánica de una asamblea del plato opresor.

Bellhousing A housing that fits over the clutch components and connects the engine and the transmission.

Concha del embrague Un cárter que encaja a los componentes del embrague y conecta al motor con la transmisión.

Bias voltage Voltage applied across a diode.

Tensión de polarización El voltaje aplicado através de un diodo.

Bolt torque The turning effort required to offset resistance as the bolt is being tightened.

Torsión del perno El esfuerzo de torsión que se requiere para compensar la resistencia del perno mientras que esté siendo apretado.

Brake horsepower (bhp) Power delivered by the engine and available for driving the vehicle; bhp = torque x rpm/5252.

Caballo indicado al freno (bhp) Potencia que provee el motor y que es disponible para el uso del vehículo; bhp = de par mortor x rpm/5252.

Brinnelling Rough lines worn across a bearing race or shaft due to impact loading, vibration, or inadequate lubrication.

Efecto brinel Lineas ásperas que aparecen en las pistas de un cojinete o en las flechas debido al choque de carga, la vibración, o falta de lubricación.

Bronze An alloy of copper and tin.

Bronce Una aleación de cobre y hojalata.

Burnish To smooth or polish by the use of a sliding tool under pressure.

Bruñir Pulir o suavizar por medio de una herramienta deslizando bajo presión.

Burr A feather edge of metal left on a part being cut with a file or other cutting tool.

Rebaba Una lima espada de metal que permanece en una parte que ha sido cortado con una lima u otro herramienta de cortar.

Bus A common connector used as an information source for the vehicle's various control units.

Bus Un conector común que se usa como un fuente de información para los varios aparatos de control del vehículo.

Bushing A cylindrical lining used as a bearing assembly made of steel, brass, bronze, nylon, or plastic.

Buje Un forro cilíndrico que se usa como una asamblea de cojinete que puede ser hecho del acero, del latón, del bronce, del nylon, o del plástico.

C-clip A C-shaped clip used to retain the drive axles in some rear axle assemblies.

Grapa de C Una grapa en forma de C que retiene a las flechas motrices en algunas asambleas de ejes traseras.

Cage A spacer used to keep the balls or rollers in proper relation to one another. In a constant-velocity joint, the cage is an open metal framework that surrounds the balls to hold them in position.

Jaula Una espaciador que mantiene una relación correcta entre los rodillos o las bolas. En una junta de velocidad constante, la jaula es un armazón abierto de metal que rodea a las bolas para mantenerlas en posición.

Cap An object that fits over an opening to stop flow.

Tapón Un objecto que tapa a una apertura para detener el flujo.

Carbon monoxide Part of the exhaust gas from an engine; an odorless, colorless, and deadly gas.

Óxido de carbono Una parte de los vapores de escape de un motor; es un gas sin olor, sin color y puede causar la muerte.

Cardan Universal Joint A nonconstant velocity universal joint consisting of two yokes with their forked ends joined by a cross. The driven yoke changes speed twice in 360 degrees of rotation.

Junta Universal Cardan Una junta universal de velocidad no constante que consiste de dos yugos cuyos extremidades ahorquilladas se unen en cruz. El yugo de arrastre cambia su velocidad dos veces en 360 grados de rotación.

Case-harden To harden the surface of steel. The carburizing method used on low-carbon steel or other alloys to make the case or outer layer of the metal harder than its core.

Cementar Endurecer la superficie del acero. El método de carburación que se emplea en el acero de bajo carbono o en otros aleaciones para que el cárter o capa exterior queda más dura que lo que esta al interior.

Case porosity Leaks caused by tiny holes which are formed by trapped air bubbles during the casting process.

Porosidad del cárter Las fugas que se causan por los hoyitos pequeños formados por burbújas de aire entrapados durante el proceso del moldeo.

Castellate Formed to resemble a castle battlement, as in a castellated nut.

Acanalado De una forma que parece a las almenas de un castillo (véa la palabra en inglés), tal como una tuerca con entallas.

Castellated nut A nut with six raised portions or notches through which a cotter pin can be inserted to secure the nut.

Tuerca con entallas Una tuerca que tiene seis porciones elevadas o muescas por los cuales se puede insertar un pasador de chaveta para retener a la tuerca.

Centrifugal clutch A clutch that uses centrifugal force to apply a higher force against the friction disc as the clutch spins faster.

Embrague centrífugo Un embrague que emplea a la fuerza centrífuga para aplicar una fuerza mayor contra el disco de fricción mientras que el embrague gira más rapidamente.

Centrifugal force The force acting on a rotating body which tends to move it outward and away from the center of rotation. The force increases as rotational speed increases.

Fuerza centrífuga La fuerza que afecta a un cuerpo en rotación moviendolo hacia afuera y alejándolo del centro de rotación. La fuerza aumenta al aumentar la velocidad de rotación.

Chamfer A bevel or taper at the edge of a hole or a gear tooth.

Chaflán Un bisél o cono en el borde de un hoyo o un diente del engranaje.

Chamfer face A beveled surface on a shaft or part that allows for easier assembly. The ends of FWD drive shafts are often chamfered to make installation of the CV joints easier.

Cara achaflanada Una superficie biselada en una flecha o una parte que facilita la asamblea. Los extremos de los árboles de mando de FWD suelen ser achaflandos para facilitar la instalación de las juntas CV.

Chase To straighten up or repair damaged threads.

Embutir Enderezar o reparar a los filetes dañados.

Chasing To clean threads with a tap.

Embutido Limpiar a los filetes con un macho.

Chassis The vehicle frame, suspension, and running gear. On FWD cars, it includes the control arms, struts, springs, trailing arms, sway bars, shocks, steering knuckles, and frame. The drive shafts, constant-velocity joints, and transaxle are not part of the chassis or suspension.

Chasis El armazón de un vehículo, la suspensión, y el engranaje de marcha. En los coches de FWD, incluye los brazos de mando, los postes, los resortes (chapas), los brazos traseros, las estabilizadoras, las articulaciones de la dirección y el armazón. Los árboles de mando, las juntas de velocidad constante, y la flecha impulsora no son partes del chasis ni de la suspensión.

Circlip A split steel snap ring that fits into a groove to hold various parts in place. Circlips are often used on the ends of FWD drive shafts to retain the constant-velocity joints.

Grapa circular Un seguro partido circular de acero que se coloca en una ranura para posicionar a varias partes. Las grapas circulares se suelen usar en las extremidades de los árboles de mando en FWD para retener las juntas de velocidad constante.

Clearance The space allowed between two parts, such as between a journal and a bearing.

Holgura El espacio permitido entre dos partes, tal como entre un muñón y un cojinete.

Clutch A device for connecting and disconnecting the engine from the transmission or for a similar purpose in other units.

Embrague Un dispositivo para conectar y desconectar el motor de la transmisión o para tal propósito en otros conjuntos.

Clutch packs A series of clutch discs and plates installed alternately in a housing to act as a driving or driven unit.

Conjuntos de embrague Una seria de discos y platos de embrague que se han instalado alternativamente en un cárter para funcionar como una unedad de propulsión o arrastre.

Clutch slippage Engine speed increases but increased torque is not transferred through to the driving wheels because of clutch slippage.

Resbalado del embrague La velocidad del motor aumenta pero la torsión aumentada del motor no se transfere a las ruedas de marcha por el resbalado del embrague.

Coefficient of friction The ratio of the force resisting motion between two surfaces in contact to the force holding the two surfaces in contact.

Coeficiente de la fricción La relación entre la fuerza que resiste al movimiento entre dos superficies que tocan y la fuerza que mantiene en contacto a éstas dos superficies.

Coil preload springs Coil springs are made of tempered steel rods formed into a spiral that resist compression; located in the pressure plate assembly.

Muelles de embrague Los muelles espirales son fabricadas de varillas de acero revenido y resisten la compresión; se ubican en el conjunto del plato opresor.

Coil spring A heavy wire-like steel coil used to support the vehicle weight while allowing for suspension motions. On FWD cars, the front coil springs are mounted around the MacPherson struts. On the rear suspension, they may be mounted to the rear axle, to trailing arms, or around rear struts.

Muelles de embrague Un resorte espiral hecho de acero en forma de alambre grueso que soporte el peso del vehículo mientras que permite a los movimientos de la suspensión. En los coches de FWD, los muelles de embrague delanteros se montan alrededor de los postes Macpherson. En la suspensión trasera, pueden montarse en el eje trasero, en los brazos traseros, o alrededor de los postes traseros.

Compound A mixture of two or more ingredients.

Compuesto Una combinación de dos ingredientes o más.

Concentric Two or more circles having a common center.

Concéntrico Dos círculos o más que comparten un centro común.

Constant-velocity joint A flexible coupling between two shafts that permits each shaft to maintain the same driving or driven speed regardless of operating angle, allowing for a smooth transfer of power. The constant-velocity joint (also called CV joint) consists of an inner and outer housing with balls in between, or a tripod and yoke assembly.

Junta de velocidad constante Un acoplador flexible entre dos flechas que permite que cada flecha mantenga la velocidad de propulsión o arrastre sin importar el ángulo de operación, efectuando una transferencia lisa del poder. La junta de velocidad constante (también llamado junta CV) consiste de un cárter interior e exterior entre los cuales se encuentran bolas, o de un conjunto de trípode y yugo.

Contraction A reduction in mass or dimension; the opposite of expansion.

Contración Una reducción en la masa o en la dimensión; el opuesto de expansión.

Control arm A suspension component that links the vehicle frame to the steering knuckle or axle housing and acts as a hinge to allow up-and-down wheel motions. The front control arms are attached to the frame with bushings and bolts and are connected to the steering knuckles with ball joints. The rear control arms attach to the frame with bushings and bolts and are welded or bolted to the rear axle or wheel hubs.

Brazo de mando Un componente de la suspención que une el armazón del vehículo al articulación de dirección o al cárter del eje y que se porta como una bisagra para permitir a los movimientos verticales de las ruedas. Los brazos de mando delanteros se conectan al armazón por medio de pernos y bujes y se conectan al articulación de dirección por medio de los articulaciones esféricos. Los brazos de mando traseros se conectan al armazón por medio de pernos y bujes y son soldados o empernados al eje trasero o a los cubos de la rueda.

Corrode To eat away gradually as if by gnawing, especially by chemical action.

Corroer Roído poco a poco, primariamente por acción químico.

Corrosion Chemical action, usually by an acid, that eats away (decomposes) a metal.

Corrosión Un acción químico, por lo regular un ácido, que corroe (descompone) un metal.

Cotter pin A type of fastener, made from soft steel in the form of a split pin, that can be inserted in a drilled hole. The split ends are spread to lock the pin in position.

Pasador de chaveta Un tipo de fijación, hecho de acero blando en forma de una chaveta que se puede insertar en un hueco tallado. Las extremidades partidas se despliegen para asegurar la posición de la chaveta.

Counterclockwise rotation Rotating in the opposite direction of the hands on a clock.

Rotación en sentido inverso Girando en el sentido opuesto de las agujas de un reloj.

Coupling A connecting means for transferring movement from one part to another; may be mechanical, hydraulic, or electrical.

Acoplador Un método de conección que transfere el movimiento de una parte a otra; puede ser mecánico, hidráulico, o eléctrico.

Coupling phase Point in torque converter operation where the turbine speed is 90% of impeller speed and there is no longer any torque multiplication.

Fase del acoplador El punto de la operación del convertidor de la torsión en el cual la velocidad de la turbina es el 90% de la velocidad del impulsor y no queda ningún multiplicación de la torsión.

Cover plate A stamped steel cover bolted over the service access to the manual transmission.

Cubrejuntas Un cubierto de acero estampado que se emperna en la apertura de servicio de la transmisión manual.

Crocus cloth A very fine polishing paper. It is designed to remove very little metal; therefore it is safe to use on critical surfaces.

Tela de óxido férrico Un papel muy fino para pulir. Fue diseñado para raspar muy poco del metal; por lo tanto, se suele emplear en las superficies críticas.

Deflection Bending or movement away from normal due to loading.

Desviación Curvación o movimiento fuera de lo normal debido a la carga.

Degree A unit of measurement equal to 1/360th of a circle.

Grado Una uneda de medida que iguala al 1/360 parte de un círculo.

Density Compactness; relative mass of matter in a given volume.

Densidad La firmeza; una cantidad relativa de la materia que ocupa a un volumen dado.

Detent A small depression in a shaft, rail, or rod into which a pawl or ball drops when the shaft, rail, or rod is moved. This provides a locking effect.

Detención Un pequeño hueco en una flecha, una barra o una varilla en el cual cae una bola o un linguete al moverse la flecha, la barra o la varilla. Esto provee un efecto de enclavamiento.

Detent mechanism A shifting control designed to hold the manual transmission in the gear range selected.

Aparato de detención Un control de desplazamiento diseñado a sujetar a la transmisión manual en la velocidad selecionada.

Diagnosis A systematic study of a machine or machine parts to determine the cause of improper performance or failure.

Diagnóstico Un estudio sistemático de una máquina o las partes de una máquina con el fín de determinar la causa de una falla o de un operación irregular.

Dial indicator A measuring instrument with the readings indicated on a dial rather than on a thimble as on a micrometer.

Indicador de carátula Un instrumento de medida cuyo indicador es en forma de muestra en contraste al casquillo de un micrómetro.

Differential A mechanism between drive axles that permits one wheel to run at a different speed than the other while turning.

Diferencial Un mecanismo entre dos semiejes que permite que una rueda gira a una velocidad distincta que la otra en una curva.

Differential action An operational situation where one driving wheel rotates at a slower speed than the opposite driving wheel.

Acción del diferencial Una situación durante la operación en la cual una rueda propulsora gira con una velocidad más lenta que la rueda propulsora opuesta.

Differential case The metal unit that encases the differential side gears and pinion gears, and to which the ring gear is attached.

Caja de satélites La unedad metálica que encaja a los engranajes planetarios (laterales) y a los satélites del diferencial, y a la cual se conecta la corona.

Differential drive gear A large circular helical gear that is driven by the transaxle pinion gear and shaft and drives the differential assembly.

Corona Un engranaje helicoidal grande circular que es arrastrado por el piñon de la flecha de transmisión y la flecha y propela al conjunto del diferencial.

Differential housing Cast iron assembly that houses the differential unit and the drive axles. Also called the rear axle housing.

Cárter del diferencial Una asamblea de acero vaciado que encaja a la unedad del diferencial y los semiejes. También se llama el cárter del eje trasero.

Differential pinion gears Small beveled gears located on the differential pinion shaft.

Satélites Engranajes pequeños biselados que se ubican en la flecha del piñon del diferencial.

Differential pinion shaft A short shaft locked to the differential case. This shaft supports the differential pinion gears.

Flecha del piñon del diferencial Una flecha corta clavada en la caja de satélites. Esta flecha sostiene a los satélites.

Differential ring gear A large circular hypoid-type gear enmeshed with the hypoid drive pinion gear.

Corona Un engranaje helicoidal grande circular endentado con el piñon de ataque hipoide.

Differential side gears The gears inside the differential case that are internally splined to the axle shafts, and which are driven by the differential pinion gears.

Planetarios (laterales) Los engranajes adentro de la caja de satélites que son acanalados a los semiejes desde el interior, y que se arrastran por los satélites.

Dipstick A metal rod used to measure the fluid in an engine or transmission.

Varilla de medida Una varilla de metal que se usa para medir el nivel de flúido en un motor o en una transmisión.

Direct drive One turn of the input driving member compared to one complete turn of the driven member, such as when there is direct engagement between the engine and driveshaft where the engine crankshaft and the driveshaft turn at the same rpm.

Mando directo Una vuelta del miembro de ataque o propulsión que se compara a una vuelta completa del miembro de arrastre, tal como cuando hay un enganchamiento directo entre el motor y el árbol de transmisión cuando el ciqueñal y el árbol de transmisión giran al mismo rpm.

Disengage When the operator moves the clutch pedal toward the floor to disconnect the driven clutch disc from the driving flywheel and pressure plate assembly.

Desembragar Cuando el operador mueva el pedal de embrague hacia el piso para desconectar el disco de embrague del volante impulsor y del conjunto del plato opresor.

Distortion A warpage or change in form from the original shape.

Distorción Abarquillamiento o un cambio en la forma original.

Dowel A metal pin attached to one object which, when inserted into a hole in another object, ensures proper alignment.

Espiga Una clavija de metal que se fija a un objeto, que al insertarla en el hoyo de otro objeto, asegura una alineación correcta.

Dowel pin A pin inserted in matching holes in two parts to maintain those parts in fixed relation one to another.

Clavija de espiga Una clavija que se inserte en los hoyos alineados en dos partes para mantener ésos dos partes en una relación fijada el uno al otro.

Downshift To shift a transmission into a lower gear.

Cambio descendente Cambiar la velocidad de una transmision a una velocidad más baja.

Driveline torque Relates to rear-wheel driveline and is the transfer of torque between the transmission and the driving axle assembly.

Potencia de la flecha motríz Se relaciona a la flecha motríz de las ruedas traseras y transfere la potencia de la torsión entre la transmisión y el conjunto del eje trasero.

Driven gear The gear meshed directly with the driving gear to provide torque multiplication, reduction, or a change of direction.

Engranaje de arrastre El engranaje endentado directamente al engranaje de ataque para proporcionar la multiplicación, la reducción, o los cambios de dirección de la potencia.

Drive pinion gear One of the two main driving gears located within the transaxle or rear driving axle housing. Together the two gears multiply engine torque.

Engranaje de piñon de ataque Uno de dos engranajes de ataque principales que se ubican adentro de la flecha de transmisión o en el cárter del eje de propulsión. Los dos engranajes trabajan juntos para multiplicar la potencia.

Drive shaft An assembly of one or two universal joints connected to a shaft or tube; used to transmit power from the transmission to the differential. Also called the propeller shaft.

Árbol de mando Una asamblea de uno o dos uniones universales que se conectan a un árbol o un tubo; se usa para transferir la potencia desde la transmisión al diferencial. También se le refiere como el árbol de propulsión.

Drop forging A piece of steel shaped between dies while hot.

Estampado Un pedazo de acero que se forma entre bloques mientras que esté caliente.

Dry friction The friction between two dry solids.

Fricción seca Fricción entre dos sólidos secos.

Dynamic In motion.

Dinámico En movimiento.

Dynamic balance The balance of an object when it is in motion; for example, the dynamic balance of a rotating drive shaft.

Balance dinámico El balance de un objeto mientras que esté en movimiento: por ejemplo el balance dinámico de un árbol de mando giratorio.

Eccentric One circle within another circle wherein both circles do not have the same center or a circle mounted off center. On FWD cars, front-end camber adjustments are accomplished by turning an eccentric cam bolt that mounts the strut to the steering knuckle.

Excéntrico Se dice de dos círculos, el uno dentro del otro, que no comparten el mismo centro o de un círculo ubicado descentrado. En los coches FWD, los ajustes de la inclinación se efectuan por medio de un perno excéntrico que fija el poste sobre el articulación de dirección

Efficiency The ratio between the power of an effect and the power expended to produce the effect; the ratio between an actual result and the theoretically possible result.

Eficiencia La relación entre la potencia de un efecto y la potencia que se gasta para producir el efecto; la relación entre un resultado actual y el resultado que es una posibilidad teórica.

Elastomer Any rubber-like plastic or synthetic material used to make bellows, bushings, and seals.

Elastómero Cualquiera materia plást parecida al hule o una materia sintética que se utiliza para fabricar a los fuelles, los bujes y las juntas.

End clearance Distance between a set of gears and their cover, commonly measured on oil pumps.

Holgura del extremo La distancia entre un conjunto de engranajes y su placa de recubrimiento, suele medirse en las bombas de aceite.

Endplay The amount of axial or end-to-end movement in a shaft due to clearance in the bearings.

Juego de las extremidades La cantidad del movimiento axial o del movimiento de extremidad a extremidad en una flecha debido a la holgura que se deja en los cojinetes.

Engage When the vehicle operator moves the clutch pedal up from the floor, this engages the driving flywheel and pressure plate to rotate and drive the driven disc.

Accionar Cuando el operador del vehículo deja subir el pedal del embrague del piso, ésto acciona la volante de ataque y el plato opresor para impulsar al disco de arrastre.

Engagement chatter A shaking, shuddering action that takes place as the driven disc makes contact with the driving members. Chatter is caused by a rapid grip and slip action.

Chasquido de enganchamiento Un movimiento de sacudo o temblor que resulta cuando el disco de ataque viene en contacto con los miembros de propulsión. El chasquido se causa por una acción rápida de agarrar y deslizar.

Engine torque A turning or twisting action developed by the engine, measured in foot-pounds or kilogram meters.

Torsión del motor Una acción de girar o torcer que crea el motor, ésta se mide en libras-pie o kilos-metros.

Essential tool kit A set of special tools designed for a particular model of car or truck.

Estuche de herramientas principales Un conjunto de herramientas especiales diseñadas para un modelo específico de coche o camión.

Etching A discoloration or removal of some material caused by corrosion or some other chemical reaction.

Grabado por ácido Una descoloración o remueva de una materia que se efectua por medio de la corrosión u otra reacción química.

Extension housing An aluminum or iron casting of various lengths that encloses the transmission output shaft and supporting bearings.

Cubierta de extensión Una pieza moldeada de aluminio o acero que puede ser de varias longitudes que encierre a la flecha de salida de la transmisión y a los cojinetes de soporte.

External gear A gear with teeth across the outside surface.

Engranaje exterior Un engranaje cuyos dientes estan en la superficie exterior.

Externally tabbed clutch plates Clutch plates that are designed with tabs around the outside periphery to fit into grooves in a housing or drum.

Placas de embrague de orejas externas Las placas de embrague que se diseñan de un modo para que las orejas periféricas de la superficie se acomoden en una ranura alrededor de un cárter o un tambor.

Extreme-pressure lubricant A special lubricant for use in hypoid-gear differentials; needed because of the heavy wiping loads imposed on the gear teeth.

Lubricante de presión extrema Un lubricante especial que se usa en las diferenciales de tipo engranaje hipóide; se requiere por la carga de transmisión de materia pesada que se imponen en los dientes del engranaje.

Face The front surface of an object.

Cara La superficie delantera de un objeto.

Fatigue The buildup of natural stress forces in a metal part that eventually causes it to break. Stress results from bending and loading the material.

Fatiga El incremento de tensiones y esfuerzos normales en una parte de metal que eventualmente causen una quebradura. Los esfuerzos resultan de la carga impuesta y el doblamiento de la materia.

Feeler gauge A metal strip or blade finished accurately with regard to thickness used for measuring the clearance between two parts; such gauges ordinarily come in a set of different blades graduated in thickness by increments of 0.001 inch.

Calibrador de laminillas Una lámina o hoja de metal que ha sido acabado precisamente con respecto a su espesor que se usa para medir la holgura entre dos partes; estas galgas típicamente vienen en un conjunto de varias espesores graduados desde el 0.001 de una pulgada.

Final drive ratio The ratio between the drive pinion and ring gear.

Relación del mando final La relación entre el piñon de ataque y la corona.

Fit The contact between two machined surfaces.

Ajuste El contacto entre dos superficies maquinadas.

Fixed-type constant-velocity joint A joint that cannot telescope or plunge tocompensate for suspension travel. Fixed joints are always found on the outer ends of the drive shafts of FWD cars. A fixed joint may be of either Rzeppa or tripod type.

Junta tipo fijo de velocidad constante Una junta que no tiene la capacidad de los movimientos telescópicos o repentinos que sirven para compensar en los viajes de suspensión. Las juntas fijas siempre se ubican en las extremidades exteriores de los árboles de mando en los coches de FWD. Una junta tipo fijo puede ser de un tipo Rzeppa o de trípode.

Flange A projecting rim or collar on an object for keeping it in place.

Reborde Una orilla o un collar sobresaliente de un objeto cuyo función es de mantenerlo en lugar.

Flexplate A lightweight flywheel used only on engines equipped with an automatic transmission. The flexplate is equipped with a starter ring gear around its outside diameter and also serves as the attachment point for the torque converter.

Placa articulada Un volante ligera que se usa solamente en los motores que se equipan con una transmisión automática. El diámetro exterior de la placa articulada viene equipado con un anillo de engranajes para arrancar y también sirve como punto de conección del convertidor de la torsión.

Fluid coupling A device in the power train consisting of two rotating members; transmits power from the engine, through a fluid, to the transmission.

Acoplamiento de fluido Un dispositivo en el tren de potencia que consiste de dos miembros rotativos; transmite la potencia del motor, for medio de un fluido, a la transmisión.

Fluid drive A drive in which there is no mechanical connection between the input and output shafts, and power is transmitted by moving oil.

Dirección fluido Una dirección en la cual no hay conecciones mecánicas entre las flechas de entrada o salida, y la potencia se transmite por medio del aceite en movimiento.

Flywheel A heavy metal wheel that is attached to the crankshaft and rotates with it; helps smooth out the power surges from the engine power strokes; also serves as part of the clutch and engine-cranking system.

Volante Una rueda pesada de metal que se fija al cigueñal y gira con ésta; nivela a los sacudos que provienen de la carrera de fuerza del motor; también sirve como parte del embrague y del sistema de arranque.

Flywheel ring gear A gear, fitted around the flywheel, that is engaged by teeth on the starting-motor drive to crank the engine.

Engranaje anular del volante Un engranaje, colocado alrededor del volante que se acciona por los dientes en el propulsor del motor de arranque y arranca al motor.

Foot-Pound (ft. lb.) A measure of the amount of energy or work required to lift 1 pound a distance of 1 foot.

Pie libra Una medida de la cantidad de energía o fuerza que requiere mover una libra a una distancia de un pie.

Force Any push or pull exerted on an object; measured in pounds and ounces, or in newtons (N) in the metric system.

Fuerza Cualquier acción empujado o jalado que se efectua en un objeto; se mide en pies y onzas, o en newtones (N) en el sistema métrico.

Four-wheel drive On a vehicle, driving axles at both front and rear, so that all four wheels can be driven.

Tracción a cuatro ruedas En un vehículo, se trata de los ejes de dirección fronteras y traseras, para que cada uno de las ruedas puede impulsar.

Frame The main understructure of the vehicle to which everything else is attached. Most FWD cars have only a subframe for the front suspension and drive train. The body serves as the frame for the rear suspension.

Armazón La estructura principal del vehículo al cual todo se conecta. La mayoría de los coches FWD sólo tiene un bastidor auxiliar para la suspensión delantera y el tren de propulsión. El carrocería del coche sirve de chassis par la suspensión trasera.

Free-wheel To turn freely and not transmit power.

Volante libre Da vueltas libremente sin transferir la potencia.

Free-wheeling clutch A mechanical device that will engage the driving member to impart motion to a driven member in one direction but not the other. Also known as an "overrunning clutch".

Embrague de volante libre Un dispositivo mecánico que acciona el miembro de tracción y da movimiento al miembro de tracción en una dirección pero no en la otra. También se conoce bajo el nombre de un "embrague de sobremarcha".

Friction The resistance to motion between two bodies in contact with each other.

Fricción La resistencia al movimiento entre dos cuerpos que estan en contacto.

Friction bearing A bearing in which there is sliding contact between the moving surfaces. Sleeve bearings, such as those used in connecting rods, are friction bearings.

Rodamientos de fricción Un cojinete en el cual hay un contacto deslizante entre las superficies en movimiento. Los rodamientos de manguitos, como los que se usan en las bielas, son rodamientos de fricción.

Friction disc In the clutch, a flat disc, faced on both sides with friction material and splined to the clutch shaft. It is positioned between the clutch pressure plate and the engine flywheel. Also called the clutch disc or driven disc.

Disco de fricción En el embrague, un disco plano al cual se ha cubierto ambos lados con una materia de fricción y que ha sido estriado a la flecha del embrague. Se posiciona entre el plato opresor del embrague y el volante del motor. También se llama el disco del embrague o el disco de arrastre.

Friction facings A hard-molded or woven asbestos or paper material that is riveted or bonded to the clutch driven disc.

Superficie de fricción Un recubrimiento remachado o aglomerado al disco de arrastre del embrague que puede ser hecho del amianto moldeado o tejido o de una materia de papel.

Front pump Pump located at the front of the transmission. It is driven by the engine through two dogs on the torque converter housing. It supplies fluid whenever the engine is running.

Bomba delantera Una bomba ubicado en la parte delantera de la transmisión. Se arrastre por el motor al través de dos álabes en el cárter del convertidor de la torsión. Provee el fluido mientras que funciona el motor.

Front-wheel drive (FWD) The vehicle has all drive train components located at the front.

Tracción de las ruedas delanteras (FWD) El vehículo tiene todos los componentes del tren de propulsión en la parte delantera.

FWD Abbreviation for front-wheeldrive.

FWD Abreviación de tracción de las ruedas delanteras.

Galling Wear caused by metal-to-metal contact in the absence of adequate lubrication. Metal is transferred from one surface to the other, leaving behind a pitted or scaled appearance.

Desgaste por fricción El desgaste causado por el contacto de metal a metal en la ausencia de lubricación adecuada. El metal se transfere de una superficie a la otra, causando una aparencia agujerado o con depósitos.

Gasket A layer of material, usually made of cork, paper, plastic, composition, or metal, or a combination of these, placed between two parts to make a tight seal.

Empaque Una capa de una materia, normalmente hecho del corcho, del papel, del plástico, de la materia compuesta o del metal, o de cualquier combinación de éstos, que se coloca entre dos partes para formar un sello impermeable.

Gasket cement A liquid adhesive material, or sealer, used to install gaskets.

Mastique para empaques Una substancia líquida adhesiva, o una substancia impermeable, que se usa para instalar a los empaques.

Gear A wheel with external or internal teeth that serves to transmit or change motion.

Engranaje Una rueda que tiene dientes interiores o exteriores que sirve para transferir o cambiar el movimiento.

Gear lubricant A type of grease or oil blended especially to lubricate gears.

Lubricante para engranaje Un tipo de grasa o aceite que ha sido mezclado específicamente para la lubricación de los engranajes.

Gear ratio The number of revolutions of a driving gear required to turn a driven gear through one complete revolution. For a pair of gears, the ratio is found by dividing the number of teeth on the driven gear by the number of teeth on the driving gear.

Relación de los engranajes El número de las revoluciones requeridas del engranaje de propulsión para dar una vuelta completa al engranaje arrastrado. En una pareja de engranajes, la relación se calcula al dividir el número de los dientes en el engranaje de arrastre por el número de los dientes en el engranaje de propulsión.

Gear reduction When a small gear drives a large gear, there is an output speed reduction and a torque increase which results in a gear reduction.

Velocidad descendente Cuando un engranaje pequeño impulsa a un engranaje grande, hay una reducción en la velocidad de salida y un incremento en la torsión que resulta en una cambio descendente de los velocidades.

Gearshift A linkage-type mechanism by which the gears in an automobile transmission are engaged and disengaged.

Varillaje de cambios Un mecanismo tipo eslabón que acciona y desembraga a los engranajes de la transmisión.

Gear whine A high-pitched sound developed by some types of meshing gears.

Ruido del engranaje Un sonido agudo que proviene de algunos tipos de engranajes endentados.

Governor pressure The transmission's hydraulic pressure which is directly related to output shaft speed. It is used to control shift points.

Regulador de presión La presión hidráulica de una transmisión se relaciona directamente a la velocidad de la flecha de salida. Se usa para controllar los puntos de cambios de velocidad.

Governor valve A device used to sense vehicle speed. The governor valve is attached to the output shaft.

Válvula reguladora Un dispositivo que se usa para determinar la velocidad de un vehículo. La válvula reguladora se monta en la flecha de salida.

Graphite Very fine carbon dust with a slippery texture used as a lubricant.

Grafito Un polvo de carbón muy fino con una calidad grasosa que se usa para lubricar.

Grind To finish or polish a surface by means of an abrasive wheel.

Amolar Acabar o pulir a una superficie por medio de una muela para pulverizar.

Heat treatment Heating, followed by fast cooling, to harden metal.

Tratamiento térmico Calentamiento, seguido por un enfriamiento rápido, para endurecer a un metal.

Horsepower A measure of mechanical power, or the rate at which work is done. One horsepower equals 33,000 ft.-lb. (foot-pounds) of work per minute. It is the power necessary to raise 33,000 pounds a distance of 1 foot in 1 minute.

Caballo de fuerza Una medida de fuerza mecánica, o el régimen en el cual se efectua el trabajo. Un caballo de fuerza iguala a 33,000 lb.p. (libras pie) de trabajo por minuto. Es la fuerza requerida para transportar a 33,000 libras una distancia de 1 pie en 1 minuto.

Hub The center part of a wheel, to which the wheel is attached.

Cubo La parte central de una rueda, a la cual se monta la rueda.

Hydraulic press A piece of shop equipment that develops a heavy force by use of a hydraulic piston-and-jack assembly.

Prensa hidráulica Una herramienta del taller que provee una fuerza grande por medio de una asamblea de gato con un pistón hidráulico.

Hydraulic pressure Pressure exerted through the medium of a liquid.

Presión hidráulica La presión esforzada por medio de un líquido.

ID Inside Diameter.

DI Diámetro Interior.

Idle Engine speed when the accelerator pedal is fully released and there is no load on the engine.

Marcha lenta La velocidad del motor cuando el pedal accelerador esta completamente desembragada y no hay carga en el motor.

Impedance The operating resistance of an electrical device.

Impedancia La resistencia operativa de un dispositivo eléctrico.

Impeller The pump or driving member in a torque converter.

Impulsor La bomba o el miembro impulsor en un convertidor de torsión.

Increments Series of regular additions from small to large.

Incrementos Una serie de incrementos regulares que va de pequeño a grande.

Index To orient two parts by marking them. During reassembly the parts are arranged so the index marks are next to each other. Used to preserve the orientation between balanced parts.

Índice Orientar a dos partes marcándolas. Al montarlas, las partes se colocan para que las marcas de índice estén alinieadas. Se usan los índices para preservar la orientación de las partes balanceadas.

Input shaft The shaft carrying the driving gear by which the power is applied, as to the transmission.

Flecha de entrada La flecha que porta el engranaje propulsor por el cual se aplica la potencia, como a la transmisión.

Inspection cover A removable cover that permits entrance for inspection and service work.

Cubierta de inspección Una cubierta desmontable que permite a la entrada para inspeccionar y mantenimiento.

Integral Built into, as part of the whole.

Integral Incorporado, una parte de la totalidad.

Internal gear A gear with teeth pointing inward, toward the hollow center of the gear.

Engranaje internal Un engranaje cuyos dientes apuntan hacia el interior, al hueco central del engranaje.

Jam nut A second nut tightened against a primary nut to prevent it from working loose. Used on inner and outer tie-rod adjustment nuts and on many pinion-bearing adjustment nuts.

Contra tuerca Una tuerca secundaria que se aprieta contra una tuerca primaria para prevenir que ésta se afloja. Se emplean en las tuercas de ajustes interiores e exteriores para las barras de acoplamiento y también en muchas de las tuercas de ajuste de portapiñones.

Journal A bearing with a hole in it for a shaft.

Manga de flecha Un cojinete que tiene un hoyo para una flecha.

Key A small block inserted between the shaft and hub to prevent circumferential movement.

Chaveta Un tope pequeño que se meta entre la flecha y el cubo para prevenir un movimiento circunferencial.

Keyway A groove or slot cut to permit the insertion of a key.

Ranura de chaveta Un corte de ranura o mortaja que permite insertar una chaveta.

Knock A heavy metallic sound usually caused by a loose or worn bearing.

Golpe Un sonido metálico fuerte que suele ser causado por un cojinete suelto o gastado.

Knurl To indent or roughen a finished surface.

Moletear Indentar o desbastar a una superficie acabada.

Lapping The process of fitting one surface to another by rubbing them together with an abrasive material between the two surfaces.

Pulido El proceso de ajustar a una superficie con otra por frotarlas juntas con una materia abrasiva entre las dos superficies.

Lash The amount of free motion in a gear train, between gears, or in a mechanical assembly, such as the lash in a valve train.

Juego La cantidad del movimiento libre en un tren de engranajes, entre los engranajes o en una asamblea mecánica, tal como el juego en un tren de vávulas.

Linkage Any series of rods, yokes, levers, and so on, used to transmit motion from one unit to another.

Biela Cualquiera serie de barras, yugos, palancas, y todo lo demás, que se usa para transferir los movimientos de una unedad a otra.

Locknut A second nut turned down on a holding nut to prevent loosening.

Contra tuerca Una tuerca segundaria apretada contra una tuerca de sostén para prevenir que ésta se afloja.

Lock pin Used in some ball sockets (inner tie-rod end) to keep the connecting nuts from working loose. Also used on some lower ball joints to hold the tapered stud in the steering knuckle.

Clavija de cerrojo Se usan en algunas rótulas (las extremidades interiores de la barra de acoplamiento) para prevenir que se aflojan las tuercas de conexión. También se emplean en algunas juntas esféricas inferiores para retener al perno cónico en la articulación de dirección.

Lockplates Metal tabs bent around nuts or bolt heads.

Placa de cerrojo Chavetas de metal que se doblan alrededor de las tuercas o las cabezas de los pernos.

Lockwasher A type of washer which, when placed under the head of a bolt or nut, prevents the bolt or nut from working loose.

Arandela de freno Un tipo de arandela que, al colocarse bajo la cabeza de un perno, previene que el perno o la tuerca se aflojan.

Low speed The gearing that produces the highest torque and lowest speed of the wheels.

Velocidad baja La velocidad que produce la torsión más alta y la velocidad más baja a las ruedas.

Lubricant Any material, usually a petroleum product such as grease or oil, that is placed between two moving parts to reduce friction.

Lubricante Cualquier substancia, normalmente un producto de petróleo como la grasa o el aciete, que se coloca entre dos partes en movimiento para reducir la fricción.

Mainline pressure The hydraulic pressure that operates apply devices and is the source of all other pressures in an automatic transmission. It is developed by pump pressure and regulated by the pressure regulator.

Línea de presión La presión hidráulica que opera a los dispositivos de applicación y es el orígen de todas las presiones en la transmisión automática. Proviene de la bomba de presión y es regulada por el regulador de presión.

Main oil pressure regulator valve Regulates the line pressure in a transmission.

Válvula reguladora de la linea de presión Regula la presión en la linea de una transmisión.

Manual control valve A valve used to manually select the operating mode of the transmission. It is moved by the gearshift linkage.

Válvula de control manual Una válvula que se usa para escojer a una velocidad de la transmisión por mano. Se mueva por la biela de velocidades.

Meshing The mating, or engaging, of the teeth of two gears.

Engrane Embragar o endentar a los dientes de dos engranajes.

Meter 1/10,000,000 of the distance from the North Pole to the Equator, or 39.37 inches.

Metro Un 1/10,000,000 de la distancia del polo del norte al ecuador.

Micrometer A precision measuring device used to measure small bores, diameters, and thicknesses. Also called a mike.

Micrómetro Un dispositivo de medida presisa que se emplea a medir a los taladros pequeños y a los espesores. También se llama un mike (mayk).

Misalignment When bearings are not on the same centerline.

Desalineamineto Cuando los cojinetes no comparten la misma linea central.

Modulator A vacuum-diaphragm device connected to a source of engine vacuum. It provides an engine load signal to the transmission.

Modulador Un dispositivo de diafragma de vacío que se conecta a un orígen de vacío en el motor. Provee un señal de carga del motor a la transmisión.

Mounts Made of rubber to insulate vibrations and noise while they support a power train part, such as engine or transmission mounts.

Monturas Hecho de hule para insular a las vibraciones y a los ruidos mientras que sujetan una parte del tren de propulsión, tal como las monturas del motor o las monturas de la transmisión.

Multiple disc A clutch with a number of driving and driven discs as compared to a single plate clutch.

Discos múltiples Un embrague que tiene varios discos de propulsión o de arraste al contraste con un embrague de un sólo plato.

Needle bearing An antifriction bearing using a great number of long, small-diameter rollers. Also known as a quill bearing.

Rodamiento de agujas Un rodamiento (cojinete) antifricativo que emplea un gran cantidad de rodillos largos y de diámetro muy pequeños.

Needle deflection Distance of travel from zero of the needle on a dial gauge.

Desviación de la aguja La distancia del cero que viaja una aguja de un indicador.

Neoprene A synthetic rubber that is not affected by the various chemicals that are harmful to natural rubber.

Neoprene Un hule sintético que no se afecta por los varios productos químicos que pueden dañar al hule natural.

Neutral In a transmission, the setting in which all gears are disengaged and the output shaft is disconnected from the drive wheels.

Neutral En una transmisión, la velocidad en la cual todos los engranajes estan desembragados y el árbol de salida esta desconectada de las ruedas de propulsión.

Neutral-start switch A switch wired into the ignition switch to prevent engine cranking unless the transmission shift lever is in neutral or the clutch pedal is depressed.

Interruptor de arranque en neutral Un interruptor eléctrico instalado en el interruptor de encendido que previene el arranque del motor al menos de que la palanca de cambio de velocidad esté en una posición neutral o que se pisa en el embrague.

Newton-meter (Nm) Metric measurement of torque or twisting force.

Metro newton (Nm) Una medida métrica de la fuerza de torsión.

Nominal shim A shim with a designated thickness.

Laminilla fina Una cuña de un espesor especificado.

Nonhardening A gasket sealer that never hardens.

Sinfragua Un cemento de empaque que no endurece.

Nut A removable fastener used with a bolt to lock pieces together; made by threading a hole through the center of a piece of metal that has been shaped to a standard size.

Tuerca Un retén removable que se usa con un perno o tuerca para unir a dos piezas; se fabrica al filetear un hoyo taladrado en un pedazo de metal que se ha formado a un tamaño especificado.

OD Outside diameter.

DE Diámetro exterior.

Oil seal A seal placed around a rotating shaft or other moving part to prevent leakage of oil.

Empaque de aciete Un empaque que se coloca alrededor de una flecha giratoria para prevenir el goteo de aceite.

One-way clutch *See* Sprag clutch.

Embrague de una via *Vea* Sprag clutch.

O-ring A type of sealing ring, usually made of rubber or a rubber-like material. In use, the O-ring is compressed into a groove to provide the sealing action.

Anillo en O Un tipo de sello anular, suele ser hecho de hule o de una materia parecida al hule. Al usarse, el anillo en O se comprime en una ranura para proveer un sello.

Oscillate To swing back and forth like a pendulum.

Oscilar Moverse alternativamente en dos sentidos contrarios como un péndulo.

Outer bearing race The outer part of a bearing assembly on which the balls or rollers rotate.

Pista exterior de un cojinete La parte exterior de una asamblea de cojinetes en la cual ruedan las bolas o los rodillos.

Out-of-round Wear of a round hole or shaft which, when viewed from an end, will appear egg-shaped.

Defecto de circularidad Desgaste de un taladro o de una flecha circular, que al verse de una extremidad, tendrá una forma asimétrica, como la de un huevo.

Output shaft The shaft or gear that delivers the power from a device, such as a transmission.

Flecha de salida La flecha o la velocidad que transmite la potencia de un dispositivo, tal como una transmisión.

Overall ratio The product of the transmission gear ratio multiplied by the final drive or rear axle ratio.

Relación global El producto de multiplicar la relación de los engranajes de la transmisión por la relación del impulso final o por la relación del eje trasero.

Overdrive Any arrangement of gearing that produces more revolutions of the driven shaft than of the driving shaft.

Sobremultiplicación Un arreglo de los engranajes que produce más revoluciones de la flecha de arrastre que los de la flecha de propulsión.

Overdrive ratio Identified by the decimal point indicating less than one driving input revolution compared to one output revolution of a shaft.

Relación del sobremultiplicación Se identifica por el punto decimal que indica menos de una revolución del motor comparado a una revolución de una flecha de salida.

Overrun coupling A free-wheeling device to permit rotation in one direction but not in the other.

Acoplamiento de sobremarcha Un dispositivo de marcha de rueda libre que permite las giraciones en una dirección, pero no en la otra dirección.

Overrunning clutch A device consisting of a shaft or housing linked together by rollers or sprags operating between movable and fixed races. As the shaft rotates, the rollers or sprags jam between the movable and fixed races. This jamming action locks together the shaft and housing. If the fixed race should be driven at a speed greater than the movable race, the rollers or sprags will disconnect the shaft.

Embrague de sobremarcha Un dispositivo que consiste de una flecha o un cárter eslabonados por medio de rodillos o palancas de detención que operan entre pistas fijas y movibles. Al girar la flecha, los rodillos o palancas de detención se aprietan entre las pistas fijas y movibles. Este acción de apretarse enclava el cárter con la flecha. Si la pista fija se arrastra en una velocidad más alta que la pista movible, los rodillos o palancas de detención desconectarán a la flecha.

Oxidation Burning or combustion; the combining of a material with oxygen. Rusting is slow oxidation, and combustion is rapid oxidation.

Oxidación Quemando o la combustión; la combinación de una materia con el oxígeno. El orín es una oxidación lenta, la combustión es la oxidación rápida.

Pascal's Law The law of fluid motion.

Ley de pascal La ley del movimiento del fluido.

Parallel The quality of two items being the same distance from each other at all points; usually applied to lines and, in automotive work, to machined surfaces.

Paralelo La calidad de dos artículos que mantienen la misma distancia el uno del otro en cada punto; suele aplicarse a las líneas y, en el trabajo automotívo, a las superficies acabadas a máquina.

Pawl A lever that pivots on a shaft. When lifted, it swings freely and when lowered, it locates in a detent or notch to hold a mechanism stationary.

Trinquete Una palanca que gira en una flecha. Levantado, mueve sín restricción, bajado, se coloca en una endentación o una muesca para mantener sín movimiento a un mecanismo.

Peen To stretch or clinch over by pounding with the rounded end of a hammer.

Martillazo Estirar o remachar con la extremidad redondeado de un martillo de bola.

Pitch The number of threads per inch on any threaded part.

Paso El número de filetes por pulgada de cualquier parte fileteada.

Pivot A pin or shaft upon which another part rests or turns.

Pivote Una chaveta o una flecha que soporta a otra parte o sirve como un punto para girar.

Planetary gear set A system of gearing that is modeled after the solar system. A pinion is surrounded by an internal ring gear and planet gears are in mesh between the ring gear and pinion around which all revolve.

Conjunto de engranajes planetarios Un sistema de engranaje cuyo patrón es el sistema solar. Un engranaje propulsor (la corona interior) rodea al piñon de ataque y los engranajes satélites y planetas se endentan entre la corona y el piñon alrededor del cual todo gira.

Planet carrier The carrier or bracket in a planetary gear system that contains the shafts upon which the pinions or planet gears turn.

Perno de arrastre planetario El soporte o la abrazadera que contiene las flechas en las cuales giran los engranajes planetarios o los piñones.

Planet gears The gears in a planetary gear set that connect the sun gear to the ring gear.

Engranages plantearios Los engranajes en un conjunto de engranajes planetario que connectan al engranaje propulsor interior (el engranaje sol) con la corona.

Planet pinions In a planetary gear system, the gears that mesh with, and revolve about, the sun gear; they also mesh with the ring gear.

Piñones planetarios En un sistema de engranajes planetarios, los engranajes que se endentan con, y giran alrededor, el engranaje propulsor (sol); también se endentan con la corona.

Plug Anything that will fit into an opening to stop fluid or airflow.

Tapón Cualquier cosa que se ajuste en una apertura para prevenir el goteo o el escape de un corriente del aire.

Pneumatic tools Power tools that rely on compressed air for power.

Herramientas neumáticas Las herramientas de motor cuyo energía proviene del aire comprimido.

Porosity A statement of how porous or permeable to liquids a material is.

Porosidad Una expresión de lo poroso o permeable a los líquidos es una materia.

Power train The mechanisms that carry the power from the engine crankshaft to the drive wheels; these include the clutch, transmission, drive line, differential, and axles.

Tren impulsor Los mecanismos que transferen la potencia desde el cigueñal del motor a las ruedas de propulsión; éstos incluyen el embrague, la transmisión, la flecha motríz, el diferencial y los semiejes.

Preload A load applied to a part during assembly so as to maintain critical tolerances when the operating load is applied later.

Carga previa Una carga aplicada a una parte durante la asamblea para asegurar sus tolerancias críticas antes de que se le aplica la carga de la operación.

Press fit Forcing a part into an opening that is slightly smaller than the part itself to make a solid fit.

Ajustamiento a presión Forzar a una parte en una apertura que es de un tamaño más pequeño de la parte para asegurar un ajustamiento sólido.

Pressure Force per unit area, or force divided by area. Usually measured in pounds per square inch (psi) or in kilopascals (kPa) in the metric system.

Presión La fuerza por unedad de una area, o la fuerza divida por la area. Suele medirse en libras por pulgada cuadrada (lb/pulg2) o en kilopascales (kPa) en el sistema métrico.

Pressure plate That part of the clutch which exerts force against the friction disc; it is mounted on and rotates with the flywheel.

Plato opresor Una parte del embraque que aplica la fuerza en el disco de fricción; se monta sobre el volante, y gira con éste.

Propeller shaft *See* Driveshaft.

Flecha de Propulsion *Vea* Flecha motríz.

Prussian blue A blue pigment; in solution, useful in determining the area of contact between two surfaces.

Azul de Prusia Un pigmento azul; en forma líquida, ayuda en determinar la area de contacto entre dos superficies.

PSI Abbreviation for pounds per square inch, a measurement of pressure.

Lb/pulg2 Una abreviación de libras por pulgada cuadrada, una medida de la presión.

Puller Generally, a shop tool used to separate two closely fitted parts without damage. Often contains a screw, or several screws, which can be turned to apply a gradual force.

Extractor Generalmente, una herramienta del taller que sirve para separar a dos partes apretadas sin incurrir daños. Suele tener una tuerca o varias tuercas, que se pueden girar para aplicar la fuerza gradualmente.

Pulsation To move or beat with rhythmic impulses.

Pulsación Moverse o batir con impulsos rítmicos.

Race A channel in the inner or outer ring of an antifriction bearing in which the balls or rollers roll.

Pista Un canal en el anillo interior o exterior de un cojinete antifricción en el cual ruedan las bolas o los rodillos.

Radial The direction moving straight out from the center of a circle. Perpendicular to the shaft or bearing bore.

Radial La dirección al moverse directamente del centro de un círculo. Perpendicular a la flecha o al taladro del cojinete.

Radial clearance Clearance within the bearing and between balls and races perpendicular to the shaft. Also called radial displacement

Holgura radial La holgura en un cojinete entre las bolas y las pistas que son perpendiculares a la flecha. También se llama un desplazamiento radial.

Radial load A force perpendicular to the axis of rotation.

Carga radial Una fuerza perpendicular al centro de rotación.

Ratio The relation or proportion that one number bears to another.

Relación La correlación o proporción de un número con respeto a otro.

Reamer A round metal-cutting tool with a series of sharp cutting edges; enlarges a hole when turned inside it.

Escariador Una herramienta redonda para cortar a los metales que tiene una seria de rebordes mordaces agudos; al girarse en un agujero lo agranda.

Rear-wheel drive A term associated with a vehicle where the engine is mounted at the front and the driving axle and driving wheels are at the rear of the vehicle.

Tracción trasera Un término que se asocia con un vehículo en el cual el motor se ubica en la parte delantera y el eje propulsor y las ruedas propulsores se encuentran en la parte trasera del vehículo.

Relief valve A valve used to protect against excessive pressure in the case of a malfunctioning pressure regulator.

Válvula de seguridad Una válvula que se usa para guardar contra una presión excesiva en caso de que malfulciona el regulador de presión.

Retaining ring A removable fastener used as a shoulder to retain and position a round bearing in a hole.

Anillo de retén Un seguro removible que sirve de collarín para sujetar y posicionar a un cojinete en un agujero.

Rivet A headed pin used for uniting two or more pieces by passing the shank through a hole in each piece and securing it by forming a head on the opposite end.

Remache Una clavija con cabeza que sirve para unir a dos piezas o más al pasar el vástago por un hoyo en cada pieza y asegurarlo por formar una cabeza en el extremo opuesto.

Roller bearing An inner and outer race upon which hardened steel rollers operate.

Cojinete de rodillos Una pista interior y exterior en la cual operan los rodillos hecho de acero endurecido.

Rollers Round steel bearings that can be used as the locking element in an overrunning clutch or as the rolling element in an antifriction bearing.

Rodillos Articulaciones redondos de acero que pueden servir como un elemento de enclavamiento en un embrague de sobremarcha o como el elemento que rueda en un cojinete antifricción.

Rotary flow A fluid force generated in the torque converter that is related to vortex flow. The vortex flow leaving the impeller is not only flowing out of the impeller at high speed but is also rotating faster than the turbine. The rotating fluid striking the slower turning turbine exerts a force against the turbine which is defined as rotary flow.

Flujo rotativo Una fuerza fluida producida en el convertidor de torsión que se relaciona al flujo torbellino. El flujo torbellino saliendo del rotor no sólo viaja en una alta velocidad sino también gira más rápidamente que el turbino. El fluido rotativo chocando contra el turbino que gira más lentamente, impone una fuerza contra el turbino que se define como flujo rotativo.

RPM Abbreviation for revolutions per minute, a measure of rotational speed.

RPM Abreviación de revoluciones por minuto, una medida de la velocidad rotativa.

RTV sealer Room-temperature vulcanizing gasket material, which cures at room temperature; a plastic paste squeezed from a tube to form a gasket of any shape.

Sellador RTV Una materia vulcanizante de empaque que cura en temperaturas del ambiente; una pasta plástica exprimida de un tubo para formar un empaque de cualquiera forma.

Runout Deviation of the specified normal travel of an object. The amount of deviation or wobble a shaft or wheel has as it rotates. Runout is measured with a dial indicator.

Corrimiento Una desviación de la carrera normal e especificada de un objeto. La cantidad de desviación o vacilación de una flecha o una rueda mientras que gira. El corrimiento se mide con un indicador de carátula.

RWD Abbreviation for rear-wheel drive.

RWD Abreviación de tracción trasera.

SAE Society of Automotive Engineers.

SAE La Sociedad de Ingenieros Automotrices.

Score A scratch, ridge, or groove marring a finished surface.

Entalladura Una raya, una arruga o una ranura que desfigure a una superficie acabada.

Scuffing A type of wear in which there is a transfer of material between parts moving against each other; shows up as pits or grooves in the mating surfaces.

Erosión Un tipo de desgaste en el cual hay una tranferencia de una materia entre las partes que estan en contacto mientras que muevan; se manifesta como hoyitos o muescas en las superficies apareadas.

Seal A material, shaped around a shaft, used to close off the operating compartment of the shaft, preventing oil leakage.

Sello Una materia, formado alrededor de una flecha, que sella el compartimiento operativo de la flecha, previniendo el goteo de aceite.

Sealer A thick, tacky compound, usually spread with a brush, which may be used as a gasket or sealant to seal small openings or surface irregularities.

Sellador Un compuesto pegajoso y espeso, comúnmente aplicado con una brocha, que puede usarse como un empaque o un obturador para sellar a las aperturas pequeñas o a las irregularidades de la superficie.

Seat A surface, usually machined, upon which another part rests or seats; for example, the surface upon which a valve face rests.

Asiento Una superficie, comúnmente maquinada, sobre la cual yace o se asienta otra parte; por ejemplo, la superficie sobre la cual yace la cara de la válvula.

Servo A device that converts hydraulic pressure into mechanical movement, often multiplying it. Used to apply the bands of a transmission.

Servo Un dispositivo que convierte la presión hidráulica al movimiento mecánico, frequentemente multiplicándola. Se usa en la aplicación de las bandas de una transmisión.

Shift lever The lever used to change gears in a transmission. Also the lever on the starting motor which moves the drive pinion into or out of mesh with the flywheel teeth.

Palanca del cambiador La palanca que sirve para cambiar a las velocidades de una transmisión. También es la palanca del motor de arranque que mueva al piñon de ataque para engranarse o desegranarse con los dientes del volante.

Shift valve A valve that controls the shifting of the gears in an automatic transmission.

Válvula de cambios Una válvula que controla a los cambios de las velocidades en una transmisión automática.

Shim Thin sheets used as spacers between two parts, such as the two halves of a journal bearing.

Laminilla de relleno Hojas delgadas que sirven de espaciadores entre dos partes, tal como las dos partes de un muñon.

Shim stock Sheets of metal of accurately known thickness which can be cut into strips and used to measure or correct clearances.

Materia de laminillas Las hojas de metal cuyo espesor se conoce precisamente que pueden cortarse en tiras y usarse para medir o correjir a las holguras.

Side clearance The clearance between the sides of moving parts when the sides do not serve as load-carrying surfaces.

Holgura lateral La holgura entre los lados de las partes en movimiento mientras que los lados no funcionan como las superficies de carga.

Sliding-fit Where sufficient clearance has been allowed between the shaft and journal to allow free-running without overheating.

Ajuste corredera Donde se ha dejado una holgura suficiente entre la flecha y el muñon para permitir una marcha libre sin sobrecalentamiento.

Snap ring Split spring-type ring located in an internal or external groove to retain a part.

Anillo de seguridad Un anillo partido tipo resorte que se coloca en una muesca interior o exterior para retener a una parte.

Spalling A condition where the material of a bearing surface breaks away from the base metal.

Escamación Una condición en la cual una materia de la superficie de un rodamiento se separa del metal base.

Spindle The shaft on which the wheels and wheel bearings mount.

Husillo La flecha en la cual se montan las ruedas y el conjunto del cojinete de las ruedas.

Spline Slot or groove cut in a shaft or bore; a splined shaft onto which a hub, wheel, gear, and so on, with matching splines in its bore is assembled so that the two must turn together.

Acanaladura (espárrago) Una muesca o ranura cortada en una flecha o en un taladro; una flecha acanalada en la cual se asamblea un cubo, una rueda, un engranaje, y todo lo demás que tiene un acanaladura pareja en el taladro de manera de que las dos deben girar juntos.

Split lip seal Typically, a rope seal sometimes used to denote any two-part oil seal.

Sello hendido Típicamente, un sello de cuerda que se usa a veces para demarcar cualquier sello de aceite de dos partes

Split pin A round split spring steel tubular pin used for locking purposes; for example, locking a gear to a shaft.

Chaveta hendida Una chaveta partida redonda y tubular hecho de acero para resorte que sirve para el enclavamiento; por ejemplo, para enclavar un engranaje a una flecha.

Spool valve A cylindrically shaped valve with two or more valleys between the lands. Spool valves are used to direct fluid flow.

Válvula de carrete Una válvula de forma cilíndrica que tiene dos acanaladuras de cañon o más entre las partes planas. Las válvulas de carrete sirven para dirigir el flujo del fluido.

Sprag clutch A member of the overrunning clutch family using a sprag to jam between the inner and outer races used for holding or driving action.

Embrague de puntal Un miembro de la familia de embragues de sobremarcha que usa a una palanca de detención trabada entre las pistas interiores e exteriores para realizar una acción de asir o marchar.

Spring A device that changes shape when it is stretched or compressed, but returns to its original shape when the force is removed; the component of the automotive suspension system that absorbs road shocks by flexing and twisting.

Resorte Un dispositivo que cambia de forma al ser estirado o comprimido, pero que recupera su forma original al levantarse la fuerza; es un componente del sistema de suspensión automotivo que absorba los choques del camino al doblarse y torcerse.

Spring retainer A steel plate designed to hold a coil or several coil springs in place.

Retén de resorte Una chapa de acero diseñado a sostener en su posición a un resorte helicoidal o más.

Squeak A high-pitched noise of short duration.

Chillido Un ruido agudo de poca duración.

Squeal A continuous high-pitched noise.

Alarido Un ruido agudo continuo.

Stall A condition where the engine is operating and the transmission is in gear, but the drive wheels are not turning because the turbine of the torque converter is not moving.

Paro Una condición en la cual opera el motor y la transmisión esta embragada pero las ruedas de impulso no giran porque no mueva el turbino del convertidor de la torsión.

Stall test A test of the one-way clutch in a torque converter.

Prueba de paro Una prueba del embrague de una vía en un convertidor de la torsión.

Static A form of electricity caused by friction.

Estático Una forma de la electridad causada por la fricción.

Stress The force to which a material, mechanism, or component is subjected.

Esfuerzo La fuerza a la cual se somete a una materia, un mecanísmo o un componente.

Sun gear The central gear in a planetary gear system around which the rest of the gears rotate. The innermost gear of the planetary gear set.

Engranaje principal (sol) El engranaje central en un sistema de engranajes planetarios alrededor del cual giran los otros engranajes. El engranaje más interno del conjunto de los engranajes planetarios.

Tap To cut threads in a hole with a tapered, fluted, threaded tool.

Roscar con macho Cortar las roscas en un agujero con una herramienta cónica, acanalada y fileteada.

Teardown A term often used to describe the process of disassembling a transmission.

Desmontaje Un término común que describe el proceso de desarmar a una transmisión.

Temper To change the physical characteristics of a metal by applying heat.

Templar Cambiar las características físicas de un metal mediante una aplicación del calor.

Tension Effort that elongates or "stretches" a material.

Tensión Un esfuerzo que alarga o "estira" a una materia.

Thickness gauge Strips of metal made to an exact thickness, used to measure clearances between parts.

Calibre de espesores Las tiras del metal que se han fabricado a un espesor exacto, sirven para medir las holguras entre las partes.

Thread chaser A device, similar to a die, that is used to clean threads.

Peine de roscar Un dispositivo, parecido a una terraja, que sirve para limpiar a las roscas.

Threaded insert A threaded coil that is used to restore the original thread size to a hole with damaged threads.

Pieza inserta roscada Una bobina roscada que sirve para restaurar a su tamaño original una rosca dañada.

Thrust bearing A bearing designed to resist or contain side or end motion as well as reduce friction.

Cojinete de empuje Un cojinete diseñado a detener o reprimir a los movimientos laterales o de las extremidades y también reducir la fricción.

Thrust load A load that pushes or reacts through the bearing in a direction parallel to the shaft.

Carga de empuje Una carga que empuja o reacciona por el cojinete en una dirección paralelo a la flecha.

Thrust washer A washer designed to take up end thrust and prevent excessive endplay.

Arandela de empuje Una arandela diseñada para rellenar a la holgura de la extremidad y prevenir demasiado juego en la extremidad.

Tolerance A permissible variation between the two extremes of a specification or dimension.

Tolerancia Una variación permisible entre dos extremos de una especificación o de un dimensión.

Torque A twisting motion, usually measured in ft.-lb. (Nm).

Torsión Un movimiento giratorio, suele medirse en pies/libra (Nm).

Torque capacity The ability of a converter clutch to hold torque.

Capacidad de la torsión La abilidad de un convertidor de embraque a sostener a la torsión.

Torque converter A turbine device utilizing a rotary pump, one or more reactors (stators), and a driven circular turbine or vane, whereby power is transmitted from a driving to a driven member by hydraulic action. It provides varying drive ratios; with a speed reduction, it increases torque.

Convertidor de la torsión Un dispositivo de turbino que utilisa a una bomba rotativa, a un reactor o más, y un molinete o turbino circular impulsado, por cual se transmite la energía de un miembro de impulso a otro arrastrado mediante la acción hidráulica. Provee varias relaciones de impulso; al descender la velocidad, aumenta la torsión.

Torque curve A line plotted on a chart to illustrate the torque personality of an engine. When the engine operates on its torque curve, it is producing the most torque for the quantity of fuel being burned.

Curva de la torsión Una linea delineada en una carta para ilustrar las características de la torsión del motor. Al operar un motor en su curva de la torsión, produce la torsión óptima para la cantidad del combustible que se consuma.

Torque multiplication The result of meshing a small driving gear and a large driven gear to reduce speed and increase output torque.

Multiplicación de la torsión El resultado de engranar a un engranaje pequeño de ataque con un engranaje más grande arrastrado para reducir la velocidad y incrementar la torsión de salida.

Torque steer An action felt in the steering wheel as the result of increased torque.

Dirección la torsión Una acción que se nota en el volante de dirección como resultado de un aumento de la torsión.

Traction The gripping action between the tire tread and the road's surface.

Tracción La acción de agarrar entre la cara de la rueda y la superficie del camino.

Transaxle Type of construction in which the transmission and differential are combined in one unit.

Flecha de transmisión Un tipo de construcción en el cual la transmisión y el diferencial se combinan en una unedad.

Transaxle assembly A compact housing most often used in front-wheel-drive vehicles that houses the manual transmission, final drive gears, and differential assembly.

Asamblea de la flecha de transmisión Un cárter compacto que se usa normalmente en los vehículos de tracción delantera que contiene la transmisión manual, los engranajes de propulsión, y la asamblea del diferencial.

Transfer case An auxiliary transmission mounted behind the main transmission. Used to divide engine power and transfer it to both front and rear differentials, either full-time or part-time.

Cárter de la transferencia Una transmisión auxiliar montada detrás de la transmisión principal. Sirve para dividir la potencia del motor y transferirla a ambos diferenciales delanteras y traseras todo el tiempo o la mitad del tiempo.

Transmission The device in the power train that provides different gear ratios between the engine and drive wheels as well as reverse.

Transmisión El dispositivo en el trén de potencia que provee las relaciones diferentes de engranaje entre el motor y las ruedas de impulso y también la marcha de reversa.

Transverse Power train layout in a front-wheel-drive automobile extending from side to side.

Transversal Una esquema del tren de potencia en un automóvil de tracción delantera que se extiende de un lado a otro.

U-joint A four-point cross connected to two U-shaped yokes that serves as a flexible coupling between shafts.

Junta de U Una cruceta de cuatro puntos que se conecta a dos yugos en forma de U que sirven de acoplamientos flexibles entre las flechas.

Universal joint A mechanical device that transmits rotary motion from one shaft to another shaft at varying angles.

Junta Universal Un dispositivo mecánico que transmite el movimiento giratorio desde una flecha a otra flecha en varios ángulos.

Upshift To shift a transmission into a higher gear.

Cambio ascendente Cambiar a la velocidad de una transmisión a una más alta.

Valve body Main hydraulic control assembly of a transmission containing the components necessary to control the distribution of pressurized transmission fluid throughout the transmission.

Cuerpo de la válvula Asamblea principal del control hidráulico de una transmisión que contiene los componentes necessarios para controlar a la distribución del fluido de la transmisión bajo presión por toda la transmisión.

Vehicle identification number (VIN) The number assigned to each vehicle by its manufacturer, primarily for registration and identification purposes.

Número de identificacion del vehículo El número asignado a cada vehículo por su fabricante, primariamente con el propósito de la registración y la identificación.

Vibration A quivering, trembling motion felt in the vehicle at different speed ranges.

Vibración Un movimiento de estremecer o temblar que se siente en el vehículo en varios intervalos de velocidad.

Viscosity The resistance to flow exhibited by a liquid. A thick oil has greater viscosity than a thin oil.

Viscosidad La resistencia al flujo que manifiesta un líquido. Un aceite espeso tiene una viscosidad mayor que un aceite ligero.

Vortex Path of fluid flow in a torque converter. The vortex may be high, low, or zero, depending on the relative speed between the pump and turbine.

Vórtice La vía del flujo de los fluidos en un convertido de torsión. El vórtice puede ser alto, bajo, o cero, depende de la velocidad relativa entre la bomba y la turbina.

Vortex flow Recirculating flow between the converter impeller and turbine that causes torque multiplication.

Flujo del vórtice El fluyo recirculante entre el impulsor del convertidor y la turbina que causa la multiplicación de la torsión.

Wet-disc clutch A clutch in which the friction disc (or discs) is operated in a bath of oil.

Embrague de disco flotante Un embrague en el cual el disco (o los discos) de fricción opera en un baño de aceite.

Wheel A disc or spokes with a hub at the center which revolves around an axle, and a rim around the outside for mounting the tire on.

Rueda Un disco o rayo que tiene en su centro un cubo que gira alrededor de un eje, y tiene un rim alrededor de su exterior en la cual se monta el neumático.

Yoke In a universal joint, the drivable torque-and-motion input and output member, attached to a shaft or tube.

Yugo En una junta universal, el miembro de la entrada y la salida que transfere a la torsión y al movimiento, que se conecta a una flecha o a un tubo.

INDEX

Note: Page numbers in **italic type** reference non-text material.

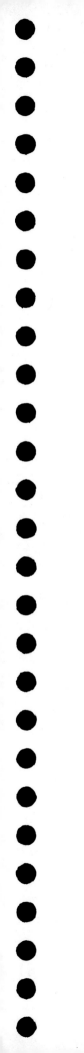

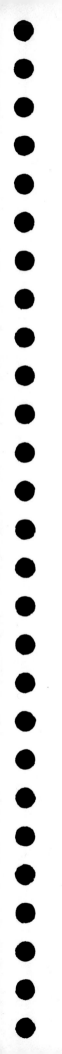

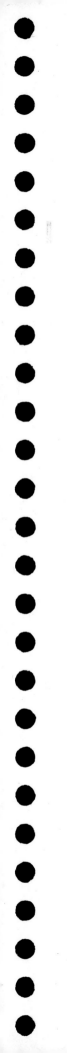

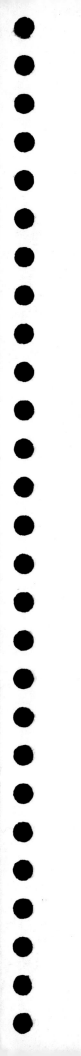